Orthogonal and Symplectic Clifford Algebras

Mathematics and Its Applications

Volume 57

Orthogonal and Symplectic Clifford Algebras
Spinor Structures

by

Albert Crumeyrolle
Faculty of Sciences,
University of Toulouse, France

KLUWER ACADEMIC PUBLISHERS
DORDRECHT / BOSTON / LONDON

Library of Congress Cataloging in Publication Data

Crumeyrolle, A. (Albert)
    Orthogonal and symplectic Clifford algebras : spinor structures /
by Albert Crumeyrolle.
        p.   cm. -- (Mathematics and its applications)
    Includes bibliographical references.
    ISBN 0-7923-0541-8
    1. Clifford algebras.   2. Spinor analysis.   I. Title.
II. Series: Mathematics and its applications (Kluwer Academic
Publishers)
    QA199.C79   1990
    512'.57--dc20   89-48828

ISBN 0–7923–0541–8

Published by Kluwer Academic Publishers,
P.O. Box 17, 3300 AA Dordrecht, The Netherlands.

Kluwer Academic Publishers incorporates
the publishing programmes of
D. Reidel, Martinus Nijhoff, Dr W. Junk and MTP Press.

Sold and distributed in the U.S.A. and Canada
by Kluwer Academic Publishers,
101 Philip Drive, Norwell, MA 02061, U.S.A.

In all other countries, sold and distributed
by Kluwer Academic Publishers Group,
P.O. Box 322, 3300 AH Dordrecht, The Netherlands.

Printed on acid-free paper

SERIES EDITOR'S PREFACE

Mathematics is a tool for thought. A highly necessary tool in a world where both feedback and non-linearities abound. Similarly, all kinds of parts of mathematics serve as tools for other parts and for other sciences.

Applying a simple rewriting rule to the quote on the right above one finds such statements as: 'One service topology has rendered mathematical physics ...'; 'One service logic has rendered computer science ...'; 'One service category theory has rendered mathematics ...'. All arguably true. And all statements obtainable this way form part of the raison d'être of this series.

This series, *Mathematics and Its Applications*, started in 1977. Now that over one hundred volumes have appeared it seems opportune to reexamine its scope. At the time I wrote

"Growing specialization and diversification have brought a host of monographs and textbooks on increasingly specialized topics. However, the 'tree' of knowledge of mathematics and related fields does not grow only by putting forth new branches. It also happens, quite often in fact, that branches which were thought to be completely disparate are suddenly seen to be related. Further, the kind and level of sophistication of mathematics applied in various sciences has changed drastically in recent years: measure theory is used (non-trivially) in regional and theoretical economics; algebraic geometry interacts with physics; the Minkowsky lemma, coding theory and the structure of water meet one another in packing and covering theory; quantum fields, crystal defects and mathematical programming profit from homotopy theory; Lie algebras are relevant to filtering; and prediction and electrical engineering can use Stein spaces. And in addition to this there are such new emerging subdisciplines as 'experimental mathematics', 'CFD', 'completely integrable systems', 'chaos, synergetics and large-scale order', which are almost impossible to fit into the existing classification schemes. They draw upon widely different sections of mathematics."

By and large, all this still applies today. It is still true that at first sight mathematics seems rather fragmented and that to find, see, and exploit the deeper underlying interrelations more effort is needed and so are books that can help mathematicians and scientists do so. Accordingly MIA will continue to try to make such books available.

If anything, the description I gave in 1977 is now an understatement. To the examples of interaction areas one should add string theory where Riemann surfaces, algebraic geometry, modular functions, knots, quantum field theory, Kac-Moody algebras, monstrous moonshine (and more) all come together. And to the examples of things which can be usefully applied let me add the topic 'finite geometry'; a combination of words which sounds like it might not even exist, let alone be applicable. And yet it is being applied: to statistics via designs, to radar/sonar detection arrays (via finite projective planes), and to bus connections of VLSI chips (via difference sets). There seems to be no part of (so-called pure) mathematics that is not in immediate danger of being applied. And, accordingly, the applied mathematician needs to be aware of much more. Besides analysis and numerics, the traditional workhorses, he may need all kinds of combinatorics, algebra, probability, and so on.

In addition, the applied scientist needs to cope increasingly with the nonlinear world and the extra mathematical sophistication that this requires. For that is where the rewards are. Linear models are honest and a bit sad and depressing: proportional efforts and results. It is in the non-linear world that infinitesimal inputs may result in macroscopic outputs (or vice versa). To appreciate what I am hinting at: if electronics were linear we would have no fun with transistors and computers; we would have no TV; in fact you would not be reading these lines.

There is also no safety in ignoring such outlandish things as nonstandard analysis, superspace and anticommuting integration, p-adic and ultrametric space. All three have applications in both electrical engineering and physics. Once, complex numbers were equally outlandish, but they frequently proved the shortest path between 'real' results. Similarly, the first two topics named have already provided a number of 'wormhole' paths. There is no telling where all this is leading - fortunately.

Thus the original scope of the series, which for various (sound) reasons now comprises five subseries: white (Japan), yellow (China), red (USSR), blue (Eastern Europe), and green (everything else), still applies. It has been enlarged a bit to include books treating of the tools from one subdiscipline which are used in others. Thus the series still aims at books dealing with:

- a central concept which plays an important role in several different mathematical and/or scientific specialization areas;
- new applications of the results and ideas from one area of scientific endeavour into another;
- influences which the results, problems and concepts of one field of enquiry have, and have had, on the development of another.

Spinors are important in theoretical physics, specifically in quantum theory. Usually one finds the word at least once in the table of contents of, say, the Journal of Mathematical Physics, or Letters of Mathematical Physics. In addition they have an important (and far from unrelated) role to play in differential geometry and topology. Even so, the concept of a spinor, and the word, tend to be used somewhat loosely (and inaccurately), possibly due, until the present volume, to the lack of a precise, thorough, and accessible, book on the topic.

Mathematically, spinors are (the elements of) a representation space of a Clifford algebra, themselves also objects of considerable current interest, and that is the point of view, originally due to Chevalley (1954), that is taken in this book, that is both rigorous enough for mathematicians and concrete enough for physicists.

The shortest path between two truths in the
real domain passes through the complex
domain.

J. Hadamard

La physique ne nous donne pas seulement
l'occasion de résoudre des problèmes ... elle
nous fait pressentir la solution.

H. Poincaré

Never lend books, for no one ever returns
them; the only books I have in my library
are books that other folk have lent me.

Anatole France

The function of an expert is not to be more
right than other people, but to be wrong for
more sophisticated reasons.

David Butler

Bussum, January 1990   Michiel Hazewinkel

Contents

Appendices :

Introduction.

This book has been conceived as a treatise on the theory of Clifford algebras and spinors, from the viewpoint of applications to differential geometry and mathematical physics, which would be both elementary and as complete as possible.

These algebras, sometimes called 'geometric' because of their relation to orthogonal transformations, were discovered by W. K. Clifford in 1878, as a generalization of W. R. Hamilton's quaternions (1858). A whole school of researchers now use them to generalize the theory of functions of a complex variable : this is the 'Clifford analysis', which will not be considered in this book [1].

In the literature, the notion of 'spinor' has remained very ambiguous. Its origins lie in the work of E. Cartan on group representation (1913), but it was given its current name only in 1930, after the discovery of spin by Pauli and Dirac. And only in 1935-38 were the Clifford algebras used by R. Brauer, H. Weyl and—again— E. Cartan to construct spinors. For many years, the book 'Leçons sur la théorie des spineurs' was the standard reference for a whole school of theorists. The clear and rigorous definition of the spinors as irreducible representation spaces for Clifford algebras—which holds in all dimensions, all signatures and all characteristics—was given by Chevalley in 1954. In spite of its depth and rigor, Chevalley's book proved too abstract for most physicists and the notions explained in it have not been applied much until recently, which is a pity. In a parallel way, the notion of spinor structure, derived by A. Haefliger (1956) from the work of Ch. Ehresmann was successfully applied by A. Lichnerowicz to field theory (1964) and by many authors in algebraic topology to differential geometry and to the theory of fiber bundles; without, surprisingly, relying on the resources of Clifford algebras and their representations.

Today, very few mathematicians and physicists have a sound knowledge of spinor theory. Spinors are used in field theory with such abuses of linear algebra that they are no longer useful for the study of global problems. Often, authors consider only one given dimension and signature, using special methods which cannot be applied to the general case. The term 'spinor' itself is not always clear, and a 'spin representation' might either be a true representation or a covering, depending on the author. Some authors ignore the modern definition of spinors as elements of minimal left (or right) ideals of a Clifford algebra; if it is used, the algebra representations are sometimes insufficiently distinguished from the representations of the associated

---

[1]cf. F. Brackx, R. Delanghe, F. Sommen, *Clifford analysis*, Research Notes in Math., Pitman and D. Hestenes, G. Sobczyk, *Clifford Algebra to Geometric Calculus*, p. 1–136, Reidel.

groups, which appear in the theory of fiber bundles : the consistency may be greatly compromised by such confusions.

We think that many researchers would like to have a systematic treatise offering the detailed exposition of these notions, which have become indispensable both in geometry and in quantum mechanics.

This book is equally meant for mathematicians, being rigorous, and for physicists, being concrete. Starting from a relatively elementary level (corresponding to the third year at university), we recall the main results from orthogonal and symplectic geometry—we will explain later why the symplectic case is included here—and some notions on tensor, exterior and symmetric algebras; the advanced reader may browse the first two chapters, noticing the influence of E. Artin, and find the main definitions and results on the Clifford algebras and their associated groups in Chapters 3 and 4. Chapter 5 is of central importance : we recall some basic facts on the representations of semi-simple algebras which explain the notion of spin representation; some properties of primitive idempotents and minimal left ideals are given, but we concentrate on the complex case, which is the most important one for physics, and introduce the essential notion of isotropic r-vector, which will prove fundamental for the sequel. Chapter 6 assumes that the reader is somewhat familiar with the classical Lie groups. Next, in Chapter 7, we give a survey of matrix methods and link our approach with an old but very widely used method. Chapters 8 and 9 are concerned with the spinors in maximal index. In these two chapters, we are closest to Chevalley's methods, but, since we are not interested in the exceptional characteristics, our methods are, we think, simpler and more concrete. The first part, which has the tools needed further on by the reader (researcher or student), ends there : the other part requires a thorough knowledge of Chapters 3, 4, 5 and 8.

Chapter 10 is one of the most important for applications to physics. It offers a new approach to 'charge conjugation' and its related notions (Majorana spinors), and has new and exhaustive results. Chapter 11 is short, but essential for the understanding of the cohomological methods; it contains the definition of new groups introduced by the author around 1970, as 'groupes de spinorialité'. Chapters 12 and 13 offer a new approach to problems which have been tackled by many different methods in recent publications : the twistors, reduced to the rank of spinors, appear in the setting of conformal geometry, and certain graded Lie algebras and supersymmetries are related to a generalized triality principle or, if one prefers, to an extension of the well-known Hurwitz problem. Chapters 14 and 15 form, along with Chapter 16 (an applications chapter for physicists), an essential part of this book, concerned with the differential geometry of spinor fields. We use an original approach to the problem of spin fibration, which relies on the algebraic setting of Chapters 5 and 8 and explains the spinor frames which, often, seem to appear from nowhere; the existence of spinor structures is that of pseudo-fields of isotropic r-vectors, an essentially geometric notion which can serve as an algorithmic tool for applications. Spheres and projective spaces are given as examples and are, for the first time, solved using elementary methods.

This formalism is applied in Chapters 15 and 16, where the principal operators of quantum mechanics are given, both in the local and in the global setting.

Chapters 17, 18 and 19 form a separate entity, only depending on Chapter 1. We define the deformations of the algebras associated to a symplectic manifold and geometrize the Maslov index. By analogy with the orthogonal case, we introduce 'symplectic Clifford algebras' which, generalizing the so-called Weyl algebras, have a similar rôle, in the symplectic case, to that of the ordinary Clifford algebras in the orthogonal case. We construct 'spinoplectic' groups (analogous to the spin groups) and symplectic spinor bundles, using a procedure which is identical to that in Chapter 14. The three-sided index and the Maslov index are defined geometrically and we show that the deformation problem can be solved in this way, our symplectic Clifford product being a generalization of the Moyal product.

The book ends with a few articles. One of them is concerned with idempotent fibrations in the Clifford algebra of a manifold, a problem which has proved very interesting in the past few years. We introduce the notion of amorphic spinor bundle, explain how it differs from that defined in Chapter 15 and warn against possible confusions and unwarranted identifications.

The second article presents the so-called Penrose transform in a purely spinorial setting (no twistors are used), and defines instantons.

Finally, the third article describes the geometrization of the Fourier transform, acting on the fields of symplectic spinors.

We think that these articles are very significant for the philosophy and the spirit of this work, and therefore thought it useful to include them, even though this further increases its size.

The author wishes to express his gratitude towards Y. Panabière for the careful typing of the main part of the manuscript.

He warmly thanks all his colleagues who have encouraged and helped him for the writing and the publication of this book.

Chapter 1

ORTHOGONAL AND SYMPLECTIC GEOMETRIES.

1.1 Background on bilinear forms.

1.1.1

The vector spaces under consideration will generally be of finite dimension over the real number field $\mathbf{R}$ or over the complex number field $\mathbf{C}$; we write $\mathbf{K}$ for either $\mathbf{R}$ or $\mathbf{C}$ when reasonings and results are true for both.

Let E and F be vector spaces of dimension n and p respectively and let B be a bilinear form :

$$B : (x, y) \in E \times F \to B(x, y) \in \mathbf{K},$$

we will write

$$\gamma_x : y \to B(x, y), \quad \text{where } x \text{ is fixed,}$$
$$\delta_y : x \to B(x, y), \quad \text{where } y \text{ is fixed.}$$

for its partial functions. Let (e_i), $i = 1, \ldots, n$ and (f_j), $j = 1, \ldots, p$ be bases for E and F respectively, then we put

$$B(e_i, f_j) = \alpha_{ij}$$

and the matrix (α_{ij}) will be denoted by α.

If we change the bases :

$$e_{i'} = A_{i'}^k e_k, \quad f_{j'} = C_{j'}^l f_l$$

where the Einstein convention on the summation of repeated indices is assumed, as it will be for the rest of this book, we know that

$$\alpha_{i'j'} = A_{i'}^k C_{j'}^l \alpha_{kl},$$

or in matrix form :

$$\alpha' = A^T \alpha C, \tag{1}$$

where A^T is the transpose matrix of A.

1

1.1.2   The rank of a bilinear form.

Let E^* and F^* be the dual spaces of E and F respectively. Define mappings

$$\gamma : x \to \gamma_x, \quad \gamma : E \to F^*,$$
$$\delta : y \to \delta_y, \quad \delta : F \to E^*.$$

The kernel of δ is a subspace E' of F called the conjugate of E in F, i.e. :

$$E' \subseteq F, \text{ and if } y \in E', B(x,y) = 0 \quad \forall x \in E.$$

We define the conjugate F' of F in E in a similar way :

$$F' \subseteq E, \text{ and if } x \in F', B(x,y) = 0 \quad \forall y \in F.$$

B is said to be non-degenerate if $E' = F' = \{0\}$, and degenerate otherwise.

If $E' = F' = \{0\}$, γ and δ are injective and clearly $n \le p$ and $p \le n$ so $n = p$. Otherwise, we write E/F' for the quotient space of E by F' (i.e. the set of classes modulo F' of elements in E). F/E' is defined analogously, and we have the following proposition :

Proposition 1.1.1 E/F' and F/E' have the same dimension. This dimension is called the rank of B (or $\mathrm{rk}(B)$ for short).

Proof. $B(x,y) = B(x,y_1)$ if $y - y_1 \in E'$ and $B(x,y) = B(x_1,y_1)$ if $x = x_1 \,(\mathrm{mod}\, F')$ and $y = y_1 \,(\mathrm{mod}\, E')$. Setting

$$\bar{x} = \mathrm{Cl}(x)\,(\mathrm{mod}\, F'), \quad \bar{y} = \mathrm{Cl}(y)\,(\mathrm{mod}\, E'),$$

we get

$$\bar{B}(\bar{x}, \bar{y}) = B(x,y),$$

where $\bar{B}$ is bilinear on $E/F' \times F/E'$ and $\bar{B}(\bar{x}, \bar{y}) = 0$, $\forall \bar{x} \in E/F'$ is equivalent to $B(x,y) = 0$, $\forall x \in E$, so $y \in E'$, and $\bar{y} = 0$. Similarly $\bar{B}(\bar{x}, \bar{y}) = 0$, $\forall \bar{y} \in F/E'$ is equivalent to $\bar{x} = 0$. Hence $\bar{B}$ is non-degenerate and $\dim E/F' = \dim F/E'$.   ∎

Note that if (φ^j) and (ϵ^i) are the dual bases of (f_j) and (e_i) respectively, $\gamma(e_i) = \alpha_{ij}\varphi^j$ and $\delta(f_j) = \alpha_{ij}\epsilon^i$ since $\gamma_{e_i}(y) = B(e_i, y) = \alpha_{ij}y^j = \alpha_{ij}\varphi^j(y)$. γ being the transpose of δ, both matrices have equal rank, which also proves Proposition (1.1.1) : if this rank is r,

$$r = \dim \gamma(E) = \dim E/F' = \dim \delta(F) = \dim F/E'.$$

We will remember that r is the common rank of γ and δ and that it is equal to the rank of the matrix α.

1.1.3 The dimension of conjugate subspaces.

Given that the rank of B is r, we know that a matrix α' can be found such that it is equivalent to α and has zero entries except for $\alpha'_{ii} = 1$ when $i = 1, \ldots, r$.

1. If B is non-degenerate we can consider bases for which the matrix of B has the reduced form we just indicated; then

$$B(x, y) = \sum_{i=1}^{n} x^i y^i.$$

   This is the classical situation of duality. Hence

   Proposition 1.1.2 *Let B be non-degenerate, let V be a subspace of E and V' its conjugate in F, then*

$$\dim V + \dim V' = n \tag{2}$$

   where $n = \dim E = \dim F$. Furthermore, if V'' is the conjugate of V' in E, $V'' = V$.

2. If B is degenerate and if $\dim E = \dim F = n$, $\mathrm{rk}(B) = r$

$$\dim V + \dim V' \geq n. \tag{3}$$

   Indeed, $\bar{V}' = V'/E' \subseteq F/E'$, $\bar{V} = V/F' \subseteq E/F'$. Also $\bar{B}$ is non-degenerate and $\bar{y} \in (\bar{V})'$ implies $\bar{y} \in \overline{V'}$, hence

$$\dim \bar{V} + \dim(\bar{V})' = r,$$
$$\dim V + \dim \overline{V'} \geq r,$$
$$\dim V + \dim V' - \dim E' \geq r,$$
$$\dim V + \dim V' - \dim E' \geq \dim F - \dim E',$$
$$\dim V + \dim V' \geq n.$$

1.1.4 Definitions.

1. If $E = F$, the determinant of α is called the *discriminant* of B for the basis (e_i). As $\alpha' = A^T \alpha A$ if $e_{i'} = A_{i}^{k} e_k$,

$$\det \alpha' = \det \alpha (\det A)^2. \tag{4}$$

   In this special case, E' is called the *right conjugate* of E and F' is called its *left conjugate*.

2. Let E and F be two vector spaces over $\mathbf{K}$ with bilinear forms B and C. An isomorphism $\sigma : E \to F$ is said to be an *isometry* iff

$$B(x,y) = C(\sigma x, \sigma y), \quad \forall x, y \in E. \tag{5}$$

If $E = F$ and $B = C$, the set of isometries is a group.

Note that if B is non-degenerate and if $\sigma x = 0$, then $C(\sigma x, \sigma y) = 0$, $\forall y \in E$ and (5) implies that $B(x,y) = 0$, so x is an element of the left conjugate of E, hence $x = 0$ and σ is injective. Also, if $\sigma : E \to F$ is an isometry,

$$B(e_i, e_j) = \alpha_{ij} = B(\sigma e_i, \sigma e_j) = \alpha_{i'j'}$$

where $e_{i'} = \sigma e_i$, so from (4) it follows that $(\det \sigma)^2 = 1$. The isometries for which $\det \sigma = 1$ are called *rotations* and these for which $\det \sigma = -1$ are called *reversions* The set of rotations is a normal subgroup of the group of isometries and its index is at most 2. $\sigma \to \det \sigma$ is a homomorphism whose kernel is the group of rotations. In certain cases, all isometries are rotations.

1.1.5   The two fundamental geometries.

Assume $E = F$. It is especially interesting to consider choices of B for which $E' = F'$, so that left and right conjugates coincide. This is the case if for all $(x, y) \in E \times E$, $B(x,y) = 0$ is equivalent to $B(y,x) = 0$, which clearly holds if B is symmetric or antisymmetric. Now we prove the converse :

Proposition 1.1.3 *If for all $(x, y) \in E \times E$, $B(x,y) = 0$ is equivalent to $B(y, x) = 0$, then B must either be symmetric or antisymmetric.*

Proof. If $x, y, z \in E$ we have

$$
\begin{aligned}
B(x, B(x, z)y) - B(x, B(x, y)z) &= B(x,y)B(x, z) - B(x, z)B(x,y) \\
&= 0.
\end{aligned}
$$

But $B(x, B(x, z)y - B(x, y)z) = 0$ is equivalent to $B(B(x, z)y - B(x, y)z, x) = 0$ and we obtain :

$$B(x, z)B(y, x) - B(x, y)B(z, x) = 0. \tag{6}$$

For $x = z$, $B(x, x)B(y, x) = B(x, x)B(x, y)$. If $B(x, y) \neq B(y, x)$, $B(x, x) = 0$ and, similarly, $B(y, y) = 0$. Assume now that for some $(x, y) \in E \times E$, $B(x, y) \neq B(y, x)$, then $B(z, z) = 0$, $\forall z \in E$. Indeed :

1. If $B(x, z) \neq B(z, x)$, the previous reasoning leads to $B(z, z) = 0$ if we replace y by z.

2. If $B(x, z) = B(z, x)$, then by (6) and because $B(y, x) \neq B(x, y)$, $B(x, z) = B(z, x) = 0$.

We may assume that $B(y, z) = B(z, y)$, otherwise we reobtain case 1 for y. Replacing x by y in 2, $B(y, z) = B(z, y) = 0$, hence

$$B(x + z, y) = B(x, y) \neq B(y, x) = B(y, x + z)$$

and if we replace z by $x + z$ and x by y in 1, we see that

$$B(x + z, x + z) = 0.$$

But from $B(x, x) = 0$ (since $B(x, y) \neq B(y, x)$ and $B(x, z) = B(z, x) = 0$), $B(z, z) = 0$. Clearly there are only two cases to consider : either $B(x, x) = 0$, $\forall x \in E$, or for some $z \in E$, $B(z, z) \neq 0$ and then $B(x, y) = B(y, x)$, $\forall x, y \in E$ since if for $x, y \in E$, $B(x, y) \neq B(y, x)$, $B(z, z) = 0$ would follow by the previous reasoning. If $B(x, x) = 0$, $\forall x \in E$, replacing x by $x + y$, we have that $B(x, y) = -B(y, x)$. If for some $z \in E$, $B(z, z) \neq 0$, $B(x, y) = B(y, x)$. ∎

If B is symmetric the geometry is called *orthogonal*. If B is antisymmetric, the geometry is called *symplectic*. No other case can occur if left and right conjugates are to coincide. From now on $E = F$ is assumed, unless stated otherwise.

1.2 Common properties of the orthogonal and the symplectic geometries.

Let B be a bilinear form on $E \times E$ which is either symmetric or antisymmetric. The left and right conjugates of V are now called orthogonal to V and V' will be written $V^\perp$. In the orthogonal case, $B(x, x)$ is written $Q(x)$ for short.

1.2.1 The radical.

Definition 1.2.1 *The set of $x \in E$ for which $B(x, y) = 0$, $\forall y \in E$, is a subspace of E called its* radical *and denoted by* rad E *or* $E^\perp$.

Clearly rad $E = 0$ if B is non-degenerate, E is then said to be *non-isotropic*. Let U be a subspace of E such that the restriction $B|_U$ of B to U induces a geometry of the same type (orthogonal or symplectic) as the geometry on E.

Definition 1.2.2 *The radical* rad U *of U is defined as $U \cap U^\perp$.*

Note that although rad $E = E^\perp$, rad $U \neq U^\perp$ in general. U is said to be *isotropic* if rad $U \neq 0$.

Definition 1.2.3 *If $E = E_1 \oplus E_2 \oplus \ldots \oplus E_r$ where the sums are direct and the E_i are pairwise orthogonal spaces (for B), we say that E is the direct orthogonal sum of the E_i and write*

$$E = E_1 \perp E_2 \perp \ldots \perp E_r.$$

Here two subspaces E_i, E_j are *orthogonal* if

$$B(x, y) = 0, \quad \forall x \in E_i, \quad \forall y \in E_j.$$

Lemma 1.2.1 *If $E = E_1 + E_2 + \ldots + E_r$ where the sum is not necessarily direct, but the E_i are pairwise orthogonal, then*

$$\text{rad } E = \text{rad } E_1 + \text{rad } E_2 + \ldots + \text{rad } E_r. \tag{1}$$

Proof. It is sufficient to note that if $x = \sum_{i=1}^r x_i$, $x_i \in E_i$, and if $x \in \text{rad } E$, $B(x, y_i) = 0$, $\forall y_i \in E_i$, so $B(x_i, y_i) = 0$ and $x_i \in \text{rad } E_i$. Conversely if every $x_i \in \text{rad } E_i$, $\sum_{i=1}^r B(x_i, y_i) = 0$ and $B(x, y) = 0$, $\forall y \in B$, so $x \in \text{rad } E$. ∎

Corollary 1.2.1 *If each B_i is non-degenerate, B is non-degenerate.*

Proof. Trivial. ∎

Lemma 1.2.2 *There exists a $U \subseteq E$ such that $E = \text{rad } E \perp U$ and $B|_U$ is non-degenerate.*

Proof. If we set $E = E^\perp \oplus U$, the sum is orthogonal :

$$E = E^\perp \perp U = \text{rad } E \perp U.$$

But by Lemma (1.2.1),

$$\text{rad } E = \text{rad}(\text{rad } E) \perp \text{rad } U = \text{rad } E \perp \text{rad } U$$

since $\text{rad}(\text{rad } E) = \text{rad } E \cap E = \text{rad } E$. Hence $\text{rad } U = 0$ and $B|_U$ is non-degenerate. ∎

Proposition 1.2.1 *Let E be a non-isotropic space and let U be one of its subspaces, then*

1. $U^{\perp\perp} = U$, $\dim U + \dim U^\perp = \dim E$,
2. $\text{rad } U = \text{rad } U^\perp = U \cap U^\perp$,
3. U *is non-isotropic if and only if $U^\perp$ is non-isotropic,*
4. *For non-isotropic U, $E = U \perp U^\perp$,*
5. *If $E = U \perp W$, U and W are non-isotropic and $W = U^\perp$.*

Proof. By formula (2) of (1.1.3) we have

$$\dim U + \dim U^\perp = \dim E.$$

That $U \subseteq (U^\perp)^\perp$ is obvious; if $\dim U = p$ and $\dim E = n$, $\dim U^\perp = n - p$ and $(U^\perp)^\perp$ has dimension p, so (1) is proved. $\text{rad } U = U \cap U^\perp$, $\text{rad } U^\perp = U^\perp \cap U^{\perp\perp} = U \cap U^\perp$. If U is non-isotropic, $\text{rad } U = 0$, hence $U^\perp$ is non-isotropic and the converse holds too. $U + U^\perp$ being a direct sum, $U \perp U^\perp = E$. If $E = U \perp W$, $W \subseteq U^\perp$, $\dim W = n - \dim U = \dim U^\perp$, so $W = U^\perp$, $U \cap U^\perp = 0 = \text{rad } U$. ∎

Definition 1.2.4 *E is said to be* totally isotropic *if $B \equiv 0$ on E. A subspace $U \subseteq E$ is said to be* totally isotropic *if $B|_U \equiv 0$.*

Both $\{0\}$ and $E^\perp$ are totally isotropic. A vector $x \in E$ is called *isotropic* if $B(x, x) = 0$. (Generally, when we say that some vector is isotropic, we will also mean that it is non-zero.)

Lemma 1.2.3 *If, in an orthogonal geometry, the subspace $U \subseteq E$ is such that $B(x, x) = 0$, $\forall x \in U$, then U is totally isotropic.*

Proof. $B(x + y, x + y) = 0$ implies $2B(x, y) = 0$ and hence $B(x, y) = 0$. ∎

In this case, the geometry defined on U by B is also symplectic.

1.2.2 Decomposition in lines and in hyperbolic planes.

Let U be a two-dimensional non-isotropic subspace of E, containing an isotropic vector $x \neq 0$. Let $y \in E$, $y \neq \lambda x$ so $(x, y) = U$.

We try to find an isotropic vector $z \in U$ for which $B(x, z) = 1$. Setting $z = \alpha x + \beta y$,

$$B(x, z) = \beta B(x, y).$$

Now $x \notin \operatorname{rad} U$ and there exists a y for which $B(x, y) \neq 0$, U being non-isotropic, hence β is uniquely defined by $B(x, z) = 1$. In the symplectic case, $B(z, z) = 0$. In the orthogonal case, to have $B(z, z) = 0$ requires

$$B(\alpha x + \beta y, \alpha x + \beta y) = 0,$$

i.e.

$$2\alpha\beta B(x, y) + \beta^2 B(y, y) = 0.$$

As $\beta B(x, y) \neq 0$, α can be determined and z follows. If we change the notations we have that

$$U = (x, y), \quad B(x, x) = B(y, y) = 0, \quad B(x, y) = 1. \tag{2}$$

Conversely if $\dim U = 2$ and if (2) holds, let $U \in \operatorname{rad} U$, where $u = \alpha x + \beta y$, then

$$B(u, y) = 0 \text{ implies } \alpha = 0,$$
$$B(x, u) = 0 \text{ implies } \beta = 0,$$

so U must be non-isotropic. In the orthogonal case the vectors collinear with x or y are the only isotropic vectors, since $B(u, u) = 0$ implies $\alpha\beta = 0$.

Definition 1.2.5 *A non-isotropic plane that contains a non-zero isotropic vector is called* hyperbolic. *An orthogonal sum of hyperbolic planes $P_1, P_2, \ldots, P_r$ is said to be a* hyperbolic space H_{2r} :

$$H_{2r} = P_1 \perp P_2 \perp \ldots \perp P_r.$$

By Lemma (1.2.1), H_{2r} must be non-isotropic. The previous results lead to :

Proposition 1.2.2 *A hyperbolic plane is generated by a pair (x, y) for which*

$$B(x, x) = B(y, y) = 0$$

and

$$B(x, y) = 1.$$

Such a pair (x, y) is called *hyperbolic*.

Definition 1.2.6 *E is said to be* irreducible *if it cannot be decomposed as an orthogonal sum of proper subspaces.*

As $E = E^\perp \perp U$ an irreducible space must be non-isotropic ($E^\perp = 0$) or totally isotropic ($E^\perp = E$), but in the latter case its dimension must be 1. (Any decomposition of a totally isotropic space as a direct sum being orthogonal.)

If E is non-isotropic,

- if the geometry is orthogonal : by Lemma (1.2.3), E contains a non-zero non-isotropic vector x, (x) is a non-isotropic space, $\mathrm{rad}(x) = 0$ and $x \perp (x)^\perp = E$ implies $\dim E = 1$.

- if the geometry is symplectic : let $x \neq 0$, $x \in E$, $\mathrm{rad}\, E$ being 0, there is an a for which $B(x, a) \neq 0$. (x, a) is a non-isotropic plane U ($B|_U$ is represented by a matrix of rank 2). $E = U \perp U^\perp$ so $U^\perp = 0$ and $E = U$, $\dim E = 2$.

We have proved :

Proposition 1.2.3

- *If the geometry is orthogonal, the space E is a direct orthogonal sum of lines :*

$$E = (a_1) \perp (a_2) \perp \ldots \perp (a_n),$$

$\{a_1, a_2, \ldots, a_n\}$ is an orthogonal basis of E and none of the a_i is isotropic if and only if E is non-isotropic.

- *If the geometry is symplectic and if B is non-degenerate, E is the direct orthogonal sum of hyperbolic planes, E is a hyperbolic space and $\dim(E)$ is even. If B is degenerate, a totally isotropic space must be added to the orthogonal sum of hyperbolic spaces.*

Proposition 1.2.4 *Let the (x_i, y_i) be hyperbolic pairs, if*

$$E = (x_1, y_1) \perp (x_2, y_2) \perp \ldots \perp (x_r, y_r)$$

is a hyperbolic space, if σ is an isometry from E to itself, and if σ fixes each of the x_i, then σ is a rotation and

$$\sigma(x_i) = x_i, \quad \sigma(y_i) = a_i^j x_j + y_i,$$

where the matrix $((a_j^i))$ is symmetric in the symplectic case and antisymmetric in the orthogonal case.

Proof.

$$\sigma(y_i) = a_i^j x_j + b_i^j y_j,$$
$$B(\sigma(x_i), \sigma(y_i)) = B(x_i, y_i)$$
$$\text{and} \quad B(\sigma(y_i), \sigma(y_j)) = B(y_i, y_j)$$

immediately lead to these conditions. ∎

Definition 1.2.7 An isometry σ of a non-isotropic space is called an involution if $\sigma^2 = \text{Id}$.

Proposition 1.2.5 Every involution has the form $-\text{Id}_U + \text{Id}_V$ where $E = U \perp V$. (Here U and V are non-isotropic, and one of them may be 0 if $\sigma = \text{Id}$ or $\sigma = -\text{Id}$).

Proof. If $\sigma^2 = \text{Id}$,

$$B(\sigma(x), \sigma(y)) = B(x, y),$$
$$B(\sigma(x), y) = B(x, \sigma(y)),$$
$$B(\sigma(x) - x, \sigma(y) + y) = 0.$$

$U = (\sigma - \text{Id})E$ and $V = (\sigma + \text{Id})E$ are orthogonal subspaces; $\sigma(\sigma(x) - x) = -\sigma(x) + x$, $\sigma(\sigma(y) + y) = \sigma(y) + y$ so $U \cap V = 0$. Any vector $x \in E$ equals

$$-\tfrac{1}{2}(\sigma(x) - x) + \tfrac{1}{2}(\sigma(x) + x)$$

so $E = U \perp V$ and $\sigma = -\text{Id}_U + \text{Id}_V$. ∎

From this proposition we deduce that if $p = \dim U$:

$$\det \sigma = (-1)^p.$$

If U is non-isotropic, p is even in the symplectic case.

Proposition 1.2.6 If U is non-isotropic and if $\sigma : E \to E$ is an isometry fixing all lines of E, then $\sigma = \pm \text{Id}_E$.

Proof. Let (x) be the space generated by $x \in E$, where $x \neq 0$.

$$\sigma(x) = ax, \quad \forall y \in (x),$$
$$\sigma(y) = \sigma(bx) = b\sigma(x) = abx = ay.$$

The coefficient a is independent of x since if (x, y) has dimension 2,

$$\sigma(x + y) = c(x + y),$$
$$\sigma(x + y) = ax + by,$$

hence $a = b = c$. Choosing x, y for which $B(x, y) \neq 0$,

$$B(x, y) = B(\sigma(x), \sigma(y)) = a^2 B(x, y),$$

so $a^2 = 1$. ∎

1.2.3 Witt decomposition and the Witt theorem.

Let B be non-degenerate, i.e. E is non-isotropic, let U be a subspace of E, then by the result in (1.2.1), $U = \operatorname{rad} U \perp W$ and W is non-isotropic. Let $(e_1, e_2, \ldots, e_r)$ be a basis of $\operatorname{rad} U$, $e_1, e_2, \ldots, e_r$ being isotropic vectors.

Proposition 1.2.7 *There are isotropic vectors $f_1, f_2, \ldots, f_r \in E$ such that every (e_i, f_i) is a hyperbolic pair, all planes $P_i = (e_i, f_i)$ are pairwise orthogonal and $P_1 \perp P_2 \perp \ldots \perp P_r \perp W$ is a non-isotropic space containing U.*

Proof. For $r = 0$ there is nothing to prove.

We use induction on r. Let $U = (e_1, e_2, \ldots, e_r) \perp W$ be assumed and let $U_0 = (e_1, e_2, \ldots, e_{r-1}) \perp W$, $e_r \in U_0^\perp$. $U_0^\perp$ contains a vector a for which $B(e_r, a) \neq 0$, otherwise e_r would be orthogonal to $U_0^\perp$, implying that $e_r \in U_0^\perp \cap U_0^{\perp\perp} = \operatorname{rad} U_0^\perp = U_0 \cap U_0^\perp$; (e_r, a) is a non-isotropic subspace of $U_0^\perp$, generated by some hyperbolic pair (e_r, f_r). Let us define $P_r = (e_r, f_r)$. $P_r \subseteq U_0^\perp$ implies $U_0 \subseteq P_r^\perp$, $P_r^\perp$ is non-isotropic, just as P_r is. The radical of U_0, $U_0 \cap U_0^\perp$, has dimension $r - 1$, $U_0 = \operatorname{rad} U_0 \perp W$. By the induction hypothesis for $P_r^\perp$, there are hyperbolic (e_i, f_i) in $P_r^\perp$ for $1 \leq i \leq r-1$ such that the P_i are pairwise orthogonal and orthogonal to W. $P_1 \perp P_2 \perp \ldots \perp P_{r-1} \perp W$ is non-isotropic and contains U_0; adding P_r to this sum we obtain the required result, since P_r is orthogonal to $P_1, P_2, \ldots, P_{r-1}$ and to the non-isotropic space W. $P_1 \perp P_2 \perp \ldots \perp P_r \perp W$ is non-isotropic by Lemma (1.2.1). ∎

Proposition 1.2.8 *(Witt decomposition.) Let U be a maximal totally isotropic subspace ('m.t.i.s.' for short) where $U = (e_1, e_2, \ldots, e_r)$. Then there exist isotropic vectors $(f_1, f_2, \ldots, f_r)$ such that the (e_i, f_i) are hyperbolic pairs, $U' = (f_1, f_2, \ldots, f_r)$ is totally isotropic and*

$$E = U \oplus U' \perp G$$

where $G = (U \oplus U')^\perp$. In the orthogonal case, if $z \in G$, $Q(z) \neq 0$ if and only if $z \neq 0$.

Proof. If U is a m.t.i.s. we will have $\operatorname{rad} U = U$ and $W = 0$ in the previous proof. So we only need to establish the second part of the proposition.

If $z \neq 0$, $z \in G$ satisfies $B(z, z) = 0$ then $B(x, z) = 0$.

$\forall x \in U$, $U + (kz)$ is totally isotropic, contradicting the maximality hypothesis on U. ∎

Proposition 1.2.9 *(Witt theorem.)* *Let E be a vector space with a non-degenerate symmetric or antisymmetric bilinear form. Let F and F' be subspaces of E for which $\sigma : F \to F'$ is an isometry. Then σ extends to an isometry of E.*

Proof. If F is isotropic, let $F = U$ as in Proposition (1.2.7), where $\operatorname{rad} U \neq 0$.

If $\sigma : U \to V'$ is a isometry, it extends to an isometry from the non-isotropic space

$$P_1 \perp P_2 \perp \ldots \perp P_r \perp W$$

to V'. Indeed, using the notations of Proposition (1.2.7), if

$$\sigma(e_i) = e_i', \quad W' = \sigma(W), \quad \sigma(U) = (e_1', e_2', \ldots, e_r') \perp W' \subseteq V',$$

we can find f_i' in V' such that $(e_i', f_i') = P_i'$ is a hyperbolic pair $(i = 1, 2, \ldots, r)$ where the P_i' are orthogonal to W' (by Proposition (1.2.4)). We extend σ by requiring that $\sigma(f_i) = f_i'$. This way we can assume that F is non-isotropic.

- The symplectic case :

$$E = F \perp F^\perp, \quad E = F' \perp F'^\perp.$$

We only need to show that $F^\perp$ and $F'^\perp$ are isometric; but $F^\perp$ and $F'^\perp$ are of equal dimension, hyperbolic and non-isotropic, so they are isometric by Proposition (1.2.3).

- The orthogonal case :

  1. If $F = F'$ we can extend σ by keeping all elements of F fixed.
  2. If $\dim F = \dim F' = 1$ and $F \neq F'$: let $F = (v)$, $F' = (v')$, $v' = \sigma(v)$, $Q(v) = Q(v') \neq 0$ and (v, v') is two-dimensional.
     - If (v, v') is non-isotropic, an isometry can be found such that it extends σ and maps v to v' and v' to v; one then applies 1.
     - If (v, v') is isotropic, $B|_{(v,v')}$ has a one-dimensional radical. Let w be a basis of this radical, $v \notin (w)$,

$$v' = av + bw.$$

$Q(v') = a^2 Q(v)$, so $a = \pm 1$; taking $a = 1$ and replacing bw by w :

$$w = v' - v.$$

Setting $z = v + v'$, $(w, z) = (w) \oplus (z)$, and (w, z) takes the part of U in Proposition (1.2.7), w that of $\mathrm{rad}(U)$ and z that of W.

By Proposition (1.2.7) there exists a $y \in E$ for which $B(y, z) = 0$, $Q(y) = 0$, $B(w, y) = 1$. $(z) \oplus (y) \oplus (w)$ is non-isotropic, there exists an isometry mapping z to z, w to $-w$ and y to $-y$. $v = (z - w)/2$ is mapped to $v' = (z + w)/2$ as should be. Invoking 1 completes this part of the proof.

3. We use induction on $\dim F$. Let $F = F_1 \perp F_2$, where $\dim(F_1)$ and $\dim(F_2)$ are greater than or equal to 1. $F_2 \subseteq F_1^\perp$, so F_1 and F_2 must be non-isotropic. $\sigma(F) = \sigma(F_1) \perp \sigma(F_2)$, $\sigma_1 = \sigma|_{F_1}$, $\sigma_2 = \sigma|_{F_2}$. $\sigma(F_2)$ is orthogonal to $\sigma(F_1) = \sigma_1(F_1)$ hence

$$\sigma(F_2) \subseteq (\sigma_1(F_1))^\perp = (\sigma(F_1))^\perp.$$

σ_2 extends to an isometry $\hat{\sigma}_2$ (by the induction hypothesis) which applies $F_1^\perp$ into $(\sigma_1(F_1))^\perp$, since $F_2 \subseteq F_1^\perp$ and $\sigma(F_2) \subseteq (\sigma_1(F_1))^\perp$. $\sigma_1 \perp \hat{\sigma}_2$ (where the notation is self-explanatory) applies $F_1 \perp F_1^\perp = E$ onto itself, hence it is an isometry.

∎

Proposition 1.2.10 *Let E be non-isotropic, then all the maximal totally isotropic subspaces of E are of the same dimension r, called the* index *of E; $2r \leq \dim E$.*

Proof. If F and F' are m.t.i.s. and if $\dim F \leq \dim F'$, F can be mapped linearly and injectively into F' by an isometry which extends to an isometry σ of E to E.

$\sigma(F) \subseteq F'$, but $\sigma^{-1}(F')$ is totally isotropic, so $F = \sigma^{-1}(F')$, $F' = \sigma(F)$, $\dim F = \dim F'$. As $F \subseteq F^\perp$, if F is totally isotropic :

$$\dim F + \dim F^\perp = n, \quad \dim F^\perp \geq \dim F = r', \quad 2r' \leq n, \quad r' \leq n/2.$$

Hence if F is a m.t.i.s. $\dim F = r$ with $2r \leq n$. ∎

Remarks.

- Two totally isotropic spaces V and V' of equal dimension can be mapped to each other by an isometry (since it is sufficient to extend any linear map of V onto V' to an isometry).

- If two totally isotropic r-dimensional spaces F and F' satisfy $F \cap F' = 0$ and $F \oplus F'$ is non-isotropic, and if furthermore $Q(z) \neq 0$, $\forall z \neq 0$ in $(F \oplus F')^\perp$ (we are in the orthogonal case), r is the index of E.

If F would not be a m.t.i.s. there would exist a totally isotropic $F_1 \supseteq F$. Then a $x_1 \in F_1$ could be found for which, if $x_1 = y + z$,

$$y \in F', \quad z \in (F \oplus F')^\perp, \quad x_1 \neq 0$$

but $B(y + z, y + z) = Q(z) = 0$, so $z = 0$. As $y \in F_1$, y is orthogonal to all elements of $F \subseteq F_1$. Furthermore $y \in F'$ implies that y is orthogonal to all elements of F', and $F \oplus F'$ would be isotropic.

1.3 Special properties of the orthogonal geometry.

1.3.1

We will assume that E is non-isotropic. Our goal is to prove that any isometry can be factored as a product of orthogonal symmetries. We will write $Q(x)$ for $B(x,x)$, so

$$2B(x,y) = Q(x+y) - Q(x) - Q(y).$$

Q is the quadratic form associated to B.

Definition 1.3.1 *Let H be a non-isotropic subspace of E. The orthogonal symmetry u_H with respect to H is defined by*

$$u_H(x + y) = x - y, \quad x \in H^\perp, \quad y \in H.$$

Clearly u_H is an isometry and

$$\det u_H = (-1)^{\dim H}.$$

Lemma 1.3.1 *If a is a non-isotropic vector, the symmetry u_a with respect to $(a)^\perp$ satisfies :*

$$u_a(x) = x - \frac{2B(x,a)a}{Q(a)}. \tag{1}$$

Proof. This is a consequence of $u_a(a) = -a$ and $u_a(x) = x$ if $x \in (a)^\perp$. ∎

Lemma 1.3.2 *If a and b are orthogonal non-isotropic vectors, the symmetries u_a and u_b commute.*

Proof.

$$
\begin{aligned}
u_b(u_a(x)) &= u_a(x) - \frac{2B(u_a(x), b)}{Q(b)} b \\
&= x - \frac{2B(x,a)a}{Q(a)} - \frac{2B(x,b)b}{Q(b)}.
\end{aligned}
$$

∎

Lemma 1.3.3 *If H is a non-isotropic subspace of E and if $(x_1, x_2, \ldots, x_h)$ is an orthogonal basis of non-isotropic vectors in H, the symmetry u_H satisfies :*

$$u_H(x) = x - \sum_{i=1}^{h} \frac{2B(x, x_i)x_i}{Q(x_i)}.$$

Proof. If $x \in H^\perp$, $u_H(x) = x$, whereas if $x \in H$, say $x = x_i$, $u_H(x_i) = -x_i$, so $u_H(x) = -x$.   ∎

Theorem 1.3.1 *(Fundamental Theorem, E. Cartan, J. Dieudonné.) Every isometry σ of E is the product of symmetries with respect to non-isotropic hyperplanes. If $\dim E = n$, any $\sigma \neq \mathrm{Id}$ can be factored as a product of at most n symmetries.*

Proof. To prove the first part is relatively easy, since it holds for $n = 1$.

1. Assume that $\sigma(x) = x$, x being a non-isotropic vector. Then

$$E = (x) \oplus (x)^\perp$$

   and $(x)^\perp$ is non-isotropic and invariant under σ. Now use induction, assuming that the restriction of σ to $(x)^\perp$ is a product of k symmetries with respect to non-isotropic $(n-2)$-dimensional spaces $F_1, F_2, \ldots, F_k$. Then u is the product of the symmetries with respect to the non-isotropic hyperplanes

$$(x) \oplus F_1, (x) \oplus F_2, \ldots, (x) \oplus F_k.$$

2. Assume that $\sigma(x) = -x$, x being a non-isotropic vector. If s is the symmetry with respect to $(x)^\perp$ we can set $v = s\sigma$ and apply the previous reasoning.

3. Assume that for all non-isotropic non-zero vectors x, $\sigma(x) \neq x$ and $\sigma(x) \neq -x$. Setting $y = \sigma(x)$, $Q(y) = Q(x)$ and $x + y$ and $x - y$ cannot both be isotropic, since $Q(x+y) - Q(x-y) = 0$ leads to $4Q(x) = 0$, i.e. $Q(x) = 0$ (if $\mathbf{K} = \mathbf{R}$ or $\mathbf{K} = \mathbf{C}$).

   - If $x - y$ is non-isotropic and s is the symmetry with respect to $(x-y)^\perp$, x and y are mapped to each other by s.

$$u_{(x-y)}(x) = x - \frac{2B(x, x-y)}{Q(x-y)}(x-y) = y.$$

   Now we can set $v = s\sigma$ to obtain the first case.

   - If $x + y$ is non-isotropic and s' is the symmetry with respect to $(x+y)^\perp$, $(-x)$ and y are mapped to each other by s'. We then set $v = s'\sigma$, $v(x) = -x$ and obtain the second case.

This completes the first part of the proof.

Now we will give a more complete proof (which holds for any field $\mathbf{K}$ not of characteristic 2.)

1. 1 need not be modified : an $x \in E$ exists for which $Q(x) \neq 0$ and $\sigma(x) = x$, we assume $k \leq n - 1$.

2. For some x, $Q(x) \neq 0$ and $\sigma(x) - x$ is non-isotropic. Then let $H = (\sigma(x) - x)^{\perp}$ and let u be the symmetry with respect to H. $B(\sigma(x) + x, \sigma(x) - x) = 0$ implies that $\sigma(x) + x \in H$. $u(\sigma(x) + x) = \sigma(x) + x$, $u(\sigma(x) - x) = -\sigma(x) + x$ so

$$(u \circ \sigma)(x) = x.$$

$u \circ \sigma$ satisfies the assumption in 1, choosing $k \leq n - 1$, σ is a product of at most n symmetries.

3. If $n = 2$. 1 and 2 allow us to assume that E has non-zero isotropic elements (for if all vectors in E are non-isotropic, either 1 or 2 applies). We may assume that $E = (x, y)$ where (x, y) is a hyperbolic pair (cf. (1.2.2)).

   - $\sigma(x) = ay$, $\sigma(x + ay) = ay + x$ is non-isotropic, apply 1.
   - $\sigma(x) = ax$, $\sigma(y) = a^{-1}y$ (we can assume that $a \neq 1$). Then for $z = x + y$, $\sigma(z) - z = (a - 1)x + (a^{-1} - 1)y$ is non-isotropic and 2 applies.

4. $n \geq 3$. Given 1 and 2 we may assume that no non-isotropic vector is fixed by σ and that $\sigma(x) - x$ is isotropic, $\forall x \in E$ for which $Q(x) \neq 0$. There are three steps in the proof :

   - First step : we can assume that for all non-isotropic elements x of E, $\sigma(x) - x$ is isotropic and prove that for all isotropic $x \in E$, $\sigma(x) - x$ is isotropic, so that $(\sigma + \mathrm{Id})E$ is a totally isotropic subspace $W \subseteq E$.
     Indeed, let x satisfy $Q(x) = 0$, then $(x)^{\perp}$ is at least two-dimensional since $n \leq 3$ and $(x)^{\perp}$ contains a non-isotropic vector y (otherwise $(x)^{\perp}$ would be totally isotropic, and as $\mathrm{rad}(x) = \mathrm{rad}(x)^{\perp} = (x)$, $\mathrm{rad}(x)^{\perp}$ would contain $x^{\perp}$). $Q(y) \neq 0$ and if $\epsilon = \pm 1$, $\sigma(y) - y$ must be isotropic (by assumption) and so must be $u = \sigma(y + \epsilon x) - (y + \epsilon x) = (\sigma(y) - y) + \epsilon(\sigma(x) - x)$. Then $Q(u) = 2\epsilon B(\sigma(y) - y, \sigma(x) - x) + \epsilon^2 Q(\sigma(x) - x) = 0$. Putting first $\epsilon = 1$, then $\epsilon = -1$ and adding the results, one obtains $Q(\sigma(x) - x) = 0$.

   - Second step : if $v' \in W^{\perp}$ then $v' = \sigma(v')$ since, W being totally isotropic : $\forall v \in E$,

     $$0 = B(\sigma(v) - v, \sigma(v') - v') = B(\sigma(v), \sigma(v')) - B(v, \sigma(v')) - B(\sigma(v) - v, v')$$

     and $\sigma(v) - v \in W$, $v' \in W^{\perp}$ imply that the last term vanishes.

     $$B(\sigma(v), \sigma(v')) = B(v, v')$$

     leads to $B(v, v' - \sigma(v')) = 0$, $v' - \sigma(v') \in \mathrm{rad}\, E = 0$, $v' = \sigma(v')$.

   - Third step : having assumed that no non-isotropic vector is invariant under σ, all vectors of $W^{\perp}$ must be isotropic and W is totally isotropic. Now $\dim W + \dim W^{\perp} = n$, $\dim W \leq n/2$ and $\dim W^{\perp} \leq n/2$ imply

     $$\dim W^{\perp} = \dim W = r$$

and $n = 2r$. Hence $E = W \oplus W^\perp$ is hyperbolic : $E = H_{2r}$. σ fixes all vectors in $W^\perp$, hence by Proposition (1.2.4), σ is a rotation. Therefore case 4 cannot occur if σ is a reversion, and the proof is complete for a reversion. If σ is a rotation in H_{2r}, and s is any symmetry, $s\sigma$ is a reversion. $s\sigma = s_1 s_2 \ldots s_k$, $k \leq 2r$ and k is odd, so $\sigma = s s_1 s_2 \ldots s_k$ with at most $2r = n$ symmetries.

∎

1.3.2 The real case.

Definition 1.3.2 Q is said to be positive definite if $Q(x) > 0$ for all non-zero $x \in E$. Q is said to be positive if $Q(x) \geq 0$ for all non-zero $x \in E$.

The definition of *negative definite* and *negative* are similar. Note that Q is positive definite if and only if Q is positive and B is non-degenerate. Indeed, if Q is positive, let $y \in E$ satisfy $Q(y) = 0$, then $\forall x \in E$, $\forall a \in \mathbf{R}$, $Q(x + ay) = Q(x) + 2aB(x, y)$; if $B(x, y) \neq 0$ the second term could have any real value, which is impossible, so $B(x, y) = 0$ and rad E is the set of $y \in E$ such that $Q(y) = 0$.

Theorem 1.3.2 *(Sylvester Theorem.)* If E is a real n-dimensional space and if Q is non-degenerate, a basis $(e_1, e_2, \ldots, e_n)$ *(said to be* orthonormal*)* can be found such that :

$$Q(x) = (x^1)^2 + \ldots + (x^r)^2 - (x^{r+1})^2 - \ldots - (x^n)^2.$$

The index of Q is r if $r \leq n - r$, otherwise $n - r$; r is independent of the choice of the orthonormal basis.

Proof. By Proposition (1.2.3) and since $\mathbf{K} = \mathbf{R}$, $Q(x)$ can be written as in the theorem. Suppose, for instance, that $r \leq n - r$, then taking

$$F = (e_1 + e_n, e_2 + e_{n-1}, \ldots, e_r + e_{n-r+1}),$$
$$F' = (e_1 - e_n, e_2 - e_{n-1}, \ldots, e_r - e_{n-r+1})$$
$$\text{and} \quad G = (e_{r+1}, \ldots, e_{n-r}),$$

a Witt decomposition is obtained and by remark 2 after Proposition (1.2.10), r is the index of Q. Hence r does not depend on the choice of the orthonormal basis. ∎

The Sylvester Theorem is thus seen to be an easy consequence of the Witt decomposition theorem.

1.3.3 The orthogonal groups.

If E is non-isotropic, i.e. B is non-degenerate, the isometry group of E is called the *orthogonal group* $O(E, B)$.

This group is most often denoted by $O(Q)$, where Q is the associated quadratic form.

If $\dim E = n$, we also write $O(n, \mathbf{K})$.

In the real case, if the Sylvester decomposition has p plus signs and q minus signs (or *signature* (p, q) for short), we will write $O(p, q, \mathbf{R})$ or $O(p, q)$.

If $q = 0$, $p = n$ we write $O(n, \mathbf{R})$ or $O(n)$.

$SO(Q)$ stands for the rotation group. The notations $SO(p, q)$, $SO(n)$ should be clear.

1.4 Special properties of the symplectic geometry.

Recall that any symplectic space E with a non-degenerate antisymmetric bilinear form B is a hyperbolic space :

$$E = H_{2r} = P_1 \perp P_2 \perp \ldots \perp P_r = (x_1, y_1) \perp (x_2, y_2) \perp \ldots \perp (x_r, y_r);$$

the hyperbolic planes P_i contain the hyperbolic pairs (x_i, y_i) and these $2r$ vectors form a basis of E which will be called *symplectic*.

Clearly the discriminant of B relative to this basis is 1.

The isometry group of E is called the *symplectic group*, denoted by $Sp(2r, \mathbf{K})$.

1.4.1 Symplectic transvections.

Let $a \in E$, $a \neq 0$. We will try to find a $\sigma \in Sp(2r, \mathbf{K})$ such that

$$\sigma(x) = x + \varphi(x)a, \quad \varphi(x) \in \mathbf{K}.$$

If such a σ exists, $B(\sigma(x), \sigma(y)) = B(x, y)$ so

$$\varphi(x)B(a, y) = \varphi(y)B(a, x).$$

Fix a y_0 for which $B(a, y_0) \neq 0$, then

$$\varphi(x) = \frac{\varphi(y_0)}{B(a, y_0)} B(a, x) = cB(a, x), \quad c \in \mathbf{K}$$

so

$$\sigma(x) = x + cB(a, x)a \tag{1}$$

where c is constant. It is easily verified that every homomorphism of the form (1) is an isomorphism (σ^{-1} being obtained if we replace c by $-c$) and that every such σ preserves B. σ is called a *symplectic transvection* of 'direction' a.

If $c = 0$, $\sigma = \mathrm{Id}$.

If $c \neq 0$, σ fixes x if and only if $x \in (a)^{\perp}$, which is analogous to the orthogonal case, where σ fixes the points of some hyperplane. Conversely if a symplectic isometry σ fixes the points of a hyperplane H, then σ is a symplectic transvection. Indeed, $\forall x \in E, \forall y \in H$:

$$\begin{aligned} B(\sigma(x) - x, y) &= B(\sigma(x), y) - B(x, y) = B(\sigma(x), y) - B(\sigma(x), \sigma(y)) \\ &= B(\sigma(x), y - \sigma(y)) = 0. \end{aligned}$$

Then $\sigma(x) - x \in H^{\perp}$, $H^{\perp}$ is a line (a) and $\sigma(x) = x + \varphi(x)a$.

Remark. One can immediately verify that the symplectic transvections of direction a make up a commutative group isomorphic to the additive group of **K**.

If σ is written σ_a then $\sigma_c \sigma_d = \sigma_{c+d}$ and $\sigma_c = \mathrm{Id}$ if and only if $c = 0$.

1.4.2 Factorization.

We will now prove a factorization property in terms of symplectic transvections, for elements of the symplectic group.

Lemma 1.4.1 *A non-zero vector x can be mapped to another non-zero vector y by a product of at most two symplectic transvections.*

Proof. σ being a symplectic transvection, if $\sigma(x) = y$, $x + cB(a, x)a - x = \sigma(x) - x = cB(a, x)a = y - x$, hence $y - x$ is collinear with a.

- If $y = x$, take $\sigma = \mathrm{Id}$.

- If $y \neq x$ and $B(x, y) \neq 0$, as

$$\sigma(X) = X + cB(x - y, X)(x - y)$$

  and

$$\sigma(x) = x - cB(y, x)(x - y),$$

  the choice of c for which $cB(x, y) = -1$ ensures that $\sigma(x) = y$.

  This reasoning fails if $B(x, y) = 0$.

  Then we look for a z such that $B(x, z) \neq 0$ and $B(y, z) \neq 0$, which we can find :

    – if $x^{\perp} = y^{\perp}$, we take a z outside of $x^{\perp}$, then x can be mapped to z, then z to y by two transvections.

    – If $x^{\perp} \neq y^{\perp}$, let $u \in x^{\perp}$, $u \notin y^{\perp}$ and $v \in y^{\perp}$, $v \notin x^{\perp}$, $w = u + v$, $B(x, w) = B(x, v) \neq 0$ and $B(y, w) = B(y, v) \neq 0$. We can map x to w, and then w on y, by two transvections.

∎

Lemma 1.4.2 *Given two hyperbolic pairs* (x_1, y_1) *and* (x_2, y_2) *there exists a product of at most four transvections mapping* (x_1, y_1) *to* (x_2, y_2).

Proof. Using at most two transvections x_1 can be mapped to x_2 and the pair (x_1, y_1) will be mapped to a hyperbolic pair (x_2, y_3). Now it is sufficient to show that (x_2, y_3) can be mapped to (x_2, y_2) using at most two transvections. Indeed, if (x, y) and (x, y') are two hyperbolic pairs :

- If $B(y, y') \neq 0$ we map y to y' by a transvection of direction $y - y'$, and since $B(x, y' - y) = 0$, x is fixed.

- If $B(y, y') = 0$, $(x, x + y)$ is a hyperbolic pair. $B(y, x + y) = -1$ so (x, y) can be mapped to $(x, x + y)$ and as $B(x + y, y') = B(x, y') \neq 0$, $(x, x + y)$ can be mapped to (x, y').

∎

Theorem 1.4.1 *(Factorization Theorem.) Every symplectic transformation σ is the product of symplectic transvections; if $n = 2r = \dim E$, σ can be factored as a product of at most $4r$ transvections. σ is always a rotation (det $\sigma = 1$).*

Proof. Let U be a non-isotropic subspace of E,

$$E = U \perp U^\perp.$$

If τ is a transvection of $U^\perp$, $\mathrm{Id}_U \perp \tau$ is a transvection of E, so if $(x, y), (x', y')$ are two hyperbolic pairs of $U^\perp$, (x, y) can be applied to (x', y') by at most four transvections of E, each of which fixes all vectors in U.

Let G be the subgroup of $\mathrm{Sp}(2r, \mathbf{K})$ generated by all transvections.

If $\sigma \in \mathrm{Sp}(2r, \mathbf{K})$ and if $(x_1, y_1), (x_2, y_2), \ldots, (x_r, y_r)$ is a symplectic basis of E, we set

$$\sigma(x_j) = x'_j, \quad \sigma(y_j) = y'_j, \quad j = 1, 2, \ldots, r.$$

If there exists a $\tau_i \in G$ mapping the first i symplectic pairs (x_α, y_α) on the (x'_α, y'_α), τ_i transforms the given basis into

$$(x'_1, y'_1), (x'_2, y'_2), \ldots (x'_i, y'_i), (x''_{i+1}, y''_{i+1}), \ldots, (x''_r, y''_r),$$

the $2i$ first elements generating U and the other elements generating $U^\perp$. Choose a $\tau \in G$ fixing all vectors of U and mapping (x''_{i+1}, y''_{i+1}) to (x'_{i+1}, y'_{i+1}), then $\tau\tau_i$ applies the $i + 1$ first pairs (x_α, y_α) on the (x'_α, y'_α), $\alpha = 1, \ldots, r + 1$. Repeating this procedure we construct an element of G which maps the (x_i, y_i) to the (x'_i, y'_i), $i = 1, 2, \ldots, r$. So $G = \mathrm{Sp}(2r, \mathbf{K})$.

At most $4r$ transvections were used in this construction.

Finally, $\sigma_c = \sigma^2_{c/2}$, so det $\sigma = 1$.

∎

Proposition 1.4.1 *The center of* $\mathrm{Sp}(2r, \mathbf{K})$ *is* $\pm \mathrm{Id}_E$.

Proof. If τ is a non-identity transvection of direction a, $\tau(x) - x \in (a)$, $\forall x \in V$.

If $\sigma \in \mathrm{Sp}(2r, \mathbf{K})$, then $\sigma\tau\sigma^{-1}(x) - x = \sigma(\tau\sigma^{-1}(x) - \sigma^{-1}(x)) \in (\sigma(a))$, so $\sigma\tau\sigma^{-1}$ is a transvection of direction $\sigma(a)$. If σ is in the center, $\sigma\tau\sigma^{-1} = \tau$ for any transvection τ, so $(\sigma(a)) = (a)$; σ fixing all lines in E is Id_E (by Proposition (1.2.6)).   ∎

1.5   Selected references.

- E. Artin, *Geometric algebra*, Interscience, New-York, 1957.

- C. Chevalley, *The algebraic theory of spinors*, Columbia University Press, 1954.

- J. Dieudonné, *Sur les groupes classiques*, Act. Sc. et Ind. 1040, troisième éd., Hermann, Paris, 1967.

- S. Lang, *Algebra*, Addison-Wesley, Reading, 1965.

Chapter 2

TENSOR ALGEBRAS, EXTERIOR ALGEBRAS AND SYMMETRIC ALGEBRAS.

2.1 Background on tensor algebras.

We will consider real or complex vector spaces ($\mathbf{K} = \mathbf{R}$ or $\mathbf{C}$), $E_1, E_2, \ldots, E_r$ of dimensions $n_1, n_2, \ldots, n_p$. In this first part, we will only recall some results, omitting all proofs, only to specify our definitions, our notations and our methods. We will assume that the reader is already somewhat familiar with tensor algebras.

2.1.1

Definition 2.1.1 *A p-linear mapping of $E_1 \times E_2 \times \ldots \times E_p$ to $\mathbf{K}$, i.e. a mapping which is linear in any of its p arguments, when the $p - 1$ other ones are fixed, is called a p-linear form.*

$$\varphi_p : (x_1, x_2, \ldots, x_p) \in E_1 \times E_2 \times \ldots E_p \to \varphi_p(x_1, x_2, \ldots, x_p) \in \mathbf{K}.$$

The set of linear p-forms is a vector space on $\mathbf{K}$ and will be called the *space of covariant tensors of order p.*

2.1.2 The tensor product of two multilinear forms.

If φ_p and ψ_p are multilinear forms of order p and q on $E_1 \times \ldots \times E_p$ and $F_1 \times \ldots \times F_q$, we define

$$(\varphi_p \otimes \psi_q)(x_1, x_2, \ldots, x_p, y_1, y_2, \ldots, y_q) = \varphi_p(x_1, x_2, \ldots, x_p)\psi_q(y_1, y_2, \ldots, y_q) \quad (1)$$

One can then verify that

$$\begin{cases} \alpha(\varphi_p \otimes \psi_q) = (\alpha\varphi_p) \otimes \psi_q = \varphi_p \otimes (\alpha\psi_q), \quad \forall \alpha \in \mathbf{K} \\ (\varphi_p + \varphi'_p) \otimes \psi_q = (\varphi_p \otimes \psi_q) + (\varphi'_p \otimes \psi_q) \\ \psi_q \otimes (\varphi_p + \varphi'_p) = (\psi_q \otimes \varphi_p) + (\psi_q \otimes \varphi'_p) \end{cases} \quad (2)$$

and that the tensor product is non-commutative.

21

The space of linear p-forms contains the tensor product of linear forms on $E_1, E_2,$ $\ldots, E_p$, which explains the notation

$$E_1^* \otimes E_2^* \otimes \ldots \otimes E_p^*$$

or $\otimes^p E^*$ if $E_1 = E_2 = \ldots = E_p$.

2.1.3 Dimension and special bases for $E_1^* \otimes E_2^* \otimes \ldots \otimes E_p^*$.

Let (e_{i_1}), $i_1 = 1, \ldots, n_1$ be a basis of E_i, let (θ^{i_1}) be its dual basis, and let similarly $(e_{i_2}), \ldots, (e_{i_p})$ and $(\theta^{i_2}), \ldots, (\theta^{i_p})$ be the bases of $E_2, \ldots, E_p$ and their duals. Define

$$\theta^{i_1 i_2 \ldots i_p} = \theta^{i_1} \otimes \theta^{i_2} \otimes \ldots \otimes \theta^{i_p} \tag{3}$$

then

$$\theta^{i_1}(e_{j_1}) = \delta^{i_1}_{j_1},$$

where $\delta^{i_1}_{j_1}$ is the Kronecker symbol; this implies that

$$\theta^{i_1 i_2 \ldots i_p}(e_{j_1}, e_{j_2}, \ldots, e_{j_p}) = \delta^{i_1 i_2 \ldots i_p}_{j_1 j_2 \ldots j_p} \tag{4}$$

where the right hand side is 1 if both sequences are equal, 0 otherwise.

$$\theta^{i_1 i_2 \ldots i_p}(x_1, x_2, \ldots, x_p) = x_1^{i_1} x_2^{i_2} \ldots x_p^{i_p} \tag{5}$$

if the $x_1^{i_1}, \ldots, x_p^{i_p}$ are the components of $(x_1, \ldots, x_p)$. The elements $\theta^{i_1 i_2 \ldots i_p}$ form a basis of $E_1^* \otimes E_2^* \otimes \ldots \otimes E_p^*$ which therefore is a vector space of dimension $n_1 n_2 \ldots n_p$. If we set $\theta^{i_1} = A^{i_1}_{j_1'} \theta^{j_1'}$, $e_{i_1} = A^{j_1'}_{i_1} e_{j_1'}$, we have the change of basis formula :

$$\begin{cases} \theta^{i_1 i_2 \ldots i_p} = A^{i_1}_{j_1'} A^{i_2}_{j_2'} \ldots A^{i_p}_{j_p'} \theta^{j_1' j_2' \ldots j_p'} \\ \theta^{j_1' j_2' \ldots j_p'} = A^{j_1'}_{i_1} A^{j_2'}_{i_2} \ldots A^{j_p'}_{i_p} \theta^{i_1 i_2 \ldots i_p}. \end{cases} \tag{6}$$

2.1.4 Contravariant and mixed tensors.

Identifying E^{**} with E, the set of p-linear forms on $E_1^* \times E_2^* \times \ldots \times E_p^*$ will be written as

$$E_1 \otimes E_2 \otimes \ldots \otimes E_p$$

or $\otimes^p E$ if all $E_i = E$. This product is called the vector space of *contravariant tensors*. The $e_{i_1} \otimes e_{i_2} \otimes \ldots \otimes e_{i_p}$ form a basis for it, denoted by $(e_{i_1 i_2 \ldots i_p})$. Then

$$e_{i_1 i_2 \ldots i_p}(\varphi^1, \varphi^2, \ldots, \varphi^p) = \varphi^1_{i_1} \varphi^2_{i_2} \ldots \varphi^p_{i_p},$$

if the $\varphi^1_{i_1}, \ldots, \varphi^p_{i_p}$ are the components of the forms $\varphi^1, \ldots, \varphi^p$. Under a coordinate change we will have

$$e_{j_1' j_2' \ldots j_p'} = A^{i_1}_{j_1'} A^{i_2}_{j_2'} \ldots A^{i_p}_{j_p'} e_{i_1 i_2 \ldots i_p}.$$

In general, for the sequence of spaces

$$E_1^*, \ldots, E_p^*, E_{p+1}, \ldots, E_{p+q}$$

the space of $(p+q)$-linear forms on $E_1^* \times \ldots \times E_p^* \times E_{p+1} \times \ldots \times E_{p+q}$ is denoted by

$$E_1 \otimes \ldots \otimes E_p \otimes E_{p+1}^* \otimes \ldots \otimes E_{p+q}^*,$$

or $\otimes^p E \otimes^q E^*$ if all $E_i = E$. A basis for this tensor product is given by $(e_{i_1 \ldots i_p} \otimes \theta^{i_{p+1} \ldots i_{p+q}})$. Under a coordinate change it transforms according to

$$e_{j_1' \ldots j_p'} \otimes \theta^{j_{p+1}' \ldots j_{p+q}'} = A_{j_1'}^{i_1} \ldots A_{j_p'}^{i_p} A_{i_{p+1}}^{j_{p+1}'} \ldots A_{i_{p+q}}^{j_{p+q}'} e_{i_1 \ldots i_p} \otimes \theta^{i_{p+1} \ldots i_{p+q}}.$$

A mixed tensor in $E \otimes E^*$ or $E^* \otimes E$ can be linearly identified with an endomorphism of E :

$$\text{End } E = E \otimes E^* = E^* \otimes E.$$

2.1.5 The contracted product.

Let $x \in E$ and let $\varphi_p \in \otimes^p E^*$ where $p > 1$. To the element φ_p we can associate the element $C_x^1(\varphi_p) \in \otimes^{p-1} E^*$ for which

$$C_x^1(x_1, x_2, \ldots, x_{p-1}) = \varphi_p(x, x_1, \ldots, x_{p-1}).$$

C_x^1 is linear. If the components of φ_p are $\lambda_{ii_1 \ldots i_{p-1}}$, the components of $C_x^1(\varphi_p)$ are $x^i \lambda_{ii_1 \ldots i_{p-1}}$. The definition of C_x^2, C_x^3 etc. are similar.

If $p = 1$, we associate the scalar $x^i \lambda_i$ to φ_1. Generally, if we take an element of $\otimes^p E \otimes^q E^*$, choose a contravariant and a covariant index and sum the components relative to these two indices, a tensor of order $(p-1)$, $(q-1)$ is obtained. If the tensor T has components t_j^i, the trace of the endomorphism to which it is identified can be obtained from T by a contraction.

2.1.6 The tensor algebra of a finite-dimensional vector space.

Define the notations $\otimes^0 E = \mathbf{K}$ and $\otimes E = \oplus_{p=0}^\infty \otimes^p (E)$, which means that every element $z \in \otimes E$ is the direct sum of its components $z_p \in \otimes^p E$, so

$$z = \sum_{p=0}^\infty z_p$$

uniquely, where only a finite number of terms do not vanish. If $z' = \sum_{q=0}^\infty z_q'$ we define

$$z \otimes z' = \sum_{p,q=0}^\infty (z_p \otimes z_q).$$

Similar definitions hold for $\otimes E^*$ and for the algebra of mixed tensors $\mathcal{T}(E) = \oplus_{p,q=0}^\infty \otimes^p E \otimes^q E^*$.

2.1.7   The universal property.

Let f_p be a p-linear mapping of $E_1 \times \ldots E_p$ into a vector space F. Then there exists a unique linear mapping g of $\otimes^p E_i$ to F such that

$$g(x_1 \otimes x_2 \otimes \ldots \otimes x_p) = f_p(x_1, x_2, \ldots, x_p), \quad \forall (x_1, x_2, \ldots, x_p) \in E_1 \times E_2 \times \ldots \times E_p$$

and $f_p \to g$ is a vector space isomorphism.

Proof. Such a g must satisfy

$$g(e_{i_1} \otimes e_{i_2} \otimes \ldots \otimes e_{i_p}) = f_p(e_{i_1}, e_{i_2}, \ldots, e_{i_p})$$

which determines it uniquely, by linearity, in $E_1 \otimes E_2 \otimes \ldots \otimes E_p$. Then g clearly satisfies the requirements, $f_p = g \circ \theta_p$ where $\theta_p(x_1, x_2, \ldots, x_p) = x_1 \otimes x_2 \otimes \ldots \otimes x_p$. ∎

In particular, if $F = \mathbf{K}$ then $f_p \in E_1^* \otimes \ldots \otimes E_p^*$ so $g \in (E_1 \otimes \ldots \otimes E_p)^*$ and we can identify both spaces. In this identification a $g^{i_1 i_2 \ldots i_p}$ corresponds to $\theta^{i_1 i_2 \ldots i_p}$,

$$g^{i_1 i_2 \ldots i_p}(e_{j_1} \otimes e_{j_2} \otimes \ldots \otimes e_{j_p}) = \delta_{j_1 j_2 \ldots j_p}^{i_1 i_2 \ldots i_p},$$

so the $g^{i_1 i_2 \ldots i_p}$ are the elements of the dual basis of $e_{j_1 j_2 \ldots j_p}$, which leads us to write

$$\langle e_{j_1 j_2 \ldots j_p}, \theta^{i_1 i_2 \ldots i_p} \rangle = \delta_{j_1 j_2 \ldots j_p}^{i_1 i_2 \ldots i_p}$$

and $\langle x_1 \otimes x_2 \otimes \ldots \otimes x_p, \varphi^1 \otimes \varphi^2 \otimes \ldots \otimes \varphi^p \rangle = \prod_{i=1}^{p} \langle \varphi^i, x_i \rangle$, expressing the duality between $\otimes_{i=0}^{p} E_i^*$ and $\otimes_{i=0}^{p} E_i$.

Now consider homomorphisms $u_i : E_i \to F_i$, $i = 1, 2$ and the bilinear mapping

$$E_1 \times E_2 \to F_1 \otimes F_2 : (x_1, x_2) \to u_1(x_1) \otimes e_2(x_2).$$

By the universal property there exists a unique linear mapping $u_1 \otimes u_2$ from $E_1 \otimes E_2$ to $F_1 \otimes F_2$ such that

$$(u_1 \otimes u_2)(x_1 \otimes x_2) = u_1(x_1) \otimes u_2(x_2).$$

This can be generalized to more than two spaces.

2.2   Essential properties of exterior algebras.

E is an n-dimensional vector space over $\mathbf{K}$ and E^* is its dual.

2.2.1   The space of exterior forms of order p.

Definition 2.2.1 A p-form $\varphi_p \in \otimes^p E^*$, $(p > 1)$ is called antisymmetric if for all $x_1, x_2, \ldots, x_p \in E$,

$$\varphi_p(x_1, x_2, \ldots, x_i, \ldots, x_j, \ldots, x_p) = -\varphi_p(x_1, \ldots, x_j, \ldots, x_i, \ldots, x_p). \tag{1}$$

If the p-form φ_p is zero whenever $x_i = x_j$, it is said to be *alternating*; it is obvious that an alternating form is antisymmetric (just take $x_i = x_j = \xi_i + \eta_i$ in (1)).

If α is a permutation of $(1, 2, \ldots, p)$, written $\alpha = \begin{pmatrix} 1 & 2 & \ldots & p \\ i_1 & i_2 & \ldots & i_p \end{pmatrix}$ of signature

$$\epsilon_\alpha = \epsilon_{i_1 i_2 \ldots i_p}^{1\ 2\ \ldots p} = \epsilon_{1\ 2\ \ldots p}^{i_1 i_2 \ldots i_p},$$

we deduce from (1) that

$$\varphi_p(x_{i_1}, x_{i_2}, \ldots, x_{i_p}) = \epsilon_\alpha \varphi_p(x_1, x_2, \ldots, x_p). \tag{2}$$

From (1) it also follows that φ_p vanishes if $x_1, x_2, \ldots, x_p$ are linearly dependent.

Definition 2.2.2 *The antisymmetrized p-form φ_p of the tensor $T_p \in \otimes^p E^*$, $p > 1$, is defined by*

$$\varphi_p(x_1, x_2, \ldots, x_p) = \frac{1}{p!} \sum \epsilon_{1\ 2\ \ldots p}^{k_1 k_2 \ldots k_p} T_p(x_{k_1}, x_{k_2}, \ldots, x_{k_p}) \tag{3}$$

for all $x_1, x_2, \ldots, x_p \in E$, the sum extending over the $p!$ terms corresponding to the $p!$ permutations of the symmetric group S_p.

(It will be more convenient to define $\epsilon_{1\ 2\ \ldots p}^{k_1 k_2 \ldots k_p} = 0$ whenever the index sequences are different and to sum independently over each index, from 1 to n).

$AT_p = \varphi_p$ is obviously antisymmetric and $AAT_p = AT_p$ (i.e. $A^2 = A$). $AT_p = T_p$ if T_p is antisymmetric. A can be extended by linearity. Then the antisymmetrization A is linear and maps $\otimes^p E^*$ into a subspace of $\otimes^p E^*$, which will be written $\wedge^p(E^*)$, the space of antisymmetric p-forms.

We also define the notations $\wedge^1(E^*) = E^*$ and $\wedge^0(E^*) = \mathbf{K}$.

Proposition 2.2.1 $\dim \wedge^p(E^*) = C_n^p$.

Proof. Any $\varphi_p \in \otimes^p E^*$ is of the form

$$\varphi_p = \alpha_{k_1 k_2 \ldots k_p} \theta^{k_1 k_2 \ldots k_p}$$

where

$$\alpha_{k_1 k_2 \ldots k_p} = \varphi_p(e_{k_1}, e_{k_2}, \ldots, e_{k_p}) \quad 1 \leq k_i \leq n.$$

If $k_i = k_j$, the corresponding component will vanish, so there are at most A_n^p non-zero terms left. Let $\chi_1, \chi_2, \ldots, \chi_p$ stand for the naturally re-ordered sequence $(k_1, k_2, \ldots, k_p)$, then

$$\alpha_{k_1 k_2 \ldots k_p} = \epsilon_{k_1 k_2 \ldots k_p}^{\chi_1 \chi_2 \ldots \chi_p} \alpha_{\chi_1 \chi_2 \ldots \chi_p}$$

(where there is no implicit summation). Grouping the $p!$ terms including any permutation of $(\chi_1, \chi_2, \ldots, \chi_p)$, we obtain :

$$\theta^{k_1 k_2 \ldots k_p} \epsilon_{k_1 k_2 \ldots k_p}^{\chi_1 \chi_2 \ldots \chi_p} \alpha_{\chi_1 \chi_2 \ldots \chi_p}$$

(where one should sum over the $k_1, k_2, \ldots, k_p$ but not over $\chi_1, \chi_2, \ldots, \chi_p$). The $\theta^{k_1 k_2 \ldots k_p} \epsilon^{\chi_1 \chi_2 \ldots \chi_p}_{k_1 k_2 \ldots k_p}$ are antisymmetric. These C_n^p elements are linearly independent, for any vanishing linear combination of them would lift to the $\theta^{k_1 k_2 \ldots k_p}$, which is impossible. ∎

The $\alpha_{\chi_1 \chi_2 \ldots \chi_p}$ where $\chi_1 < \chi_2 < \ldots < \chi_p$ will be called the *strict components* of φ_p to distinguish them from the antisymmetric tensor components $\alpha_{k_1 k_2 \ldots k_p}$.

2.2.2 The graded algebra $\wedge(E^*)$.

Now we define

$$\wedge (E^*) = \oplus_{p=0}^n \wedge^p (E^*). \tag{4}$$

Note that, by a remark of (2.2.1), $\wedge^p(E) \equiv 0$ if $p > n$. The dimension of $\wedge(E^*)$ must then be

$$\sum_{p=0}^n C_n^p = 2^n.$$

Lemma 2.2.1 *The set of all $\theta \in \otimes(E^*)$ such that $A(\theta) = 0$ is a two-sided ideal of $\otimes(E^*)$.*

Proof. Let $J(E^*)$ be the set of these elements. Then $J(E^*)$ obviously is a vector space and contains all θ for which

$$\theta(x_1, \ldots, x_k, \ldots, x_l, \ldots, x_p) = \theta(x_1, \ldots, x_l, \ldots, x_k, \ldots, x_p).$$

The lemma will be proved if it can be shown that

$$A(\theta \otimes T) = A(T \otimes \theta) = 0, \quad \forall T \in \otimes^p(E^*), \quad \forall \theta \in \otimes^q(E^*)$$

whenever $A(\theta) = 0$.

This follows from a property of the permutations : any $\alpha \in S_{p+q}$ can be factored as $\alpha_0 \circ \beta$, where β is an element of the subgroup H of all elements in S_{p+q} fixing $p+1, p+2, \ldots, p+q$. Now if $\varphi_{p+q} = A(\theta \otimes T)$ and writing $\alpha(x_i)$ for $x_{\alpha(i)}$,

$$\varphi_{p+q}(x_1, \ldots, x_{p+q}) = \frac{1}{(p+q)!} \sum_{\alpha \in S_{p+q}} \epsilon_\alpha \theta(\alpha(x_1), \ldots, \alpha(x_p)) T(\alpha(x_{p+1}), \ldots, \alpha(x_{p+q})).$$

First keeping α_0 fixed and letting β vary, we obtain :

$$\frac{1}{(p+q)!} T(\alpha_0(x_{p+1}), \ldots, \alpha_0(x_{p+q})) \sum_{\beta \in H} \epsilon_{\alpha_0 \circ \beta} \theta(\alpha_0 \circ \beta(x_1), \ldots, \alpha_0 \circ \beta(x_p))$$

and the sum yields 0 by the antisymmetry of θ. ∎

Proposition 2.2.2 *The quotient of the algebra $\otimes(E^*)$ by the two-sided ideal $J(E^*)$ is an associative algebra whose underlying vector space is $\wedge(E^*)$.*

Proof. The first part of the proposition follows from a well-known fact concerning algebras.

Note that every $T \in \otimes(E^*)$ can be uniquely decomposed as

$$T = \varphi + \theta, \quad \varphi \in \wedge(E^*), \quad \theta \in J(E^*),$$

since $T = AT + (T - A(T))$ and $A(T - A(T)) = 0$.

Conversely, if $T = \varphi + \theta$ as above, $A(T) = A\varphi = \varphi$ and $\theta = T - A(T)$. Hence the restriction of

$$\varphi : \otimes(E^*) \to \otimes(E^*)/J(E^*)$$

to $\wedge(E^*)$ is an isomorphism, therefore allowing the identification in the proposition and the definition of a multiplication $\wedge$, called the *exterior product*, turning $\wedge(E^*)$ into an associative algebra. ∎

Definition 2.2.3 *The exterior product $\wedge$ is defined by*

$$\psi \wedge \varphi = A(\psi \otimes \varphi), \quad \forall \psi, \varphi \in \wedge(E^*) \tag{5}$$

Note that if $T, T' \in \otimes(E^*)$,

$$A(T \otimes T') = A(T) \wedge A(T'),$$

and hence

$$A(T \otimes T' \otimes T'' \otimes \ldots) = A(T) \otimes A(T') \otimes A(T'') \otimes \ldots$$

If T, T' and T'' are antisymmetric :

$$T \wedge T' \wedge T'' = A(T \otimes T' \otimes T'').$$

Because of the decomposition in (4), $\wedge(E^*)$ is a *graded algebra* on $\mathbf{Z}$ with the property

$$\psi_p \wedge \varphi_q = (-1)^{p+q} \varphi_q \wedge \psi_p \tag{6}$$

if ψ_p and φ_q are of order p and q.

In fact, (6) means that $\wedge(E^*)$ is *graded commutative*. To prove (6), note that if $\pi_{p+q} = \psi_p \wedge \varphi_q$ and $\pi'_{p+q} = \varphi_q \wedge \psi_p$,

$$(p+q)! \pi_{p+q}(x_1, \ldots, x_{p+q}) = \epsilon^{i_1 \ldots i_p i_{p+1} \ldots i_{p+q}}_{1 \ldots p(p+1) \ldots (p+q)} \psi_p(x_{i_1}, \ldots, x_{i_p}) \varphi_q(x_{i_{p+1}}, \ldots, x_{i_{p+q}}),$$

$$(p+q)! \pi'_{p+q}(x_1, \ldots, x_{p+q}) = \epsilon^{i_{p+1} \ldots i_{p+q} i_1 \ldots i_p}_{(p+1) \ldots (p+q) 1 \ldots p} \varphi_q(x_{i_{p+1}}, \ldots, x_{i_{p+q}}) \psi_p(x_{i_1}, \ldots, x_{i_p})$$

and it takes pq transpositions to bring both expressions to the same order.

2.2.3   Exterior product of linear forms.

Note that if $\theta^i, \theta^j, \ldots$ are linear p-forms,

$$\theta^i \wedge \theta^j = A(\theta^i \otimes \theta^j) = \tfrac{1}{2}(\theta^i \otimes \theta^j - \theta^j \otimes \theta^i)$$
$$\text{and} \quad \theta^i \wedge \theta^i = 0.$$

Generally, if $\chi_1, \chi_2, \ldots, \chi_p$ is a strictly increasing sequence of indices,

$$\theta^{\chi_1} \wedge \theta^{\chi_2} \wedge \ldots \wedge \theta^{\chi_p} = A(\theta^{\chi_1} \otimes \theta^{\chi_2} \otimes \ldots \otimes \theta^{\chi_p})$$

so that

$$\theta^{\chi_1} \wedge \theta^{\chi_2} \wedge \ldots \wedge \theta^{\chi_p} = \frac{1}{p!} \epsilon_{k_1 k_2 \ldots k_p}^{\chi_1 \chi_2 \ldots \chi_p} \theta^{k_1 k_2 \ldots k_p}. \tag{7}$$

By the reasoning used to compute the dimension of $\wedge^p(E^*)$, the C_n^p exterior products obtained in (7) from all sequences $(\chi_1, \chi_2, \ldots, \chi_p)$ form a basis of $\wedge^p(E^*)$ if the $\theta^1, \theta^2, \ldots, \theta^n$ form a basis of E^*.

In particular, if we set $p = n$, if $\varphi^1, \varphi^2, \ldots, \varphi^n$ are n forms of E^*,

$$\varphi^1 \wedge \varphi^2 \wedge \ldots \wedge \varphi^n = \sigma \theta^1 \wedge \theta^2 \wedge \ldots \wedge \theta^n, \quad \sigma \in \mathbf{K},$$

since $\dim \wedge^n(E^*) = 1$. $\sigma = \det(\varphi^1, \varphi^2, \ldots, \varphi^n)$, the determinant of the n elements $\varphi^1, \varphi^2, \ldots, \varphi^n$ relative to the basis $(\theta^1, \theta^2, \ldots, \theta^n)$.

Proposition 2.2.3 *Two linear forms are linearly independent if and only if their exterior product does not vanish.*

Proof. If $\theta^1, \theta^2, \ldots, \theta^p$ are linearly dependent linear forms, and, for instance, $\theta^1 = \lambda_j^1 \theta^j$, then

$$\lambda_j^1 \theta^j \wedge \theta^2 \wedge \ldots \wedge \theta^p$$

consists only of zero terms.

Conversely, if $\theta^1, \theta^2, \ldots, \theta^p$ are linearly independent, we can add some forms $\theta^{p+1}, \ldots, \theta^n$, completing them to a basis of E^*. The C_n^p exterior products of p of these forms, ordered naturally on the indices, form a basis for $\wedge^p(E^*)$, hence $\theta^1 \wedge \theta^2 \wedge \ldots \wedge \theta^p$, being an element of this basis, cannot vanish. ∎

All the facts we have established concerning the exterior product on E^* are equally true for E, we will henceforth rather consider $\wedge(E)$.

2.2.4   The universal property.

Let f_p be a multilinear antisymmetric mapping of $\prod^p E$ into the vector space F. Then there exists a unique linear mapping g from $\wedge^p(E)$ to F such that

$$g(x_1 \wedge x_2 \wedge \ldots \wedge x_p) = f_p(x_1, x_2, \ldots, x_p), \quad \forall x_1, x_2, \ldots, x_p \in E$$

and $f_p \to g$ is a vector space isomorphism.

The proof is similar to that of (2.1.7).

Some consequences. If $F = \mathbf{K}$, $f_p \in \wedge^p(E^*)$ by definition and $g \in (\wedge^p(E))^*$, establishing a duality between $\wedge(E)^*$ and $\wedge^*(E)$, if we specify that for $p \neq q$, $\wedge^p(E^*)$ and $\wedge^q(E)$ are orthogonal under the duality. Considering $\wedge^p(E^*)$ and $\wedge^p(E)$ as subspaces of $\otimes^p(E^*)$ and $\otimes^p(E)$, the duality introduced before for the tensor spaces leads to

$$\langle \theta^{x_1} \wedge \theta^{x_2} \wedge \ldots \wedge \theta^{x_p}, e_{\lambda_1} \wedge e_{\lambda_2} \wedge \ldots \wedge e_{\lambda_p} \rangle = \frac{1}{p!} \epsilon^{x_1 x_2 \ldots x_p}_{\lambda_1 \lambda_2 \ldots \lambda_p}$$

if (θ^α) and (e_β) $(\alpha, \beta = 1, \ldots, n)$ are dual bases of E^* and E; more generally, if $\varphi^i \in E^*$ and $x_j \in E$, $(i, j = 1, 2, \ldots, p)$:

$$\langle \varphi^1 \wedge \varphi^2 \wedge \ldots \wedge \varphi^p, x_1 \wedge x_2 \wedge \ldots \wedge x_p \rangle = \frac{1}{p!} \det \langle \varphi^i, x_j \rangle. \tag{8}$$

Exterior powers of a linear mapping. Let $u : E \to F$ be a linear mapping. If to $(x_1, x_2, \ldots, x_p) \in \prod^p E$ we associate $u(x_1) \wedge u(x_2) \wedge \ldots \wedge u(x_p) \in \wedge^p F$, the universal property can be applied and there exists a unique linear map $\wedge^p(u) : \wedge^p(E) \to \wedge^p(F)$, for which :

$$\wedge^p(u)(x_1 \wedge x_2 \wedge \ldots \wedge x_p) = u(x_1) \wedge u(x_2) \wedge \ldots \wedge u(x_p). \tag{9}$$

$\wedge^p(u)$ has the following elementary properties :

1. If u maps E into F and v maps F into G, $\wedge^p(v \circ u)$ maps $\wedge^p(E)$ linearly into $\wedge^p(G)$ and

$$\wedge^p(v \circ u) = \wedge^p(v) \circ \wedge^p(u).$$

2. If u is surjective, so is $\wedge^p(u)$.

3. If u is an isomorphism from E to F, $\wedge^p(u)$ is an isomorphism from $\wedge^p(E)$ to $\wedge^p(F)$ and the inverse isomorphism is $\wedge^p(u^{-1})$.

Proof. 1 is straightforward, for 2 one should note that every element of $\wedge^p F$ is a linear combination of elements $y_1 \wedge y_2 \wedge \ldots \wedge y_p$, $y_i \in F$, the y_i being of the form $y_i = u(x_i)$, $x_i \in E$. 3 is obvious. $\blacksquare$

2.2.5 Another definition of the exterior algebra.

To facilitate the transition to Clifford algebras, let us consider the two-sided ideal $\dot{J}$ in $\otimes(E)$ generated by the elements of the form $x \otimes x$, $x \in E$. $\dot{J}$ consists of all sums of the form

$$\sum_i t_i \otimes x_i \otimes x_i \otimes s_i, \quad x_i \in E, \quad t_i, s_i \in \otimes(E).$$

In fact, replacing x by $x + y$, $\dot{J}$ could equally well be generated by the elements of the form $x \otimes y + y \otimes x$ where $x, y \in E$.

Proposition 2.2.4 *The exterior algebra* $\wedge(E)$ *is the quotient algebra of the tensor algebra* $\otimes(E)$ *by the two-sided ideal* $\dot{J}$ *generated by the* $x \otimes x$, $x \in E$.

Proof. A multiplication $\overset{.}{\wedge}$ can be defined as follows on the quotient $\overset{.}{\wedge}(E)$: if $\mathrm{Cl}(u)$ stands for the quotient class of u, $u \in \otimes(E)$, we set

$$\mathrm{Cl}(u) \overset{.}{\wedge} \mathrm{Cl}(v) = \mathrm{Cl}(u \otimes v).$$

$E \rightarrow \overset{.}{\wedge}(E) : x \rightarrow \mathrm{Cl}(x)$ being injective, $x \in E$ and $\mathrm{Cl}(x)$ can be identified : if $x, x' \in E$ and $\mathrm{Cl}(x) = \mathrm{Cl}(x')$, $x - x' \in \dot{J}$ and the elements of $\dot{J}$ being at least of order 2, $x = x'$. Similarly, the elements of $\mathbf{K}$ can be identified with their image.

$x \otimes x \in \dot{J}$ implies $x \overset{.}{\wedge} x = 0$ and $x \overset{.}{\wedge} y = -y \overset{.}{\wedge} x$, $\forall x, y \in E$. These properties, along with the associativity, imply that if the sequence of elements $x_1, x_2, \ldots, x_p \in E$ has one or more repeated members,

$$x_1 \overset{.}{\wedge} x_2 \overset{.}{\wedge} \ldots \overset{.}{\wedge} x_p = 0$$

and hence

$$x_{\alpha(1)} \overset{.}{\wedge} x_{\alpha(2)} \overset{.}{\wedge} \ldots \overset{.}{\wedge} x_{\alpha(p)} = \epsilon_\alpha(x_1 \overset{.}{\wedge} x_2 \overset{.}{\wedge} \ldots \overset{.}{\wedge} x_p), \quad \alpha \in S_p.$$

$\overset{.}{\wedge}(E)$ is generated by the

$$e_{\chi_1} \overset{.}{\wedge} e_{\chi_2} \overset{.}{\wedge} \ldots \overset{.}{\wedge} e_{\chi_p}, \quad 1 \leq \chi_1 < \chi_2 < \ldots < \chi_p \leq n$$

if the (e_i) form a basis of E. The multiplication table of these elements is the same as that of $\wedge(E)$, as indicated before. The algebras $\overset{.}{\wedge}(E)$ and $\wedge(E)$ can therefore be identified. ∎

2.2.6   Decomposable elements of $\wedge(E)$, p-vectors.

Definition 2.2.4 *An element* $x_1 \wedge x_2 \wedge \ldots \wedge x_p \in \wedge^p(E)$ *is said to be decomposable; it is also called a* p-*vector.*

Lemma 2.2.2 *Let* $z \neq 0$, $z \in \wedge^p(E)$, $p < n$, *then the elements* $x \in E$ *for which* $z \wedge x = 0$ *form a* q-*dimensional subspace* V_z *of* E, $q \leq p$. *If* (x_i), $i = 1, \ldots, q$ *is a basis of* V_z, $z = v \wedge x_1 \wedge \ldots \wedge x_q$ *for some* $v \in \wedge^{p-q}(E)$.

Proof. V_z is obviously a subspace and if q is its dimension, a basis of E can be obtained from $x_1, x_2, \ldots, x_q$ and $(n - q)$ other elements.

$$z = a^{\alpha_1 \alpha_2 \ldots \alpha_p} x_{\alpha_1} \wedge x_{\alpha_2} \wedge \ldots \wedge x_{\alpha_p},$$

where $\alpha_1, \alpha_2, \ldots, \alpha_p$ is an increasing sequence of indices.

$$z \wedge x_{\alpha_i} = 0$$

for $\alpha_i = 1, 2, \ldots, q$, by the assumption, hence the $a^{\alpha_1 \alpha_2 \cdots \alpha_p}$ where none of the indices $1, 2, \ldots, q$ occur must vanish. Furthermore, if one of the indices α_j from 1 to q is absent from certain sequences, exterior multiplication by x_{α_j} will, through the vanishing of the left-hand side, prove that all the coefficients of such sequences must be zero : all non-vanishing terms therefore contain $x_1 \wedge x_2 \wedge \ldots \wedge x_q$, $q < p$ and

$$z = v \wedge x_1 \wedge x_2 \wedge \ldots \wedge x_q.$$

∎

Note that obviously $V_z = E$ if $p = n$.

Lemma 2.2.3 z *is decomposable if and only if* $\dim(V_z) = p$.

Proof. If $\dim(V_z) = p$, v is scalar. Conversely, if $z = y_1 \wedge y_2 \wedge \ldots \wedge y_p \neq 0$, $y_i \in E$, then the $y_1, y_2, \ldots, y_p$ are linearly independent elements of V_z, so $\dim(V_z) = p$ since $\dim(V_z)$ cannot exceed p. ∎

Proposition 2.2.5 *Let* $\{x_i\}$ *and* $\{y_i\}$ *be two linearly independent sets,* $1 \leq i \leq p$. *Then* $y_1 \wedge y_2 \wedge \ldots \wedge y_p = \lambda(x_1 \wedge x_2 \wedge \ldots \wedge x_p)$, $\lambda \in \mathbf{K}^*$, *if and only if the* (x_i) *and the* (y_i) *generate the same subspace of* E.

Proof. If these subspaces coincide and if $(f_1, f_2, \ldots, f_p)$ is a basis of this subspace F,

$$x_1 \wedge x_2 \wedge \ldots \wedge x_p = \alpha(f_1 \wedge f_2 \wedge \ldots \wedge f_p), \qquad \alpha \in \mathbf{K}^*$$
$$y_1 \wedge y_2 \wedge \ldots \wedge y_p = \beta(f_1 \wedge f_2 \wedge \ldots \wedge f_p), \qquad \beta \in \mathbf{K}^*,$$

since $\dim \wedge^p(F) = 1$. Conversely if $y_1 \wedge y_2 \wedge \ldots \wedge y_p = \lambda x_1 \wedge x_2 \wedge \ldots \wedge x_p \neq 0$, $\lambda \in \mathbf{K}^*$, $z = y_1 \wedge y_2 \wedge \ldots \wedge y_p$ satisfies $\dim(V_z) = p$ since V_z contains $y_1, y_2, \ldots, y_p$. As it also contains the $x_1, x_2, \ldots, x_p$, it is also the subspace generated by them. ∎

Consequences. To each decomposable element $z \in \wedge^p(E)$ a unique p-dimensional subspace can be associated, viz. V_z. Conversely, to each p-dimensional subspace F there corresponds a p-vector which is unique up to a non-zero scalar factor. $(x_1, x_2, \ldots, x_p)$ being a basis of F and $e_1, e_2, \ldots, e_n$ a basis of E,

$$x_1 \wedge x_2 \wedge \ldots \wedge x_p = a^{\alpha_1 \alpha_2 \cdots \alpha_p} e_{\alpha_1} \wedge e_{\alpha_2} \wedge \ldots \wedge e_{\alpha_p},$$

where the $\alpha_1, \alpha_2, \ldots, \alpha_p$ are a strictly increasing sequence of indices.

The C_n^p coefficients $a^{\alpha_1 \alpha_2 \cdots \alpha_p}$, defined up to a non-zero factor, are called the *Plücker* or *Grassmann coordinates* of F.

2.3   Essential properties of symmetric algebras.

2.3.1   The space of symmetric forms of order p.

Definition 2.3.1 *A p-form $\varphi_p \in \otimes^p(E^*)$, $p > 1$, is said to be symmetric if for every sequence $x_1, x_2, \ldots, x_p \in E$,*

$$\varphi_p(x_1, \ldots, x_i, \ldots, x_j, \ldots, x_p) = \varphi_p(x_1, \ldots, x_j, \ldots, x_i, \ldots, x_p), \tag{1}$$

so that for any $\alpha \in S_p : (1, 2, \ldots, p) \to (i_1, i_2, \ldots, i_p)$

$$\varphi_p(x_{i_1}, x_{i_2}, \ldots, x_{i_p}) = \varphi_p(x_1, x_2, \ldots, x_p). \tag{2}$$

For $p = 1$, φ_1 is a linear form. For $p = 0$, φ_0 is a scalar.

Definition 2.3.2 *The symmetrized p-form φ_p of the tensor $T_p \in \otimes^p E^*$, $p > 1$, is defined by*

$$\varphi_p(x_1, x_2, \ldots, x_p) = \frac{1}{p!} \sum_{\alpha \in S_p} T_p(\alpha(x_1), \alpha(x_2), \ldots, \alpha(x_p)). \tag{3}$$

(Here $\alpha(x_i)$ stands for $x_{\alpha(i)}$).

We write $S(T_p) = \varphi_p$. If T_p is symmetric, $S(T_p) = T_p$, so $S^2 = S$. We extend S by linearity. S is a linear mapping from $\otimes^p(E^*)$ into the subspace of symmetric forms in $\otimes^p(E^*)$ which is denoted by $\vee^p(E^*)$. Setting $\vee^0(E^*) = \mathbf{K}$, $\vee^1(E^*) = E^*$, we define

$$\vee(E^*) = \oplus_{p=0}^{\infty} \vee^p(E^*). \tag{4}$$

$\vee(E^*)$ is an infinite-dimensional subspace of $\otimes(E^*)$.

Proposition 2.3.1 $\dim(\vee^p(E^*)) = C^p_{n+p-1}$.

Proof. An element $\varphi_p \in \vee^p(E^*)$ can be written uniquely as

$$\varphi_p = a_{k_1 k_2 \ldots k_p} \theta^{k_1} \otimes \theta^{k_2} \otimes \ldots \otimes \theta^{k_p} \in \otimes^p(E^*)$$

where the $a_{k_1 k_2 \ldots k_p}$ are symmetric. Let $k_1, k_2, \ldots, k_p$ be a sequence of not necessarily distinct numbers from 1 to n and let $\chi_1, \chi_2, \ldots, \chi_p$ be the same sequence, but ordered, so that

$$1 \leq \chi_1 \leq \chi_2 \leq \ldots \leq \chi_q \leq n. \tag{5}$$

Regrouping the terms of φ_p involving some permutation of $k_1, k_2, \ldots, k_p$, a linearly independent set is obtained. Its cardinal number equals the number of sequences satisfying (5).

Now if we note that the p numbers $\chi_1, \chi_2 + 1, \chi_3 + 2, \ldots, \chi_p + p - 1$ are pairwise different and form an arbitrary sequence of numbers between 1 and $n + p - 1$, the proposition is proved.   ∎

2.3.2   The symmetric algebra $\vee(E^*)$.

The set of $\theta \in \otimes(E^*)$ such that $S(\theta) = 0$ is a two-sided ideal of $\otimes(E^*)$, denoted by $\mathcal{J}(E^*)$.

The proof is similar to the antisymmetric case, considering $S(T \otimes \theta)$ and $S(\theta \otimes T)$ and factoring $\alpha \in S_{p+q}$ as $\alpha_0 \circ \beta$, $\beta \in H$. (cf. subsection 2.2).

An element T of $\otimes(E^*)$ can be uniquely split as $T = \varphi + \theta$, $\varphi \in \vee(E^*)$, $\theta \in \mathcal{J}(E^*)$. $\vee(E^*)$ and $\otimes(E^*)/\mathcal{J}(E^*)$ are identified. We define

$$\psi_p \vee \varphi_q = S(\psi_p \otimes \varphi_q) = \varphi_q \vee \psi_p. \tag{6}$$

As before we deduce that $S(T \otimes T' \otimes T'') = S(T) \vee S(T') \vee S(T'')$. A basis for $\vee^p(E^*)$ is given by

$$\theta^{\chi_1} \vee \theta^{\chi_2} \vee \ldots \vee \theta^{\chi_p}, \quad 1 \leq \chi_1 \leq \chi_2 \leq \ldots \leq \chi_p \leq n \tag{7}$$

if the θ^i form a basis of E^*. It is common practice to write $\theta^{\chi_1} \vee \theta^{\chi_1} = (\theta^{\chi_1})^2$ etc.

Clearly any element of this basis can be written in the form

$$(\theta^1)^{\alpha_1} \vee (\theta^2)^{\alpha_2} \vee \ldots \vee (\theta^n)^{\alpha_n} \tag{8}$$

where $\alpha_1, \alpha_2, \ldots, \alpha_n$ are non-negative integers satisfying $\alpha_1 + \alpha_2 + \ldots + \alpha_n = p$. Therefore we can identify $\vee(E^*)$ with the algebra of polynomials in n variables with coefficients in $\mathbf{K}$,

$$\vee(E^*) = \mathbf{K}[\theta^1, \theta^2, \ldots, \theta^n]. \tag{9}$$

Finally we can consider a universal property and consequences which parallel the antisymmetric case. The formula (8) can now be written

$$\langle \varphi^1 \vee \varphi^2 \vee \ldots \vee \varphi^p, x_1 \vee x_2 \vee \ldots \vee x_p \rangle = \frac{1}{p!} \operatorname{perm} \langle \varphi^i, x_j \rangle$$

where perm stands for the permanent of the matrix $(\langle \varphi^i, x_j \rangle)$.

As a final remark, $\vee(E)$ can be defined as the quotient of $\otimes(E)$ by the two-sided ideal $\dot{\mathcal{J}}$ generated by all $x \otimes y - y \otimes x$, where $x, y \in E$.

2.4   The inner product and the annihilation operators.

2.4.1   In the exterior algebra.

Let $x \in E$, we define for all $a \in \otimes(E)$,

$$e_x(a) = x \otimes a, \tag{1}$$

e_x is called a *creation operator*.

Proposition 2.4.1 *Given $\varphi \in E^*$, there is a unique linear mapping $i_\varphi : \otimes(E) \to \otimes(E)$ for which*

$$i_\varphi(1) = 0 \qquad (2)$$

$$i_\varphi \circ e_x + e_x \circ i_\varphi = \langle x, \varphi \rangle, \quad \forall x \in E. \qquad (3)$$

Proof. (2) determines i_φ on $\mathbf{K} = \otimes^0(E)$, (3) means that

$$i_\varphi(x \otimes a) + x \otimes i_\varphi(a) = \langle x, \varphi \rangle a = i_\varphi(x)a$$

since $i_\varphi(x \otimes 1) = \langle x, \varphi \rangle = i_\varphi(x)$.

Assume that i_φ is defined on $\otimes^0(E), \otimes^1(E), \ldots, \otimes^{h-1}(E)$, then (3) yields, for $x \otimes a \in \otimes^h(E)$,

$$i_\varphi(x \otimes a) = -x \otimes i_\varphi(a) + i_\varphi(x)a.$$

The existence and uniqueness of i_φ on $\otimes^h(E)$ follow from the bilinearity of the right-hand side and from the universal property of the tensor algebra. ∎

Proposition 2.4.2 *On $\otimes^h(E)$,*

$$i_\varphi = \sum_{j=1}^h (-1)^{j+1} C_\varphi^j \qquad (4)$$

where C_φ^j is the contraction with φ, using the jth index.

Proof. (4) holds for $h = 1$. Assume that $x \otimes a \in \otimes^h(E)$, then

$$
\begin{aligned}
i_\varphi(x \otimes a) &= -x \otimes i_\varphi(a) + C_\varphi^1(x)a \\
&= -x \otimes \sum_{j=1}^{h-1}(-1)^{j+1}C_\varphi^j(a) + C_\varphi^1(x \otimes a) \\
&= -\sum_{j=1}^{h-1}(-1)^{j+1}C_\varphi^{j+1}(x \otimes a) + C_\varphi^1(x \otimes a) \\
&= \sum_{j=1}^{h}(-1)^{j+1}C_\varphi^j(x \otimes a).
\end{aligned}
$$

∎

Proposition 2.4.3 *If φ and ψ are elements of E^*,*

$$(i_\varphi)^2 = 0 \quad \text{and} \quad i_\varphi \circ i_\psi + i_\psi \circ i_\varphi = 0. \qquad (5)$$

Proof. $(i_\varphi)^2$ and e_x commute, since

$$(i_\varphi)^2 \circ e_x = -i_\varphi \circ e_x \circ i_\varphi + \langle x, \varphi \rangle i_\varphi = e_x \circ (i_\varphi)^2.$$

As $(i_\varphi)^2 = 0$ on $\mathbf{K}$, $(i_\varphi)^2 = 0$ on $\otimes^h(E)$ by induction. The second formula can be obtained from $(i_\varphi + i_\psi)^2 = (i_{\varphi+\psi})^2 = 0$. ∎

Proposition 2.4.4 i_φ is an anti-derivation of the tensor algebra, its degree is -1:

$$i_\varphi(a \otimes b) = i_\varphi(a) \otimes b + (-1)^h a \otimes i_\varphi(b), \tag{6}$$

if a is homogeneous of order h.

Proof. This is an immediate consequence of (4). ∎

Proposition 2.4.5 i_φ has a natural extension to an antiderivation on the exterior algebra and

$$i_\varphi(x_1 \wedge x_2 \wedge \ldots \wedge x_h) = \sum_{j=1}^{h} (-1)^{j+1} \langle x_j, \varphi \rangle (x_1 \wedge \ldots \wedge \widehat{x_j} \wedge \ldots \wedge x_h). \tag{7}$$

(Here $\frown$ indicates a missing factor.)

Proof. It is sufficient to verify that, using the definitions of (2.2.5), the two-sided ideal J is invariant under i_φ, so that we can divide by J in order to define the action of i_φ on the exterior algebra. Assuming that a has order h, $i_\varphi(a \otimes x \otimes x \otimes b) = i_\varphi(a) \otimes x \otimes x \otimes b + (-1)^h a \otimes x \otimes x \otimes i_\varphi(b)$ together with $i_\varphi(x \otimes x) = 0$ yields the result.

(7) is a simple consequence of (4). ∎

Definition 2.4.1 Considered as an endomorphism of the exterior algebra, i_φ defines an inner product or annihilation operator.

2.4.2 In the symmetric algebra.

Using the same notations, but replacing i_φ by j_φ, an analog to Proposition (2.4.1), is obtained, where

$$j_\varphi(1) = 0 \tag{8}$$

$$j_\varphi \circ e_x - e_x \circ j_\varphi = \langle x, \varphi \rangle. \tag{9}$$

Then

$$j_\varphi = \sum_{k=1}^{h} C_\varphi^k \tag{10}$$

on $\otimes^h(E)$, from which

$$j_\varphi \circ j_\psi - j_\psi \circ j_\varphi = 0 \tag{11}$$

and j_φ is a derivation of degree (-1) on the tensor algebra :

$$j_\varphi(a \otimes b) = j_\varphi(a) \otimes b + a \otimes j_\psi(b). \tag{12}$$

Finally :

Proposition 2.4.6 j_φ has a natural extension to a derivation on the symmetric algebra and

$$j_\varphi(x_1 \vee x_2 \vee \ldots \vee x_h) = \sum_{k=1}^{h} \langle x_k, \varphi \rangle (x_1 \vee \ldots \vee \widehat{x_k} \vee \ldots \vee x_h). \tag{13}$$

Proof. It is sufficient to verify, using (2.2.5), that the ideal $\dot{\mathcal{J}}$ is invariant under the action of j_φ :

$$j_\varphi(a \otimes (x \otimes y - y \otimes x) \otimes b) = j_\varphi(a) \otimes (x \otimes y - y \otimes x) \otimes b + a \otimes (x \otimes y - y \otimes x) \otimes j_\varphi(b)$$

since $j_\varphi(x \otimes y - y \otimes x) = 0$.   ∎

Definition 2.4.2 *Considered as an endomorphism of the symmetric algebra, j_φ defines an inner product or annihilation operator.*

2.5   Selected references.

- C. Chevalley, *Theory of Lie groups*, Princeton University Press, 1946.

- A. Lichnerowicz, *Eléments de calcul tensoriel*, A. Colin, Paris, 1958.

- D. Kastler, *Introduction a l'électrodynamique quantique*, Dunod, Paris, 1961.

Chapter 3

ORTHOGONAL CLIFFORD ALGEBRAS.

3.1 Definition and general properties.

Let E be an n-dimensional vector space on $\mathbf{K}$, ($\mathbf{R}$ or $\mathbf{C}$), Q a quadratic form and B its associated symmetric bilinear form :

$$B(x,y) = \frac{Q(x+y) - Q(x) - Q(y)}{2}, \quad x,y \in E.$$

3.1.1

Definition 3.1.1 *The orthogonal Clifford algebra is defined as the quotient of the tensor algebra* $\otimes(E)$ *by the two-sided ideal* $\mathcal{N}(Q)$ *generated by the elements of the form*

$$x \otimes x - Q(x), \quad x \in E \subset \otimes(E).$$

We will generally speak of 'Clifford algebra', as long as the symplectic case is not considered. If $Q(x) = 0$, the orthogonal Clifford algebra coincides with $\wedge(E)$ (cf. Chapter 2, 2.5).

The Clifford algebra will be written as $C(E,Q)$, but often this will be shortened to $C(Q)$.

Let ρ_Q be the canonical mapping $\otimes(E) \rightarrow \otimes(E)/\mathcal{N}(Q) = C(Q)$, then just as $\otimes(E)$ is generated by E, $C(Q)$ will be generated by $\rho_Q(E)$. We will soon establish that $\rho_Q(E)$ and E can be identified.

The definition immediately leads to

$$(\rho_Q(x))^2 = Q(x), \tag{1}$$

from which, expanding $(\rho_Q(x+y))^2 = Q(x+y)$, we obtain

$$\rho_Q(x)\rho_Q(y) + \rho_Q(y)\rho_Q(x) = 2B(x,y). \tag{2}$$

The gradation on $C(Q)$. The tensor algebra is the direct sum of two spaces :
$\otimes(E) = T^+ \oplus T^-$, where

$$T^+ = \oplus_{h=0}^{\infty} \otimes^{2h}(E)$$

37

and
$$T^- = \oplus_{h=0}^{\infty} \otimes^{2h+1}(E).$$

The generators of $C(Q)$ are elements of T^+, so $\mathcal{N}(Q) = \mathcal{N}^+(Q) \oplus \mathcal{N}^-(Q)$, where

$$\mathcal{N}^+(Q) = \mathcal{N}(Q) \cap T^+$$

and

$$\mathcal{N}^-(Q) = \mathcal{N}(Q) \cap T^-.$$

It follows that

$$C(Q) = C^+(Q) \oplus C^-(Q),$$

where $C^+(Q) = \rho_Q(T^+)$ and $C^-(Q) = \rho_Q(T^-)$.

Obviously $C^+(Q)$ is a subalgebra of $C(Q)$ and the properties

$$C^+(Q)C^+(Q) \subseteq C^+(Q),$$
$$C^-(Q)C^-(Q) \subseteq C^+(Q),$$
$$C^+(Q)C^-(Q) \subseteq C^-(Q)$$

prove that $C(Q)$ is graded over the group $\mathbf{Z}_2 = \{0, 1\}$, i.e. it is semi-graded.

3.1.2   The universal property of $C(Q)$.

Theorem 3.1.1 *(Fundamental Theorem.) Let F be an algebra over* **K** *and let u be a linear mapping from E to F such that*

$$(u(x))^2 = Q(x), \quad \forall x \in E,$$

then there exists a unique homomorphism $\hat{u}$ from $C(Q)$ to F such that

$$u = \hat{u} \circ \rho_Q.$$

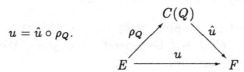

Proof. $\hat{u}$ must be unique since the set of all $\rho_Q(x)$, $x \in E$, generates $C(Q)$. u has a unique natural extension to a homomorphism v from $\otimes(E)$ to F,

$$v(x_1 \otimes x_2 \otimes \ldots \otimes x_h) = v(x_1)v(x_2)\ldots v(x_h) = u(x_1)u(x_2)\ldots u(x_h).$$

From $(u(x))^2 = Q(x)$ it follows that

$$v(x \otimes x - Q(x)) = v(x \otimes x) - Q(x) = (u(x))^2 - Q(x) = 0$$

so v vanishes on $\mathcal{N}(Q)$ and $v(t) = v(t')$ whenever $t - t' \in \mathcal{N}(Q)$.

Therefore, taking the quotient, a mapping $\hat{u}$ can be deduced from v; $\hat{u}$ is a homomorphism from $C(Q)$ to F. Setting $\hat{t} = \rho_Q(t)$, $\hat{u}(\hat{t}) = v(t)$ by definition, so $\hat{u}(\hat{x}) = u(x)$ and $\hat{u} \circ \rho_Q = u$ on E. ∎

Consequences : main involution and anti-involution.

- Let F be the reverse algebra $C(Q)^\circ$ of $C(Q)$ and $u = \rho_Q$. The restriction of $\hat{u}$ to $\rho_Q(E) \subset C(Q)$ will be the identity since $\hat{u}$ must map $\rho_Q(x) \in C(Q)$ to $\rho_Q(x) \in C(Q)^\circ$: therefore a unique β exists such that

$$\beta(zz') = \beta(z')\beta(z)$$

and that its restriction to $\rho_Q(E)$ is the identity. Clearly $\beta^2 = \mathrm{Id}$, since

$$\beta(\hat{x}_1\hat{x}_2\ldots\hat{x}_h) = \hat{x}_h\ldots\hat{x}_2\hat{x}_1, \quad \hat{x}_i = \rho_Q(x_i), \quad x_i \in E.$$

$\beta^{-1} = \beta$ and β must be injective.

β is called the main antiautomorphism (or main anti-involution) of $C(Q)$; we stress the fact that its restriction to ρ_Q is the identity.

- Let us consider a vector space E' with quadratic form Q' and let $f : E \to E'$ be linear. $C(Q')$ being the Clifford algebra of E', we will assume that $Q' \circ f = Q$ so that

$$(\rho_{Q'}(f(x)))^2 = Q'(f(x)) = Q(x),$$

and we may consider $u = \rho'_Q \circ f : E \to C(Q')$.

Then there exists a homomorphism C_f from $C(Q)$ to $C(Q')$ such that

$$C_f \circ \rho_Q = \rho_{Q'} \circ f$$

on E. In particular, for $E' = E$, $Q' = Q$, $f = -\mathrm{Id}$, we find that *there exists a unique homomorphism α from $C(Q)$ to $C(Q)$ such that $\alpha \circ \rho_Q = -\rho_Q$ on E.* As $\alpha^2 = \mathrm{Id}$, $\alpha^{-1} = \alpha$. α is called the main automorphism (or main involution) on $C(Q)$ and its restriction to $\rho_Q(E)$ is $-\mathrm{Id}$.

- The universal property also proves that if (E, Q) and (E_1, Q_1) are isometric spaces, the Clifford algebras $C(E, Q)$ and $C(E_1, Q_1)$ are isomorphic.

3.1.3 The dimension of the algebra $C(Q)$.

Proposition 3.1.1 *If n is the dimension of E, $\dim C(Q) \leq 2^n$.*

Proof. Let $(e_1, e_2, \ldots, e_n)$ be a basis of E, then, writing ρ as ρ_Q now :

$$\rho(e_i)\rho(e_j) + \rho(e_j)\rho(e_i) = 2B(e_i, e_j),$$

$$(\rho(e_i))^2 = Q(e_i)$$

imply, by an easy reasoning, that $C(Q)$ can be generated by products of the form

$$\rho(e_{i_1})\rho(e_{i_2})\ldots\rho(e_{i_p}), \quad 1 \leq i_1 < i_2 < \ldots < i_p \leq n$$

and by $\rho(1)$, whose number does not exceed 2^n. ∎

Definition 3.1.2 *Let F and G be two graded algebras, then their tensor product can be given the structure of an associative algebra if we define, for homogeneous elements,*

$$(a_1 \otimes a_2)(b_1 \otimes b_2) = (-1)^{p_2 q_1} a_1 b_1 \otimes a_2 b_2$$

where p_2 is the degree of a_2 and q_1 is the degree of b_1, and extend this multiplication by linearity.

The proof is immediate; the algebra obtained in this way is called the *graded tensor product* of F and G.

Lemma 3.1.1 *Let Q and Q' be quadratic forms on the finite-dimensional vector spaces E and E', then there exists a surjective homomorphism from $C(Q \oplus Q')$ to $C(Q) \otimes C(Q')$.*

Proof. $Q \oplus Q'$ is the quadratic form on $E \oplus E'$, considered as a direct orthogonal sum; the tensor product $C(Q) \otimes C(Q')$ has the multiplicative structure from Definition (3.1.2) and the Clifford algebras are $\mathbf{Z}_2$-graded. Now define

$$u(x, x') = \rho(x) \otimes 1 + 1 \otimes \rho(x'), \quad x \in E, \quad x' \in E'.$$

It is easily verified that this mapping $u : E \oplus E' \to C(Q) \otimes C(Q')$ satisfies

$$(u(x, x'))^2 = Q(x) + Q(x') = (Q \oplus Q')(x, x').$$

Applying the Fundamental Theorem yields the existence of a homomorphism $\hat{u}$ from $C(Q \oplus Q')$ to $C(Q) \otimes C(Q')$. The latter algebra being generated by the elements of the form $\rho(x) \otimes 1$ and $1 \otimes \rho(x')$, the result is proved.   ∎

Proposition 3.1.2 *The dimension of $C(Q)$ is 2^n. If $(e_1, e_2, \ldots, e_n)$ is a basis of E, the set of*

$$\rho(1), \quad \rho(e_{i_1})\rho(e_{i_2}) \ldots \rho(e_{i_p}), \quad 1 \le i_1 < i_2 < \ldots < i_p \le n$$

is a basis for $C(Q)$.

Proof. We use induction on n.

If $\dim E = 1$, $\otimes(E)$ can be identified with the algebra of polynomials in one variable and $\mathcal{N}(Q)$ is the ideal generated by $X^2 - a$ where $Q(x) = a$; by a classical result, $C(Q)$ is two-dimensional.

If $\dim E > 1$, we set $E = E' \oplus E_1$, where $\dim E_1 = 1$, and $Q = Q' \oplus Q_1$ as in Lemma (3.1.1). By this lemma, $C(Q \oplus Q')$ can be mapped to $C(Q') \otimes C(Q_1)$, so $\dim C(Q) \ge \dim(C(Q') \otimes C(Q_1))$. The induction hypothesis being that $\dim C(Q') = 2^{n-1}$, this implies $\dim C(Q) \ge 2^n$ and the result is proved.   ∎

Consequences.

- $\wedge(E)$ and $C(Q)$ are *linearly isomorphic.*

  An isomorphism between them maps $e_{i_1} \wedge e_{i_2} \wedge \ldots \wedge e_{i_p}$ to $\rho(e_{i_1})\rho(e_{i_2})\ldots\rho(e_{i_p})$ if $1 \leq i_1 < i_2 < \ldots < i_p \leq n$ and maps 1 to $\rho(1)$.

- *The elements of* **K** *and* E *can be identified with their images under* ρ.

  As $\rho(1)$ and $\rho(e_i)$ are elements of a basis of $C(Q)$, they form a linearly independent set :

  $\rho(a) = a\rho(1) = 0$ implies that $a = 0$, and

  $\rho(x) = \sum_{i=1}^n x^i \rho(e_i) = 0$ implies $x^i = 0$, so $x = 0$.

  The restrictions of ρ to **K** and to E are therefore injective, and the result is obtained. Henceforth we will write a for $\rho(a)$, e_i for $\rho(e_i)$ and $\{e_{i_1} e_{i_2} \ldots e_{i_p}\}$ for the basis of $C(Q)$ given before; also $1 = e_0$. The relations (1) and (2) become $x^2 = Q(x)$, $xy + yx = 2B(x,y)$.

- The homomorphism of Lemma (3.1.1) is an algebra isomorphism :

$$C(Q \oplus Q') = C(Q) \otimes C(Q') \tag{3}$$

where the tensor product is graded.

3.1.4 Simple examples.

- We have already noted that if $\dim E = 1$, $C(Q)$ is the algebra of polynomials in one variable. If e is a non-zero element of E, $e^2 = Q(e)$ and $(1, e)$ is a basis of $C(Q)$. In the real case, e can be chosen so as to have $Q(e) = \pm 1$ or $Q(e) = 0$.

  - If $Q(e) = -1$, $e^2 = -1$, the real algebra obtained is the complex number field.

  - If $Q(e) = 1$, $e^2 = 1$, setting

$$e_1 = (1 + e)/2, \quad e_2 = (1 - e)/2,$$

  (e_1, e_2) is a new basis for $C(Q)$ such that

$$(e_1)^2 = e_1, \quad (e_2)^2 = e_2, \quad e_1 e_2 = e_2 e_1 = 0.$$

  $C(Q)$ is then the direct composition of 2 fields isomorphic to **K** :

$$C(Q) = \mathbf{K}e_1 \oplus \mathbf{K}e_2.$$

  The elements of $C(Q)$ are called *hyperbolic complex numbers.*

— If $Q(e) = 0$, $e^2 = 0$, $C(Q)$ is the set D of *dual numbers*. D is a commutative ring.

Remark. If F is a three-dimensional real vector space and if A is the affine space associated to F, we define

$$F' = F \otimes D$$

and call the elements of F' *dual vectors*. These dual vectors can be applied to torsor reduction : if T is a torsor whose reduction elements in $M \in A$ are S_M and g_M (vector sum and resulting moment), the dual vector

$$T_M = \vec{S}_M + e\vec{g}_M$$

is associated to T and the study of torsors can be reduced to the geometry of the space F'.

The hyperbolic complex numbers can also be applied to certain physical theories (e.g. general relativity).

- Let $\dim E = 2$ and let (e_1, e_2) be an orthogonal basis of E for which $Q(e_1) = \alpha_1$ and $Q(e_2) = \alpha_2$.

  A basis for $C(Q)$ is then given by $1, e_1, e_2, e_1e_2$ and it immediately follows that

$$e_1e_2 + e_2e_1 = 0, \text{ by (2)}$$
$$(e_1)^2 = \alpha_1 \text{ and } (e_2)^2 = \alpha_2 \text{ by (1)}$$
$$(e_1e_2)^2 = -\alpha_1\alpha_2,$$
$$(e_1e_2)e_1 = -e_1(e_1e_2) = -\alpha_1 e_2 \text{ and}$$
$$(e_1e_2)e_2 = e_1(e_2)^2 = \alpha_2 e_1.$$

$C(Q)$ is a generalized quaternion algebra; if $\alpha_1 = \alpha_2 = -1$, the skewfield $\mathbf{H}$ of Hamilton's quaternions is obtained. (In the classical notations, $e_1 = i$, $e_2 = j$, $e_1e_2 = k$.)

In Chapter 7, other examples will be given, from the work of E. Cartan and others, related to the traditional exposition of mathematical physics.

3.1.5   The rôle of the 'fermionic' creation and annihilation operators.

The vector space isomorphism obtained between $\wedge(E)$ and $C(Q)$ may depend on the choice of the base for $\wedge(E)$. The reader may verify, using formula (8) in the sequel, with $xu = e_{i_1} e_{i_2} \ldots e_{i_p}$, that two isometrically related bases define the same linear isomorphism. We will now display a linear isomorphism between $\wedge(E)$ and $C(Q)$, defined independently from any choice of bases, and related to the creation and annihilation operators introduced in Chapter 2, 4.1. Using the notations of Chapter 1, 1.1 for the symmetric bilinear form B,

$$\gamma_y : x \to B(x, y)$$

is an element of E^*, the 'covariant' vector associated to the 'contravariant' vector y. Let us write $\gamma_y = y^*$, so that $y^*(x) = B(x, y)$.

By Chapter 2, 4.1,

$$i_{y^*} \circ e_x + e_x \circ i_{y^*} = B(x, y) \tag{4}$$

and both i_{y^*} and e_x can be extended to the exterior algebra. Now set $L_x = i_{x^*} + e_x$, then L_x is an endomorphism of $\wedge(E)$ and

$$(L_x)^2 = i_{x^*} \circ e_x + e_x \circ i_{x^*} = Q(x).$$

By the Fundamental Theorem (3.1.1), there exists a unique homomorphism $\hat{u}$ from $C(Q)$ to $\mathrm{End}(\wedge(E))$ such that

$$\hat{u} \circ \rho(x) = L_x, \quad \forall x \in E.$$

Using a classical definition which will be considered in Chapter 5, we obtain a *representation* of $C(Q)$ in the exterior algebra of E.

Lemma 3.1.2 *Given a bilinear form F on E, there exists a unique linear endomorphism λ_F on $\otimes(E)$ such that :*

$$\lambda_F(1) = 1 \tag{5}$$
$$\lambda_F \circ e_x = (e_x + i_x^F) \circ \lambda_F \tag{6}$$

where $i_x^F = i_{x^*}$ and $x^*(y) = F(x, y)$.

Proof. (5) determines λ_F on **K**.

If $x \otimes a \in \otimes^h(E)$,

$$\lambda_F(x \otimes a) = x \otimes \lambda_F(a) + i_x^F(\lambda_F(a)).$$

The right-hand side being bilinear in (x, a), the existence and uniqueness of λ_F follow by induction on the degree h. ∎

Lemma 3.1.3 *λ_F and i_φ commute, $\forall \varphi \in E^*$.*

Proof. Again we use induction. A simple computation involving the formulas

$$
\begin{aligned}
i_\varphi(x \otimes a) &= -x \otimes i_\varphi(a) + \varphi(x)a, \\
i_\varphi \circ e_x + e_x \circ i_\varphi &= \varphi(x) \\
i_\varphi \circ i_x^F + i_x^F \circ i_\varphi &= 0
\end{aligned}
$$

yields the result if we apply $\lambda_F \circ i_\varphi$ to $x \otimes a$. ∎

Lemma 3.1.4 *If F and F' are bilinear forms on E,*

$$\lambda_F \circ \lambda_{F'} = \lambda_{F+F'} \tag{7}$$

so, in particular, λ_F is a bijection on $\otimes(E)$.

Proof. Reasoning by induction, we apply $\lambda_F \circ \lambda_{F'} \circ e_x = \lambda_{F+F'} \circ e_x$ to $\otimes^{h-1}(E)$. The result follows immediately. If $F = 0$, $i_x^F = 0$ and $\lambda_F \circ e_x = e_x \circ \lambda_F$. By induction, $\lambda_F = \mathrm{Id}$ when $F = 0$, hence

$$\lambda_F \circ \lambda_{F'} = \mathrm{Id} .$$

∎

$\dot{J}$ denotes the two-sided ideal of $\otimes(E)$ generated by the $x \otimes x$ (cf. Chapter 2, 2.5).

Proposition 3.1.3 *If B is the symmetric bilinear form associated to Q, (i.e. $Q(x) = B(x, x)$), the operator λ_B maps the ideal $\mathcal{N}(Q)$ into $\dot{J}$ and defines a vector space isomorphism from $C(Q)$ to $\wedge(E)$.*

Proof. λ_B being bijective in $\otimes(E)$, it is sufficient to prove that $\lambda_B(\mathcal{N}(Q)) \subseteq \dot{J}$.
    We see that

$$\lambda_B \circ (e_x)^2 = ((e_x)^2 + B(x, x)) \circ \lambda_B,$$

$$\lambda_B(x \otimes x \otimes b) = ((e_x)^2 + B(x, x)) \circ \lambda_B(b), \quad b \in \otimes(E),$$

and $\lambda_B(x \otimes x \otimes b - Q(x)b) = (x \otimes x) \otimes \lambda_B(b) \in \dot{J}$. Now if $c = x \otimes x \otimes b - Q(x)b$, we compute $\lambda_B(a \otimes c)$ where $\lambda_B(c) \in \dot{J}$. If $y \in E$, $\lambda_B(y \otimes c) = y \otimes \lambda_B(c) + i_{y^*}\lambda_B(c)$ so that by induction (and noticing that $\dot{J}$ is invariant under i_φ), $\lambda_B(y \otimes c) \in \dot{J}$ etc. ∎

Consequence. Taking the quotient in (6), applied to $u \in C(Q)$ for $F = B$,

$$\lambda_B(xu) = x \wedge \lambda_B(u) + i_{x^*}(\lambda_B(u)), \tag{8}$$

where the product in $C(Q)$ is, as always, simply indicated by the juxtaposition of the factors.

    Analogous isomorphisms λ_F could be obtained using any bilinear form F for which $F(x, x) = Q(x)$, but the choice of the (symmetric) B is natural.

    Note that both sides of (8) are elements of the exterior algebra, so that a second multiplication can be defined on $\wedge(E)$, identifying $\lambda_B(v)$ with $v \in C(Q)$, using

$$xu = x \wedge u + i_{x^*}(u), \quad x \in E, \quad u \in \wedge(E). \tag{9}$$

Some theoretical physicists call this algebra over $\wedge(E)$ a *Kähler-Atiyah* algebra; but the correspondence given by (8) or (9) is already given by Chevalley and Bourbaki, and is implicit in E. Cartan's work.

    The importance of (9) is that it expresses, step by step, computations in the Clifford algebra $C(Q)$ in terms of other computations in $\wedge(E)$. Conversely, noticing that λ_{-B} maps $\dot{J}$ into $\mathcal{N}(Q)$, one could exchange $C(Q)$ and $\wedge(Q)$ in the reasoning, having noted that i_φ can be obtained on the Clifford algebra by taking the quotient, to express computations in the exterior algebra in terms of computations in $C(Q)$. Sadly, λ_B is not an algebra isomorphism and the identifications in $\wedge(E)$ are no longer covariantly functorial in $C(Q)$, or, put more plainly, a basis change in $C(Q)$ messes up the **Z**-homogeneous components of an element that is linearly identified to its image in the exterior algebra. Hence (9) does not provide a new definition of the Clifford algebra, but merely offers a computational method, whose interest is only to yield an intrinsic linear isomorphism (having chosen $F = B$) between $C(Q)$ and $\wedge(E)$.

The most important point is that any computation in the Clifford algebra can be expressed in the exterior algebra, and vice versa.

Remark. If we compare Proposition (3.1.3), (8), (9) and the results deduced before from (4), the creation and annihilation operators are seen to generate a Clifford algebra which is isomorphic to an endomorphism subalgebra of $\wedge(E)$; $x \in E$ and $L_x \in \mathrm{End}(\wedge(E))$ can obviously be identified to yield another definition of the Clifford algebra, which is often used by theoretical physicists.

3.1.6 Extension of the scalar field.

If **K** is a subfield of the (commutative) field **K$'$**, the *amplified* vector space E' of E by **K$'$** is the tensor product $\mathbf{K}' \otimes E$. It is also a vector space over **K$'$** under the scalar multiplication $a'(a \otimes x) = (a'a) \otimes x$, where $a', a \in \mathbf{K}'$ and $x \in E$.

If $(e_1, e_2, \ldots, e_n)$ is a basis of E over **K**, the $e_1, e_2, \ldots, e_n$ are also a basis of E' over **K$'$** : if the f_i are a (possibly infinite) [1] basis of **K$'$** over **K**, if $z \in \mathbf{K}' \otimes E$ where $z \in \sum_{i,j} t^{ij}(f_i \otimes e_j)$ and only a finite number of terms do not vanish, $t^{ij} \in \mathbf{K}$,

$$z = \sum \lambda^j \otimes e_j = \sum \lambda^j (1 \otimes e_j), \quad \lambda^j \in \mathbf{K}'.$$

Furthermore, $\sum \lambda^k (1 \otimes e_k) = 0$ implies

$$\sum \lambda^k \otimes e_k = 0 \text{ and } \sum t^{ik}(f_i \otimes e_k) = 0,$$

hence $t^{ik} = 0$ and $\lambda^k = 0$; the result then follows from the identification of $1 \otimes e_j$ with e_j.

Let B be a symmetric bilinear form on $E \times E$ and let $Q(x) = B(x, x)$ be its associated quadratic form, then B has a unique extension B' to $E' \times E'$. Indeed, if B' extends B, $B'(e_i, e_j) = B(e_i, e_j)$, defining B' completely. If $Q'(x) = B'(x, x)$, Q' is an extension of Q. Conversely, if Q'_1 extends Q, the associated symmetric bilinear form extends B, since :

$$2B'_1(x, y) = Q'_1(x + y) - Q'_1(x) - Q'_1(y)$$

---

[1] This case will occur in Chapter 17; **K$'$** will then be the ring of formal power series in one indeterminate. The extension of the tensor algebra operations to this case is obvious.

and $B' = B'_1$.

Q' is therefore unique, just as B'.

If Q is non-degenerate, Q' is also non-degenerate, the discriminants of Q and Q' relative to the basis (e_i) being equal.

Proposition 3.1.4 *Let* **K** *be a subfield of* **K**'*, let* E' *and* $C'(Q)$ *be the vector space and the Clifford algebra obtained from* E *and* $C(Q)$ *by the extension of the scalar field, then* $C'(Q)$ *and* $C(Q')$ *are isomorphic.*

Proof. $C'(Q)$ is an algebra with the multiplication

$$(a \otimes v)(a' \otimes v') = aa' \otimes vv', \quad \forall a, a' \in \mathbf{K}', \quad \forall v, v' \in C(Q).$$

We associate to $1 \otimes e_i = e_i \in E'$ the element $1 \otimes \rho_Q(e_i) \in C'(Q)$ and define by linearity a $\mathbf{K}'$-linear mapping from E' to $C'(Q)$ written as $1 \otimes \rho_Q = u$.

Every element of E' being a finite sum of terms of the form $a \otimes x$, $a \in \mathbf{K}'$, $x \in E$, it is sufficient to verify that

$$(u(a \otimes x))^2 = (a \otimes \rho_Q(x))^2 = Q'(a \otimes x)$$

and to apply the universal property :

$$(a \otimes \rho_Q(x))^2 = a^2 Q(x), \quad Q'(a \otimes x) = Q'(ax) = a^2 Q(x).$$

$\hat{u}$ (cf. Theorem (3.1.1)) is a homomorphism from $C(Q')$ to $C'(Q)$ and is obviously surjective, hence bijective since the dimensions of $C'(Q)$ and $C(Q')$ are equal.   ∎

In the sequel, we will apply these results to $\mathbf{K} = \mathbf{R}$ and $\mathbf{K}' = \mathbf{C}$.

3.2   Real and complex Clifford algebras : the 'periodicity theorems'.

In the previous section, the only special property of **R** and **C** we used was the fact that their characteristic differs from 2, and all results and properties carry over to Clifford algebras over finite-dimensional modules of characteristic different from 2. In this section, however, we will fully use the special properties of the real and complex number fields.

3.2.1   The real case.

Let B be a non-degenerate symmetric bilinear form, we will assume that the associated quadratic form Q has signature (r, s) where $r + s = n$, so that in an orthonormal basis :

$$Q(x) = (x^1)^2 + \ldots + (x^r)^2 - (x^{r+1})^2 - \ldots - (x^n)^2.$$

$C(r, s)$ will stand for $C(Q)$. The tensor product used here is *over non-graded algebras*.

Proposition 3.2.1 *If $r > 0$ or $s > 0$, there are algebra isomorphisms*

$$\begin{align}
C(r+1, s+1) &\simeq C(1,1) \otimes C(r,s) \tag{1} \\
C(s+2, r) &\simeq C(2,0) \otimes C(r,s) \tag{2} \\
C(s, r+2) &\simeq C(0.2) \otimes C(r,s) \tag{3}
\end{align}$$

Proof. Let E, E' be $(n = r + s)$-dimensional spaces, let E_2 be a two-dimensional space, and let Q, Q', Q_2 be their non-degenerate quadratic forms.

Consider the direct orthogonal sums $E \oplus E_2$ and $E' \oplus E_2$. Let (e_1, e_2) be an orthonormal basis of E_2 and let

$$u : x' + y \to (e_1 e_2 \otimes x) + (y \otimes 1), \quad x' \in E', \quad y \in E_2,$$

where x is the image of x' under a linear isomorphism identifying E' and E. u is defined as a mapping from $E' \oplus E_2$ into $C(Q_2) \otimes C(Q)$.

$$(u(x' + y))^2 = -(e_1)^2 (e_2)^2 x^2 + y^2 + (y e_1 e_2 + e_1 e_2 y) \otimes x,$$

but $y = \alpha e_1 + \beta e_2$ causes the last term to vanish.

Taking $Q' = Q$ if $Q(e_1)Q(e_2) = -1$ or $Q' = -Q$ if $Q(e_1)Q(e_2) = 1$, we get

$$u(x' + y)^2 = (x')^2 + y^2 = (Q' \oplus Q_2)(x' + y).$$

The proposition then follows from the universal property of Clifford algebras and from their dimensions, since the image of u contains $y \otimes 1$, $e_1 \otimes 1$, $e_2 \otimes 1$, $e_1 e_2 \otimes x$ and hence $(e_1 e_2 \otimes x)(e_1 e_2 \otimes 1) = \pm(1 \otimes x)$; these elements generate $C(Q_2) \otimes C(Q')$, u is surjective and therefore it is bijective. ∎

These formulas imply that :

$$\begin{align}
C(p, p) &\simeq \otimes^p C(1,1) \\
C(p + k', p) &\simeq \otimes^p C(1,1) \otimes C(k', 0) \\
C(k', 0) &\simeq C(2,0) \otimes C(0,2) \otimes C(k' - 4, 0), \quad k' > 4
\end{align}$$

and if

$$\begin{align}
k' = 4q : \quad & C(4q, 0) \simeq C(2,0) \otimes C(0,2) \otimes \ldots \otimes C(0,2) \\
k' = 4q + 1 : \quad & C(4q + 1, 0) \simeq C(2,0) \otimes C(0,2) \otimes \ldots \otimes C(0,2) \otimes C(1,1) \\
k' = 4q + 2 : \quad & C(4q + 2, 0) \simeq C(2,0) \otimes C(0,2) \otimes \ldots \otimes C(0,2) \otimes C(2,0) \\
k' = 4q + 3 : \quad & C(4q + 3, 0) \simeq C(2,0) \otimes C(0,2) \otimes \ldots \otimes C(0,2) \otimes C(0,1).
\end{align}$$

$C(p, p + k')$ and $C(0, k')$ are deduced from these formulas by exchanging r and s in $C(r,s)$.

Finally, we need only compute $C(1,0)$, $C(0,1)$, $C(1,1)$, $C(2,0)$ and $C(0,2)$. Let $M_n(\mathbf{R})$, $M_n(\mathbf{C})$ and $M_n(\mathbf{H})$ be the matrix algebras with real, complex and quaternion components. Taking the basis

$$\begin{pmatrix} 0 & 1 \\ 1 & 0 \end{pmatrix}, \begin{pmatrix} 0 & -1 \\ 1 & 0 \end{pmatrix}, \begin{pmatrix} 1 & 0 \\ 0 & -1 \end{pmatrix}, \begin{pmatrix} 1 & 0 \\ 0 & 1 \end{pmatrix}$$

of $M_2(\mathbf{R})$ and bearing in mind the examples from section 1.4,

$$\begin{aligned}
C(2,0) &\simeq C(1,1) \simeq M_2(\mathbf{R}), \\
C(1,0) &\simeq \mathbf{R} \oplus \mathbf{R}, \\
C(0,1) &\simeq \mathbf{C}, \\
C(0,2) &\simeq \mathbf{H}.
\end{aligned}$$

Furthermore $M_n(\mathbf{R}) \otimes M_p(\mathbf{R}) \simeq M_{np}(R)$ and the isomorphism between $C(r,r)$ and $M_{2^r}(\mathbf{R})$ is obvious.

In the sequel we will use the classical results $M_p(\mathbf{H}) \simeq M_p(\mathbf{R}) \otimes \mathbf{H}$ and $M_p(\mathbf{C}) \simeq M_p(\mathbf{R}) \otimes \mathbf{C}$. Then

$$\begin{aligned}
C(2,0) \otimes C(0,2) &\simeq M_2(\mathbf{R}) \otimes \mathbf{H} \simeq M_2(\mathbf{H}) \simeq C(0,4), \\
C(0,2) \otimes C(2,0) &\simeq \mathbf{H} \otimes M_2(\mathbf{R}) \simeq M_2(\mathbf{H}) \simeq C(4,0).
\end{aligned}$$

Using (3) $(r = 0,\ s = 2)$ we obtain

$$\mathbf{H} \otimes \mathbf{H} \simeq M_4(\mathbf{R}).$$

Then $\mathbf{H} \otimes \mathbf{C} \simeq C(0,2) \otimes C(0,1)$ leads, using (3) and (1), to

$$\mathbf{H} \otimes \mathbf{C} \simeq C(1,2) \simeq C(1,1) \otimes C(0,1) \simeq M_2(\mathbf{R}) \otimes \mathbf{C} \simeq M_2(\mathbf{C})$$

and these results allow us to compute all real Clifford algebras. The following results are helpful :

1.

$$C(r,s) \otimes C(0,8) \simeq C(r,s+8) \text{ and } C(r,s) \otimes C(8,0) \simeq C(r+8,s)$$

(the 'periodicity' isomorphisms). Indeed, $C(r,s) \otimes C(2,0) \simeq C(s+2,r)$ by (2), $C(s+2,r) \otimes C(0,2) \simeq C(r,s+4)$ by (3) etc. $C(r,s+4) \otimes C(2,0) \simeq C(s+6,r)$. $C(s+6,r) \otimes C(0,2) \simeq C(r,s+8)$, and $C(0,8) \simeq C(0,2) \otimes C(6,0) \simeq C(0,2) \otimes C(2,0) \otimes C(0,4) \simeq C(0,2) \otimes C(2,0) \otimes C(0,2) \otimes C(2,0)$, which proves this result.

2. $C(r,s) \otimes C(p,q) \simeq C(r+p,s+q)$ if $p = q$, $(\mathrm{mod}\ 4)$.

3. $C(r,s) \otimes C(p,q) \simeq C(s+p,r+q)$ if $p = q+2$, $(\mathrm{mod}\ 4)$. These formulas are proved in a similar way to 1, using Proposition (3.2.1).

Finally, we can tabulate the Clifford algebras for the signatures (p,q), $p+q = n$, p plus signs and q minus signs, and include all cases, when the 'periodicity' is taken into account. We write $\mathbf{R}$, $\mathbf{C}$, $\mathbf{H}$ for matrix algebras and $^2\mathbf{R}$, $^2\mathbf{H}$ for the direct sum of two identical matrix algebras. Note that $^2\mathbf{C}$ does not occur here, which will be a very important result in the sequel.

$p+q$	\-7	\-6	\-5	\-4	\-3	\-2	\-1	0	1	2	3	4	5	6	7
								$p-q$							
0		R		H		H		R		R		H		H	
1	^{2}R		C		^{2}H		C		^{2}R		C		^{2}H		C
2		R		H		H		R		R		H		H	
3	^{2}R		C		^{2}H		C		^{2}R		C		^{2}H		C
4		R		H		H		R		R		H		H	
5	^{2}R		C		^{2}H		C		^{2}R		C		^{2}H		C
6		R		H		H		R		R		H		H	
7	^{2}R		C		^{2}H		C		^{2}R		C		^{2}H		C

It is known that the matrix algebras $M_n(\mathbf{R})$, $M_n(\mathbf{C})$ and $M_n(\mathbf{H})$ are simple (cf. ultra), so we have that
$C(p,q)$ is not simple, but rather the direct composition of two simple ideals, if and only if $p - q = 1$, (mod 4).

Note the isomorphisms $C(p,q) \simeq C(p+4, q-4)$ if $q \geq 4$ and $C(p,q) \simeq C(q+1, p-1)$ if $p \neq 0$.

The algebras constructed on even-dimensional spaces E are simple and can be identified with real or quaternion matrix algebras.

3.2.2 The complex case.

We have proved before that $C(Q') = C'(Q)$ if the prime stands for complexification or for a complex space. By tensor product, we see that

$$C'(2r) \simeq C(r,r) \otimes \mathbf{C} \simeq M_{2^r}(\mathbf{C}).$$

A similar reasoning to the one in the real case proves that

$$C'(r+1) \simeq C'(1) \otimes C'(r). \tag{4}$$

One considers a $(r+1)$-dimensional complex space $E' = E'_1 \oplus (e_1)$ where $(e_1)^2 = 1$, e_1 being orthogonal to E'_1. Now define $u(x_1 + y) = (e_1 \otimes x_1) + (y_1 \otimes 1)$, $y \in (e_1)$, $x_1 \in E'_1$. Then

$$(u(x_1 + y))^2 = (x_1)^2 + (y)^2 + (e_1 y + y e_1) \otimes x_1 = (x_1)^2 + y^2$$

and the conclusion is similar to the real case. Just as for $C(1,0)$, we see that $C'(1) = \mathbf{C} \oplus \mathbf{C}$, hence

$$C'(2r+1) \simeq C'(2r) \oplus C'(2r) \simeq M_{2^r}(\mathbf{C}) \oplus M_{2^r}(\mathbf{C}).$$

A periodicity of order two appears. The Clifford algebras on even-dimensional spaces E' are simple, those on odd-dimensional spaces are the direct composition of two simple ideals.

3.3 The direct study of the Clifford algebra structure.

We first assume that Q is non-degenerate, $\dim E = n > 0$. Some of the results in this section have already been obtained in section 2; nevertheless we give a direct treatment of Clifford algebras without resorting to the special properties of **R** and **C**. In fact, the methods and results in this section can be applied to any finite-dimensional vector space over a commutative field whose characteristic is different from 2.

3.3.1 If E is an even-dimensional vector space over K $(n = 2r)$.

If the index of Q equals r, Q is said to be *neutral*.

Proposition 3.3.1 *If* $\dim E = 2r$ *and* Q *is neutral :*

1. $C(Q)$ *is isomorphic to the endomorphism algebra of a* (2^r)*-dimensional vector space* S *over* **K**.

2. $C^+(Q)$ *is the direct composition of two ideals, each of which is isomorphic to the endomorphism algebra of a* (2^{r-1})*-dimensional vector space over* **K**.

Proof. Q being neutral, $E = F \oplus F'$ where F and F' are maximal totally isotropic subspaces (Chapter 1, 2.3). Hence a basis

$$x_1, x_2, \ldots, x_r, y_1, y_2, \ldots, y_r$$

can be found such that the x_i span F and the y_i span F'. Then

$$B(x_i, y_j) = \delta_{ij}.$$

Let S be the subspace of $C(Q)$ generated by the set of

$$x_{i_1} x_{i_2} \ldots x_{i_h} f, \quad f = y_1 y_2 \ldots y_r, \quad 1 \le i_1 < i_2 \ldots < i_h \le r,$$

and by $x_\emptyset f = f$.

If we notice that the set of $x_{i_1} x_{i_2} \ldots x_{i_h} y_{j_1} y_{j_2} \ldots y_{j_k}$ for $1 \le j_1 < j_2 < \ldots j_k \le r$ is a basis of $C(Q)$, it is easily seen that S is $C(Q)f$.

The dimension of S is 2^r.

Let ρ be the representation of $C(Q)$ in S :

$$\rho(u)vf = uvf, \quad u \in \mathbf{C}(Q).$$

If we set $x_{i_1} \ldots x_{i_h} = x_H$ for short, we get $x_J(\prod_{i \notin H} y_i x_i) y_H(x_H f) = x_J f$, modulo a non-zero factor. $x_J(\prod_{i \notin H} y_i x_i) y_H(x_K f) = 0$ if $K \ne H$. This proves that ρ is surjective since we can obtain every element of the canonical basis of End S. Due to the dimensions, ρ is bijective which proves 1.

To establish 2 it is sufficient to note that if $u \in C^+(Q)$, vf and uvf have the same parity.

$C^+(Q)f$ and $C^-(Q)f$ are each invariant under $C^+(Q)$, ρ being injective, the image of $C^+(Q)$ by ρ is in $\operatorname{End} S^+ \times \operatorname{End} S^-$, $(S^+ = C^+(Q) \cap S$, $S^- = C^-(Q) \cap S)$ and the dimensions guarantee that ρ is a bijection. ∎

The center Z of $C(Q)$. Q still is assumed non-degenerate, but the index of Q may now again be arbitrary.

Let $(e_1, e_2, \ldots, e_n)$ be an orthogonal basis of E, such that $Q(x) = \alpha_i(x^i)^2$, $x = x^i e_i$, $e_i e_j + e_j e_i = 0$ if $i \neq j$ and $(e_i)^2 = \alpha_i$, $\alpha_i \neq 0$.

Lemma 3.3.1 The elements $(e_{i_1} e_{i_2} \ldots e_{i_m})$, $1 \leq i_1 < i_2 < \ldots < i_m \leq n$, are not in the center.

Proof. If m is even :

$$(e_{i_1} e_{i_2} \ldots e_{i_m}) e_{i_m} = \alpha_{i_m} e_{i_1} e_{i_2} \ldots e_{i_{m-1}}$$
$$e_{i_m}(e_{i_1} e_{i_2} \ldots e_{i_m}) = -\alpha_{i_m} e_{i_1} e_{i_2} \ldots e_{i_{m-1}}.$$

If m is odd : n being even there exists some $e_{i_k} \neq e_{i_1}, e_{i_2}, \ldots, e_{i_m}$ and

$$(e_{i_1} e_{i_2} \ldots e_{i_m}) e_{i_k} = -e_{i_k}(e_{i_1} e_{i_2} \ldots e_{i_m}).$$

∎

An algebra is said to be *central* if its center is the scalar field **K**.

Proposition 3.3.2 If $\dim E = n$ is even and Q is non-degenerate, $C(Q)$ is a central algebra.

Proof. From the proof of Lemma (3.3.1) we know that in both cases an e_i can be found such that it anticommutes with an element of the basis of $C(Q)$ different from e_0 and e_i; here e_0 stands for $e_\emptyset$.

Let $u = \lambda^{i_1 i_2 \ldots i_m} e_{i_1} e_{i_2} \ldots e_{i_m} \in Z$.

Given any $v = e_{k_1} e_{k_2} \ldots e_{k_m} \neq e_0$, let e_i anticommute with v. The mappings $u \to e_i u$ and $u \to u e_i$ are bijective, $e_i(e_{i_1} \ldots e_{i_m})$ and $(e_{i_1} \ldots e_{i_m}) e_i$ are equal to the same elements in the basis of $C(Q)$, except for some coefficient.

If $e_i u - u e_i = 0$ then

$$2\lambda^{k_1 k_2 \ldots k_m} e_i e_{k_1} e_{k_2} \ldots e_{k_m} + \sum \mu^{j_1 j_2 \ldots j_p} e_{j_1} e_{j_2} \ldots e_{j_p} = 0$$

where the summation has $2^n - 1$ terms (since $e_i e_{k_1} e_{k_2} \ldots e_{k_m}$ does not appear). Hence $\lambda^{k_1 k_2 \ldots k_m} = 0$ and only $\lambda^\emptyset = \lambda^0$ need not vanish. ∎

The ideals of $C(Q)$. Let $\mathcal{A} \neq \{0\}$ be a two-sided ideal of $C(Q)$ and let $u \in \mathcal{A}$, explicitly

$$u = \sum_{p=0}^{\nu} C_p e_p, \quad \nu = 2^n - 1.$$

If $C_\alpha \neq 0$, $0 \leq \alpha \leq \nu$, then $v = e_\alpha^{-1}u \in \mathcal{A}$. The coefficient of e_0 in v cannot vanish since, up to a non-zero scalar, e_α^{-1} is equal to e_α, and $u \to e_\alpha^{-1}u$ is bijective and permutes, up to non-zero coefficients, the elements e_p. In the proof of Lemma (3.3.1), the existence of a e_i such that $e_\nu e_i = -e_i e_\nu$, $i = 1, 2, \ldots, n$, was established, $e_i e_\nu e_i = -\alpha_i e_\nu$ and $\alpha_i v + e_i v e_i \in \mathcal{A}$, but its term in e_ν has vanished, whereas the coefficient of e_0 still differs from 0. Proceeding in this fashion, an element of $\mathcal{A}$ of the form $\lambda^0 e_0$ could be found, hence $e_0 \in \mathcal{A}$ and $\mathcal{A}$ is a trivial ideal. This yields the

Proposition 3.3.3 *If* $\dim E = n$ *is even and* Q *is non-degenerate, the two-sided ideals of* $C(Q)$ *are trivial.*

Such an algebra will be called *simple* (cf. Chapter 5). Joining Propositions (3.3.1) and (3.3.3), we have :

Theorem 3.3.1 *If* $\dim E = n = 2r$ *and if* Q *is non-degenerate, the algebra* $C(Q)$ *is simple and central.*

Remarks. If $\mathbf{K} = \mathbf{R}$ we proved in section 2.1 that $C(Q)$ is isomorphic to a matrix algebra $M_l(\mathbf{R})$ or $M_{l'}(\mathbf{H})$; it is possible to prove directly, in the framework of matrix theory, that $C(Q)$ is simple and central. Similarly, if $\mathbf{K} = \mathbf{C}$, $C'(2r) \simeq M_{2^r}(\mathbf{C})$ and the same conclusion is valid.

The algebras $C(3,1)$ and $C(1,3)$, which are used in relativity theory, are $M_4(\mathbf{R})$ and $M_2(\mathbf{H})$ respectively.

Clearly the results of section 3.1 can be applied to other cases, $\mathbf{K}$ might even be a commutative ring and E a finite-dimensional module.

These remarks are meant for the mathematicians who read this book; they will also verify that the results tabulated in section 2.1 correspond, in the even-dimensional case, to a theorem of Wedderburn on simple algebras (cf. Curtis-Reiner p. 175).

3.3.2 E is an odd-dimensional space, $n = 2r+1$, and Q is non-degenerate.

Theorem 3.3.2 *If* $\dim E = 2r + 1$ *and if* Q *is non-degenerate,*

1. $C^+(Q)$ *is simple and central.*

2. *If* Q *is of maximum index* r, $C^+(Q)$ *is isomorphic to* $\operatorname{End} S$ *where* S *is a* (2^r)-*dimensional vector space over* E.

Proof. Let us consider a non-isotropic $x_0 \in E$ and let $(x_0)^\perp$ be its orthogonal complement. On $(x_0)^\perp$ we define the quadratic form Q_1 :

$$Q_1(y) = -Q(x_0)Q(y), \quad y \in (x_0)^\perp.$$

Q_1 is non-degenerate, since if $z \in E$:

$$Q(z) = Q(x + y) = \alpha_0(x_0)^2 + Q(y)$$

if $x \in (x_0)$, $y \in (x_0)^\perp$, so if Q_1 would be degenerate a decomposition of $Q(x)$ in less than n squares would be obtained, which is impossible.

Note that if $y \in (x_0)^\perp$, $x_0y = -yx_0$ and

$$(x_0y)^2 = -y^2x_0^2 = -Q(x_0)Q(y) = Q_1(y).$$

By the universal property, $y \to x_0y$ can be extended to a homomorphism v from $C(Q_1)$ to $C^+(Q)$: these two algebras having the same dimension and $C(Q_1)$ being simple, v is an isomorphism.

Hence C^+ is a simple central algebra isomorphic to the algebra $C(Q_1)$ over a space of dimension $2r$.

To prove 2 a Witt decomposition can be used :

$$E = (x_0) \oplus F \oplus F',$$

Q_1 is neutral and Proposition (3.3.1) can be applied. ∎

Corollary 3.3.1 *The Clifford algebra $C(Q)$ of an odd-dimensional space E, $n = 2r - 1$, is isomorphic to a Clifford algebra $C^+(Q_1)$ of a $(2r)$-dimensional space E_1.*

Proof. Taking $E_1 = E \oplus (x_0)$ and Q_1 such that $Q_1(x_0) = \alpha$, $\alpha Q_1(y) = -Q(y)$ for $\alpha \neq 0$, $y \in E$, and that (x_0) is orthogonal to E for Q_1, the previous result can be applied. ∎

The center Z of $C(Q)$. The reasoning is similar to the case $n = 2r$. Setting $e_N = e_1e_2 \ldots e_n$, e_N is central since it commutes with all $(e_{i_1} e_{i_2} \ldots e_{i_m})$ introduced in section 3.1 and Z contains $\mathbf{K}e_0 \oplus \mathbf{K}e_N$. From now on the proof runs along the same lines as that in 3.1, only λ^0 and λ^N need not vanish. We have :

Proposition 3.3.4 *If $\dim E = n = 2r + 1$ and Q is not degenerate the center of $C(Q)$ is $Z = \mathbf{K}e_0 + \mathbf{K}e_N$.*

Z is called the 'quadratic extension of $\mathbf{K}$', considered as a ring.

The ideals of $C(Q)$. Let $\mathcal{A} \neq \{0\}$ be a two-sided ideal of $C(Q)$.

We reason as in the case where n is even, but e_N behaves differently since $e_Ne_i = e_ie_N$. A similar computation proves that $\mathcal{A}$ contains a central element of the form $e_0 + \alpha_Ne_N$, where $\alpha_N \in \mathbf{K}$.

- If $(e_N)^2 = (-1)^{\epsilon_1} \alpha_1 \alpha_2 \ldots \alpha_n$ in the notations of 3.1, where $\epsilon_1 = n(n-1)/2$, is not a square in $\mathbf{K}$ (i.e. $(-1)^r D$, D being the discriminant of Q, is not a square), then

$$\frac{(e_0 + \alpha_N e_N)(e_0 - \alpha_N e_N)}{e_0^2 - \alpha_N^2 (e_N)^2} = e_0 \in \mathcal{A}$$

and $\mathcal{A}$ is a trivial ideal. Then obviously Z is a field.

- If for some $j \in \mathbf{K}$, $(e_N)^2 = 1/j^2$, we take

$$u = \frac{e_0 + j e_N}{2} \quad \text{and} \quad v = \frac{e_0 - j e_N}{2}.$$

Then $vu = uv = 0$, $u^2 = u$ and $v^2 = v$; $Z = \mathbf{K}u \oplus \mathbf{K}v$ and Z is the direct composition of two fields, each of which is isomorphic to $\mathbf{K}$.

Furthermore, $C(Q) = C(Q)u \oplus C(Q)v$ since if $w \in C(Q)$, $w = we_0 = w(u+v) = wu + wv$ and $w'u = wv$ implies $w'u^2 = wvu$ and $w'u = 0$.

If $w \in \mathcal{A}$, $w = yu + zv$, $y, z \in C(Q)$, $u(yu + zv) = yu \in \mathcal{A}$ and $v(yu + zv) = zv \in \mathcal{A}$. If $zv \neq 0$ we repeat the reasoning used in the case where n is even (replacing the element v of the previous proof by zv). In the formula for z, e_N is replaced by $(u - v)/j$ and we form

$$\alpha_i(zv) + e_i(zv)e_i \in \mathcal{A}$$

eventually to establish the existence of an element λv in Z, $\lambda \in \mathbf{K}^*$.

The same can be done if $yu \neq 0$.

Finally, $\mathcal{A}$ contains $u + v = e_0$ if both yu and zv are non-zero : then $\mathcal{A}$ is trivial. Hence if $\mathcal{A}$ is non-trivial, it is either of the form $C(Q)u$ or $C(Q)v$. And since u and v are elements of Z, each ideal of $C(Q)u$ is an ideal of $C(Q)$, so $C(Q)u$ and $C(Q)v$ have no proper two-sided ideals and must be simple.

This can be summarized as follows :

Proposition 3.3.5 *If* $\dim E = 2r + 1$, *if* Q *is non-degenerate and if* D *is the discriminant of* Q :

1. *If* $(-1)^r D$ *is not a square in* $\mathbf{K}$, $C(Q)$ *is simple and its center is a field.*

2. *If* $(-1)^r D$ *is a square in* $\mathbf{K}$, $C(Q)$ *is the direct composition of two simple ideals :*

$$C(Q) = C(Q)u + C(Q)v, \quad u^2 = u, \quad v^2 = v, \quad uv = 0,$$

where u *and* v *are central elements. The center is the direct composition of two fields, each of which is isomorphic to* $\mathbf{K}$: $Z = \mathbf{K}u \oplus \mathbf{K}v$.

For instance, if Q is positive definite $(\mathbf{K} = \mathbf{R})$, 1 applies if $n = 3$, $(\bmod\, 4)$ and 2 applies if $n = 1$, $(\bmod\, 4)$. If $\mathbf{K} = \mathbf{C}$, 2 always applies.

3.3.3

Some results valid for $K = C$ will be summarized here; they rely on the algebraic closedness of C. Q is assumed of maximal index.

Theorem 3.3.3

- If E is a $(n = 2r)$-dimensional complex vector space and if Q is non-degenerate :

  1. $C(Q)$ is isomorphic to the endomorphism algebra of a (2^r)-dimensional complex vector space; $C(Q)$ is simple and central, $Z = C$.

  2. $C^+(Q)$ is the direct composition of two ideals, each of which is isomorphic to the endomorphism algebra of a (2^{r-1})-dimensional complex vector space, so $C^+(Q)$ is the direct composition of two simple ideals.

- If E is a $(n = 2r + 1)$-dimensional complex vector space and if Q is non-degenerate :

  1. $C^+(Q)$ is simple and central, and it is isomorphic to the endomorphism algebra of a (2^r)-dimensional complex vector space.

  2. $C(Q)$ is the direct composition of two simple ideals, each of which is isomorphic to the endomorphism algebra of a (2^r)-dimensional complex vector space.

$$C(Q) = C(Q)u \oplus C(Q)v, \quad u^2 = u, \quad v^2 = v, \quad uv = 0$$

and $Z = Cu + Cv = Ce_0 + Ce_N$.

Both cases 1 and both cases 2 correspond by an isomorphism if the dimensions match.

3.3.4 Degenerate Clifford algebras.

Now we will assume that the rank n_1 of Q is less than n. As in Chapter 1, 2.1, we set

$$E = E_1 \oplus F, \quad \dim E_1 = n_1,$$

F is the radical of (E, Q) and Q induces on E_1 a quadratic form Q_1 of rank n_1 (Lemma (1.2.2)).

Proposition 3.3.6 *The Clifford algebra $C(Q)$ is isomorphic to the graded tensor product of $C(Q_1)$ and $\wedge(F)$*

Proof. This is a special case of formula 3 in section 1.3. ∎

Proposition 3.3.7 *If Z is the center of $C(Q)$:*

- If $\mathrm{rk}(Q) = n_1$ is even, $Z = \wedge^+(F)$.

- If $\mathrm{rk}(Q) = n_1$ is odd, $Z = \wedge^+(F) \oplus e_{N_1} \wedge^+ (F)$, where $e_{N_1} = e_1 e_2 \ldots e_{n_1}$ is the product of the elements of an orthonormal basis for (E_1, Q_1).

Proof. Every element $u \in C(Q)$ can be written in the form

$$u = \sum (a^{HL} e_H f_L + b^M e_M + c^R f_R)$$

where H, L, M, R are index sets, $e_M, f_R, e_H f_L, \ldots$ are ordered products of elements of (e_i, f_j), $e_i \in E_1$ and $f_j \in F$, $a^{HL}, b^M, \ldots$ are scalars, $|H| = h_1 + h_2 + \ldots + h_s$, $e_H = e_{h_1} e_{h_2} \ldots e_{h_s}$ etc. If u is a central element and if we multiply from the left and from the right by f_P, then subtract, we obtain as terms involving only the (f_j) :

$$\sum_R c^R f_R f_P (1 - (-1)^{|R||P|}),$$

which must vanish, so either $c^R = 0$, f_R is of maximal degree or f_R is even. But f^R cannot be of odd maximal degree since it must commute with every e_L, and $\sum_R c^R f_R \in \wedge^+(F)$ follows.

$\sum_M b^M e_M$ must be central in $C(Q_1)$; by the previous results on the center of a non-degenerate Clifford algebra,

$$\sum_M b^M e_M \in \mathbf{K} \text{ if } n_1 \text{ is even,}$$
$$\sum_M b^M e_M \in \mathbf{K}e_0 + \mathbf{K}e_{N_1} \text{ if } n_1 \text{ is odd;}$$

but the latter case cannot occur since e_{N_1} cannot commute with f_P if $|P|$ is odd.

We are left with central elements of the form

$$\sum_{H,L} a^{HL} e_H f_L.$$

Given a term with even $|L|$, we multiply by e_P from the left and from the right, and see that $a^{HL} e_H$ is central in $C(Q_1)$; if n_1 is even, it lies in $\wedge^+(F)$, if n_1 is odd, in $(\mathbf{K}e_0 + \mathbf{K}e_{N_1}) \wedge^+ (F)$.

Given a term with odd $|L|$, we multiply by e_H and see that $|H|$ must be even. If e_H does not contain e_i as a factor, the multiplication by e_i will lead to $a^{HL} = 0$, so $e_H = e_{N_1}$ and n_1 is even, but then $e_{N_1} f_L$, $|L|$ odd, does not commute with $y \in \wedge^-(F)$ and the proof is complete.   ∎

The center of $\wedge(F)$ is easily seen to be $\wedge^+(F)$ is $\dim F$ is even and $\wedge^p(F) + \wedge^+(F)$ if $\dim F = p$ is odd. Also, an element of $\wedge(F)$ is invertible if and only if its zero degree term does not vanish.

3.4   Selected references.

- Atiyah, Bott, Shapiro, *Clifford Modules*, Topology, Vol. 3 Suppl. 1, 1964.

- N. Bourbaki, *Algèbre, Chapitre 9*, Hermann, Paris, 1959.

- C. Chevalley, *Theory of Lie groups*, Princeton University Press, 1946.

- Curtis, Reiner, *Representation theory of finite groups and associative algebras*, Interscience, 1966.

- Lam, *The algebraic theory of quadratic forms*, Benjamin, 1973.

Chapter 4

THE CLIFFORD GROUPS, THE TWISTED CLIFFORD GROUPS AND THEIR FUNDAMENTAL SUBGROUPS.

E is an n-dimensional vector space over $\mathbf{K}$ with quadratic form Q. In sections 1 and 2 we will assume that Q is non-degenerate.

4.1 Clifford groups.

4.1.1

Let G be the set of invertible $g \in C(Q)$ such that

$$\varphi_g(x) = gxg^{-1} \in E, \quad \forall x \in E.$$

Then clearly if $g' \in G$, $gg' \in G$; φ_g is linear and injective, hence surjective and $g^{-1} \in G$.

Definition 4.1.1 *The multiplicative group G of invertible $g \in C(Q)$ such that*

$$\varphi_g(x) = gxg^{-1} \in E, \quad \forall x \in E$$

is called the Clifford group.

The intersection of G and $C^+(Q)$ is called the *special Clifford group G^+*.

Theorem 4.1.1

1. φ *is a representation of G in $O(Q)$ whose kernel is the set of invertible elements in the center Z of $C(Q)$.*

2. $E \cap G$ *is the set of non-isotropic vectors in E and if $a \in E \cap G$, $-\varphi(a)$ is the symmetry relative to the hyperplane $(a)^\perp$.*

3. *If $\dim E = n = 2r$ then $\varphi(G) = O(Q)$, $\varphi(G^+) = SO(Q)$ and $\ker \varphi = \mathbf{K}^*$.*

4. *If $\dim E = n = 2r + 1$ then $\varphi(G) = \varphi(G^+) = SO(Q)$ and $\ker(\varphi|_{G^+}) = \mathbf{K}^*$.*

5. G is the set of elements in $C(Q)$ of the form $\lambda x_1 x_2 \ldots x_k$, where $\lambda \in \mathbf{K}^*$ and $x_1, x_2, \ldots, x_k$ are non-isotropic vectors, if n is even. If n is odd, this is true for G^+, in which case k must be even.

Proof.

- Clearly $\varphi_g = \varphi(g)$ is an isometry, since $Q(gxg^{-1}) = (gxg^{-1})^2 = gx^2g^{-1} = x^2 = Q(x)$. φ is a homomorphism from G to $O(Q)$. If $g \in \ker\varphi$, $\varphi_g = \mathrm{Id}_E$, $gx = xg$, $g \in G \cap Z$.

- Let $a \in E$, if $Q(a) = 0$, $a^2 = 0$ and a is not invertible in $C(Q)$. If $Q(a) \neq 0$, the inverse of a is $a(Q(a))^{-1}$ and

$$\varphi_a(x) = \frac{axa}{Q(a)} = \frac{a(2B(x,a) - ax)}{Q(a)} = -x + \frac{2B(x,a)}{Q(a)}a,$$

hence $a \in E \cap G$ and $-\varphi(a)$ is the symmetry relative to $a^\perp$.

- If n is even and $u \in O(Q)$, then u is a product $u_1 u_2 \ldots u_h$, the u_i being symmetries relative to hyperplanes $(x_1)^\perp, (x_2)^\perp, \ldots, (x_h)^\perp$, where the x_i are non-isotropic vectors (Chapter 1, 3.1). Write

$$u_1 = -\varphi(x_1), \ldots, u_h = -\varphi(x_h),$$
$$g = x_1 x_2 \ldots x_h,$$
$$\varphi(g) = (-1)^h u_1 u_2 \ldots u_h = (-1)^h u.$$

If h is even, $g \in G^+$ and $\varphi(G^+) \supseteq SO(Q)$. If the dimension is even, $e_N = e_1 e_2 \ldots e_n$ anticommutes with all $x \in E$, $\varphi(e_N) = -\mathrm{Id}_E$. If h is odd, $\varphi(ge_n) = u$, so

$$\varphi(G) = O(Q).$$

We only have to verify that $\varphi(G^+) \subseteq SO(Q)$. If $u \in O(Q) \setminus SO(Q)$, h is odd and if a $t \in G^+$ would exist for which $\varphi(t) = u = \varphi(ge_N)$ then $ge_N t^{-1} \in Z$, but since $Z = \mathbf{K}$, this is impossible, as $ge_N t^{-1} \in C^-(Q)$.

- If n is odd, the previous reasoning shows that

$$\varphi(G^+) \supseteq SO(Q).$$

$(-\mathrm{Id}_E)$ is an element of $O(Q) \setminus SO(Q)$ and every element of $O(Q)$ is generated by $(-\mathrm{Id}_E)$ and $SO(Q)$. So we only have to verify that $(-\mathrm{Id}_E)$ is not an element of $\varphi(G)$.

If for some $t \in G$, $txt^{-1} = -x$, $\forall x \in E$, $te_i t^{-1} = -e_i$ would imply $te_N t^{-1} = -e_N$, but this contradicts the fact that $e_N \in Z$.

The results on the kernels are obvious. Finally, 5 is a consequence of the previous proof. ∎

4.1.2   Reduced Clifford groups.

Lemma 4.1.1 *Let β be the main anti-automorphism of $C(Q)$. Then $\forall g \in G^+$, $\beta(g)g \in K^*$ and $N : g \to \beta(g)g$ is a homomorphism from G^+ to K^*.*

Proof. If $x \in E$ and $g \in G$, $x' = gxg^{-1}$ is invariant under β :

$$\beta(x') = \beta(g^{-1})\beta(x)\beta(g)$$

so $\beta(G) = G$ and $\beta(G^+) = G^+$. $\varphi(\beta(g)g) = \mathrm{Id}_E$, so $\beta(g)g \in Z$, but if $g \in G^+$, as $Z^+ = K$, $\beta(g)g \in K^*$.

$$N(gg') = \beta(gg')gg' = \beta(g')\beta(g)gg' = N(g)N(g').$$

Note that if $n = 2r$, $Z = K$ and the lemma holds for $g \in G$.   ∎

Definition 4.1.2 *If $g \in G^+$ or if $g \in G$ with $n = 2r$, the scalar*

$$N(g) = \beta(g)g$$

is called the spin norm on g.

Definition 4.1.3 *The kernel G_0 of N in G^+ is called the special reduced Clifford group, it is the set of $g \in G^+$ such that $\beta(g)g = 1$. If n is even, the kernel G_0 of N in G is called the reduced Clifford group, $G_0^+ = G_0 \cap C^+(Q)$.*

4.1.3   Spin groups.

- Assume that $n = 2r$.

    Definition 4.1.4 *The subgroup of $g \in G$ for which $N(g) = \pm 1$ is called $\mathrm{Pin}(Q)$ and the intersection $\mathrm{Pin}(Q) \cap C^+(Q)$ is called $\mathrm{Spin}(Q)$.*

    In the real case (which is by far the most important), we have that :

    - If Q is positive definite, as $g = \lambda x_1 x_2 \dots x_k$, $\beta(g)g > 0$ hence

    $$\mathrm{Spin}(Q) = G_0^+ \subset G_0 = \mathrm{Pin}(Q).$$

    - If Q is negative definite,

    $$\mathrm{Spin}(Q) = G_0^+ = G_0 \subset \mathrm{Pin}(Q).$$

    - If Q is indefinite,

    $$G_0^+ \subset G_0 \subset \mathrm{Pin}(Q), \quad G_0^+ \subset \mathrm{Spin}(Q).$$

Clearly, $\varphi(\mathrm{Pin}(Q)) = O(Q)$, $\varphi(\mathrm{Spin}(Q)) = SO(Q)$ and $\ker \varphi = \pm 1$.

- Assume that $n = 2r + 1$.

Definition 4.1.5 *The subgroup of* $g \in G^+$ *such that* $N(g) = \pm 1$ *is called* $\mathrm{Spin}(Q)$.

In the real case :

- If Q is definite : $G_0^+ = \mathrm{Spin}(Q)$.
- If Q is indefinite : $G_0^* \subset \mathrm{Spin}(Q)$.

$\varphi(\mathrm{Spin}(Q)) = O(Q)$ and the kernel of φ is ± 1.

Whenever $\ker \varphi$ is discrete, we will say that φ is a *covering*.

4.2 Twisted Clifford groups or Clifford α-groups.

4.2.1

The groups introduced this far do not allow to reach the whole of $O(Q)$ if n is odd. Therefore new definitions are in order.

Definition 4.2.1 *If* α *is the main automorphism of* $C(Q)$, *the group of invertible elements* $g \in G$ *such that* $\alpha(g)xg^{-1} \in E$, $\forall x \in E$, *is called the* Clifford α-group Γ.

We also define $\Gamma^+ = \Gamma \cap C^+(Q)$.
  Let $p_g(x) = \alpha(g)xg^{-1} = x' \in E$, then

$$x'\alpha(x') = -x'^2, \quad \alpha(g)xg^{-1}g\alpha(x)\alpha(g^{-1}) = -\alpha(g)x^2\alpha(g^{-1}) = -x'^2$$

so $x^2 = x'^2$ and $p(g) = p_g$ is an element of $O(Q)$. Clearly p is then a representation of Γ in $O(Q)$.
  If $a \in E$, $Q(a) \neq 0$,

$$p_a(x) = -axa^{-1} = \frac{-axa}{Q(a)} = \frac{-a(2B(x,a) - ax)}{Q(a)} = x - \frac{2B(x,a)a}{Q(a)} \in E$$

and p_a is the symmetry relative to $(a)^{\perp}$. If $u = u_1 u_2 \ldots u_h$ where $u_i = p_{x_i}$, $Q(x_i) \neq 0$, $g = x_1 x_2 \ldots x_h$ satisfies $p(g) = u$. Then $p(\Gamma) = O(Q)$ and $p(\Gamma^+) = SO(Q)$.
  What is the kernel of p ? If $u \in \ker p$, $\alpha(u)x = xu$, $\forall x \in E$ and if $u = u^+ + u^-$, $u^+x = xu^+$ and $u^-x = -xu^-$.
  Let (e_i) be an orthogonal basis of E; in the basis $(e_{i_1} e_{i_2} \ldots e_{i_m})$ of $C(Q)$, we will distinguish the elements involving e_1 and those where e_1 does not occur as a factor.

$u^+ = v_0 + e_1 v_1$ where v_0 and v_1 are linear combinations of elements in which e_1 does not occur.

$u^+ e_1 = e_1 u^+$ implies that

$$v_0 + e_1 v_1 = e_1 v_0 e_1^{-1} + (e_1)^2 v_1 e_1^{-1} = v_0 - e_1 v_1$$

and v_1 must vanish. Replacing e_1 by $e_2, \ldots, e_n$, we see that $u^+ \in \mathbf{K}^*$. Similarly, $u^- = w_0 + e_1 w_1$, $u^- e_1 = -e_1 u^-$ leads to $w_0 + e_1 w_1 = w_0 - e_1 w_1$ and $w_1 = 0$ so that $u^- = 0$ since $\mathbf{K}^* \not\subset C^-(Q)$.

Finally, we see that $u \in \mathbf{K}^*$. Furthermore, Γ is the set of $(\lambda x_1 x_2 \ldots x_k)$ such that $Q(x_i) \neq 0$ and $\lambda \in \mathbf{K}^*$, whereas Γ^+ is obtained in the same way, restricting k to even values.

We can formulate the

Theorem 4.2.1 *If $p_g(x) = \alpha(g) x g^{-1}$, $x \in E$, $g \in \Gamma$:*

1. *p is a representation of Γ in $O(Q)$ whose kernel is $\mathbf{K}^*$ and $p(\Gamma^+) = SO(Q)$.*

2. *If a, $Q(a) \neq 0$, is an element of E, $p(a)$ is the symmetry relative to $(a)^\perp$.*

3. *Γ is the set of elements in $C(Q)$ of the form $\lambda x_1 x_2 \ldots x_k$, x_i being a non-isotropic vector, $i = 1, 2, \ldots, k$, $\lambda \in \mathbf{K}^*$, and Γ^+ is obtained by the even k.*

4. *$\Gamma = G$ if n is even, $\Gamma^+ = G^+$ if n is odd.*

Note that if n is odd, $\Gamma \subset G$. Indeed, $\lambda x_1 x_2 \ldots x_k \in G$ but $\lambda + \mu e_N = \gamma$ where λ and μ are non-zero and γ is invertible, is an element of G but not of Γ in general.

Spin α-groups or twisted spin groups.

Lemma 4.2.1 *For every $g \in \Gamma$, $\beta(\alpha(g))g = \tilde{N}(g)$ is a scalar and*

$$\tilde{N}(gg') = \tilde{N}(g)\tilde{N}(g').$$

Proof. This can be verified by an analogous computation to that of Lemma (4.1.1), or directly by Theorem (4.2.1, part 3).   ■

Note that when both are defined, $N(g) = \pm \tilde{N}(g)$.

Definition 4.2.2 *The subgroup of elements $g \in \Gamma$ such that $\tilde{N}(g) = \pm 1$ is called $Pin^\alpha(Q)$ and its subgroup $Pin^\alpha(Q) \cap C^+(Q)$ is called $Spin^\alpha(Q)$.*

If n is even, $Pin^\alpha(Q) = Pin(Q)$ and $Spin^\alpha(Q) = Spin(Q)$. If n is odd, $Spin(Q) = Spin^\alpha(Q)$. Although φ and p only coincide on $\Gamma^+ = G^+$ we will use the notations $Pin(Q)$ and $Spin(Q)$ regardless of the parity of n. This will not cause confusion; but if n is odd it should be noted that $Pin(Q)$ is a subgroup of Γ.

4.2.2 The torogonal spin groups.

Let us assume that E is real and let E', Q' and $C(Q')$ be the complexified versions of E, Q and $C(Q)$.

Definition 4.2.3 *The multiplicative group of invertible* $g \in C(Q')$ *such that* $p_g(x) = \alpha(g)xg^{-1} \in E$, $\forall x \in E$ *and* $|\tilde{N}(g)| = 1$ *is called the torogonal group* $\mathrm{Tor}(Q)$. *Also,* $\mathrm{Tor}^+(Q)$ *is defined as* $\mathrm{Tor}(Q) \cap C^+(Q')$.

Clearly $p(\mathrm{Tor}(Q)) = O(Q)$ and $p(\mathrm{Tor}^+(Q)) = SO(Q)$. The kernel of p can be determined using the same method as for Γ, it satisfies $\ker p = S^1 = U(1)$, and $\mathrm{Tor}(Q)$ is the set of elements in $C(Q')$ of the form

$$g = \lambda x_1 x_2 \ldots x_k, \quad \lambda \in S^1, \quad x_i \in E, \quad Q(x_i) = \pm 1, \quad i = 1, 2, \ldots, k.$$

Since we can simultaneously replace x_i by $-x_i$ and λ by $-\lambda$ in the previous factorization, we write

$$\mathrm{Tor}(Q) = S^1 \times_{\mathbf{Z}_2} \mathrm{Pin}(Q).$$

This factorization justifies the name of *reduced enlarged spin group* [1], its elements can be written as

$$g = \exp(i\theta)\gamma, \quad \gamma \in \mathrm{Pin}(Q), \quad \theta \in \mathbf{R}.$$

4.2.3 The algebras $C(Q)$ and $C(-Q)$ and their associated groups.

In Chapter 3, 2 above it was shown that the algebras corresponding to Q and $-Q$ have, in general, different structures : although $C(8k, 0)$ and $C(0, 8k)$ are both simple real matrix algebras, and $C(8k + 4, 0)$ and $C(0, 8k + 4)$ are both simple quaternion matrix algebras, all other pairs of algebras where Q is positive definite and $-Q$ therefore negative definite, have a different structure.

Definition 4.2.4 *Let* F *and* G *be graded algebras and let* $f : F \to G$ *be a degree-preserving linear mapping such that for all pairs* a_1, a_2 *of* F, *whose degrees are* p_1 *and* p_2,

$$f(a_1 a_2) = (-1)^{p_1 p_2} f(a_1) f(a_2).$$

Then f *is called a left graded algebra homomorphism.*

Proposition 4.2.1 *There exists a left isomorphism from the* $\mathbf{Z}_2$-*graded algebra* $C(Q)$ *to the* $\mathbf{Z}_2$-*graded algebra* $C(-Q)$.

---

[1]Later on, we will introduce enlarged spin groups where S^1 is replaced by $\mathbf{C}^*$. We could also define the subgroup $\mathrm{pin}(Q)$ of elements $g \in \mathrm{Tor}(Q)$ such that $N(g) = 1$ (cf. J. Popovici).

Proof. Consider $\rho_{-Q} : E \to C(-Q)$, then $(\rho_{-Q}(x))^2 = -Q(x)$. Extend ρ_{-Q} to a mapping $v : \otimes(E) \to C(-Q)$, by

$$v(x_1 \otimes x_2 \otimes \ldots \otimes x_h) = \rho_{-Q}(x_1) \otimes \rho_{-Q}(x_2) \ldots \otimes \rho_{-Q}(x_h)(-1)^{[h/2]},$$

$[h/2]$ being the integer part of $h/2$. Then v is a naturally defined left homomorphism as can be verified if the parity of $[(h + h')/2] - [h/2] - [h'/2]$ is proved equal to that of hh'.

Since $v(x \otimes x - Q(x)) = -(\rho_{-Q}(x))^2 - Q(x) = 0$, by a reasoning similar to that used to prove the universal property, v will give rise to a mapping from $C(Q)$ to $C(-Q)$ which will be a left homomorphism and, due to the dimensions, is necessarily an isomorphism. ∎

Corollary 4.2.1 $C^+(Q)$ and $C^+(-Q)$ are isomorphic (in the usual sense).

Proof. This is a direct consequence of their definition. ∎

Corollary 4.2.2 The underlying sets of $\Gamma(Q)$ and $\Gamma(-Q)$ can be identified and the groups $\Gamma(Q)$ and $\Gamma(-Q)$ are isomorphic.

Proof. This follows from Theorem (4.2.1, part 3) and the isomorphism from $\Gamma^+(Q)$ to $\Gamma^+(-Q)$ follows from Corollary (4.2.1). ∎

4.3 Degenerate Clifford groups.

4.3.1

We use the notations of Chapter 3, 3.4. The rank of Q is $n_1 < n$ and $E = E_1 \oplus F$ where $\dim E_1 = n_1$ and F is the radical of (E, Q). Q induces a quadratic form Q_1 of rank n_1 on E_1; its associated symmetric bilinear form will be written B_1.

The definition of the twisted Clifford group Γ is the same as in section (4.2).

Proposition 4.3.1 p is a homomorphism from Γ to the monoid $O(E, Q)$ of isometries of (E, Q) whose kernel is the set $\wedge(F)^*$ of invertible elements in $\wedge(F)$.

Proof. One easily sees that $\wedge(F)^*$ is a subset of $\ker p$. The computation of this kernel is completely similar to the non-degenerate case, all other statements are obvious. ∎

Proposition 4.3.2 p is a homomorphism onto the group Σ of isometries of (E, Q) whose restriction to F is the identity.

Proof. Indeed, $p(g)$ is clearly injective, so it is an isomorphism of E, which obviously has to be an element of Σ. We will prove that $p(\Gamma) = \Sigma$.

If $a \in E$ and $Q(a) \neq 0$, we can write $a = a_1 + b$ where $a_1 \in E_1$ and $b \in F$. Then

$$p_a(x) = x - \frac{2B(x, a)}{Q(a)} a$$

if $b = 0$,

$$p_{a_1} = x - \frac{2B(x, a_1)}{Q(a_1)} a_1$$

and p_{a_1} is the symmetry relative to $(a_1)^{\perp}$ in the space (E_1, Q_1).

Choosing a_1 for which $Q(a_1) \neq 0$, any element of $O(Q_1)$ can be obtained by the composition of such symmetries; more precisely, if (e_k) is an orthogonal basis of (E_1, Q_1) and if (f_j) is an arbitrary basis of F, a $g \in \Gamma$ can be found such that $p_g(e_i) = \sum_k a_i^k e_k$ and $p_g(f_j) = f_j$, where the matrix (a_i^k) corresponds to an isometry in (E_1, Q_1). Now consider $\gamma_k = 1 + c^i e_k f_i$, k fixed, the c^i being arbitrary coefficients,

$$\gamma_k^{-1} = 1 - c^i e_k f_i,$$

$p_{\gamma_k}(e_k) = e_k - 2c^i(e_k)^2 f_i$, $p_{\gamma_k}(e_l) = l$ if $l \neq k$ and $p_{\gamma_k}(f_j) = f_j$. For fixed s we therefore obtain

$$p_{\gamma_s}(e_k) = e_k - 2\sum_i g_{ks} c^{si} f_i, \quad k = 1, 2, \ldots, n_1,$$

where the g_{ks} are the components of B_1 in the chosen basis. Using $\gamma_1, \gamma_2, \ldots, \gamma_{n_1}$ and $\gamma = \gamma_1 \gamma_2 \ldots \gamma_{n_1}$ we obtain

$$p_{\gamma}(e_k) = e_k - 2\sum_{s,i} g_{ks} c^{si} f_i = e_k + \sum b_k^i f_i$$

where the b_k^i are the components of an arbitrary $(n - n_1) \times n_1$ matrix.

If we compose these transformations of g and γ, the result is obtained, since Σ is the semi-direct product of the matrix endomorphism groups $\begin{pmatrix} A & 0 \\ 0 & I_{n_2} \end{pmatrix}$ and $\begin{pmatrix} I_{n_1} & 0 \\ A' & I_{n_2} \end{pmatrix}$ where A is an isometry matrix for (E_1, Q_1), A' is an $(n - n_1) \times n_1$ matrix and I_{n_1}, I_{n_2} are unit matrices $(n_2 = n - n_1)$. ∎

Corollary 4.3.1 *When $n_1 = n - 1$, $p(\Gamma)$ is isomorphic to the group of affine transformations of E which, on E_1, induce an isometry for Q_1.*

Proof. A' can indeed be identified with an isometry of E_1. ∎

Proposition 4.3.3 *Every $g \in \Gamma$ can be factored as :*

$$g = a_1 a_2 \ldots a_k \gamma_1 \gamma_2 \ldots \gamma_{n_1} v$$

where $a_i \in E_1$, $Q(a_i) \neq 0$, $\gamma_k = 1 + \sum_i c^{ik} e_k f_i$ and $v \in \wedge(F)^*$.

We could have written $\gamma_k = \exp(\sum_i c^{ik} e_k f_i)$ and

$$\gamma = \gamma_1 \gamma_2 \ldots \gamma_{n_1} = \exp(\sum_{i,k} c^{ik} e_k f_i)$$

(the convergence does not raise any problem here). We will consider the exponential on Clifford algebras more generally later on.

4.3.2 Spin groups.

As before, considering the center of $C(Q)$ leads to $\tilde{N}(g) = \beta(\alpha(g))g \in \wedge^+(F)^*$ and

$$\tilde{N}(gg') = \tilde{N}(g)\tilde{N}(g').$$

Note that $v \in \wedge(F)^*$ need not be reduced to a scalar, even if $\tilde{N}(v) \in \mathbf{K}^*$, so that to consider only the elements for which $\tilde{N}(g) = \pm 1$ does not necessarily lead to a discrete $\ker p$. Therefore we introduce the following

Definition 4.3.1 *Let θ be the normal subgroup of elements if Γ which have the form $a + u$, where $a \in (\mathbf{K}^*)^2$, and $u \in \wedge(F)$ is of non-zero degree. Then the quotient group Γ/θ will be called $\mathrm{Pin}(Q)$.*

Taking the quotient, $p : \Gamma \to \Sigma$ gives a $\tilde{p} : \mathrm{Pin}(Q) \to \Sigma$ for which $\ker(\tilde{p}) = \{-1, 1\}$.
   By Proposition (4.3.3), every element of $\mathrm{Pin}(Q)$ is a product

$$a_1 a_2 \ldots a_k \exp(\sum_{i,k} c^{ik} e_k f_i), \quad Q(a_i) = \pm 1.$$

Here k is even if an element of $\mathrm{Spin}(Q)$ is considered.

Remark. We could define the quotient group $\hat{\Gamma} = \Gamma/\Delta$, where Δ is the invariant subgroup of elements of the form $1 + u$, $u \in \wedge(F)$ being of non-zero degree, to obtain from $\tilde{N}$ a mapping $\hat{N}$ from Γ/Δ to $\wedge(F)/\Delta$. Then $\mathrm{Pin}(Q)$ would be defined by elements $\hat{g}$ such that $\hat{N}(\hat{g}) = \pm 1$.

4.4 Selected references.

- N. Bourbaki, *Algèbre, Chapitre 9*, Hermann, Paris, 1959.

- C. Chevalley, *The algebraic theory of spinors*, Columbia University Press.

● A. Crumeyrolle, *Algèbres de Clifford dégénérées et revêtements des groupes conformes affines orthogonaux et symplectiques*, Ann. Inst. Poincaré, Vol. 33, no. 3, 1980, pp. 235–249.

● M. Karoubi, *Algèbres de Clifford et K-théorie*, Ann. Scient. Ec. Norm. Sup., 4ᵉᵐᵉ série, t. 1, 1968, pp. 161–270.

● J. Popovici, *Représentations irréductibles des fibrés de Clifford*, Ann. Inst. H. Poincaré, Vol. 25, no 1, 1976, pp. 35–39.

Chapter 5

SPINORS AND SPIN REPRESENTATIONS.

5.1 Notions on the representations of associative algebras and on semi-simple rings.

5.1.1 Representations of an associative algebra A.

Definitions.

Let A be an associative algebra on the commutative field $\mathbf{K}$ ($\mathbf{R}$ or $\mathbf{C}$), $\mathbf{K} \subseteq A$. If $E \neq 0$ is a vector space, a homomorphism ρ from A to $\operatorname{End} E$ which maps the unit element of A to Id_E is called a *representation of A in E*.

The *degree* of the representation is the dimension of E. The addition in E and the mapping $A \times E \rightarrow E : (a, x) \rightarrow \rho(a)x$ turn E into a A-module called the *representation module*. Conversely, if A is an algebra over $\mathbf{K}$ and if E is an A-module, E is a vector space over $\mathbf{K}$ and if $a \in A$, the mapping $\gamma : a \rightarrow \gamma_a$, with $\gamma_a(x) = ax$, $x \in E$, is a homomorphism from A to $\operatorname{End} E$, so it is a representation of A in E.

Therefore the study of A-modules is equivalent to that of the representations of A.

$\rho(a)(x)$ will often be written $a.x$, $a * x$ or ax.

A representation ρ is said to be *faithful* if the kernel of ρ is zero : $\rho(a)x = 0$, $\forall x \in E$ should imply that $a = 0$. In general the *kernel* of ρ, which is the set of $a \in A$ such that $a.x = 0$, $\forall x \in E$, is also called the *annihilator* of its module.

A representation ρ is said to be *simple* or *irreducible* if the only invariant subspaces of $\rho(a)$, $\forall a \in A$, are E and $\{0\}$. This also means that the representation module is simple, i.e. it has no proper submodule.

A representation is called *semi-simple* or *completely reducible* if its module is semi-simple, i.e. it is the direct sum of simple modules. This also means that E is the direct sum of subspaces which are globally invariant under $\rho(a)$, $\forall a \in A$.

Equivalence of two representations. Two A-modules E and E', whose exterior multiplications will be written . and $*$, are *isomorphic* if there exists a bijection φ from E to E' such that

68

$$\varphi(x + x') = \varphi(x) + \varphi(x') \text{ and}$$
$$\varphi(a.x) = a * \varphi(x), \quad \forall a \in A, \quad \forall x, x' \in E.$$

Accordingly, two representations are called *equivalent* if their modules are isomorphic, which amounts to the existence of a $\mathbf{K}$-linear isomorphism φ from E to E' such that $\varphi \circ \rho(a) = \rho'(a) \circ \varphi, \forall a \in A$, or

$$\rho'(a) = \varphi \circ \rho(a) \circ \varphi^{-1}.$$

If $\dim E = n$, then $\dim E' = n$.

5.1.2 Simple A-modules.

We will only assume that A is a ring with unit.

1. For the module M over A, to be simple just means that $M = Ax$ for all $x \neq 0$, since Ax is a submodule of M containing $x = 1.x$; a simple module is therefore generated by a single $x \neq 0$.

2. A_S being the module obtained from A under left multiplication, a left ideal J of A can be identified to a submodule of A_S and vice versa. This ideal is said to be maximal if it differs from A and if it is not part of any other proper left ideal in A. A_S/J is simple if and only if J is maximal, for if $J' \supseteq J$ is another left ideal, $A_S/J' \subseteq A_S/J$.

3. If M is simple, the annihilator J of $x \neq 0$ is a maximal left ideal of A. Indeed, $\varphi : a \rightarrow ax$ has the left ideal J as kernel, but $A_S/J = M$ and M is simple. Conversely, every left maximal ideal of A is the annihilator of some element of a simple A-module : A_S/J is simple if J is maximal, by 1, $A_S/J = A\bar{\alpha}_0$, denoting by $\bar{\alpha}_0$ the equivalence class of α_0 modulo J; we can take $\bar{\alpha}_0 = \bar{1}$ since if $1 \in J$, J is trivial. Then $A_S/J = A.\bar{1}$ and $a.\bar{1} = \bar{0}$ if and only if $a.1 = a \in J$.

5.1.3 Semi-simple modules.

Let us consider a module obtained as the direct sum of simple modules.

Lemma 5.1.1 *Let $M = \oplus_{i \in I} M_i$ where M_i is simple. If N is a submodule of M, there exists a subset J of I such that*

$$M = N \oplus (\oplus_{j \in J} M_j)$$

and a semi-simple supplement to N in M can therefore be found.

Proof. We have that $M = N + \sum_{i \in I} M_i$. Now define $H_1 = N$, $H_2 = N + M_1 = H_1 + M_1$, $H_{k+1} = H_k + M_k$. Then $H_k \cap M_k$ is a submodule of M_k so it is either zero or identical to M_k. If the intersection is M_k, $H_{k+1} = H_k + M_k = H_k$ and H_{k+1} can be deleted; if the intersection is zero, H_{k+1} is kept. ∎

Lemma 5.1.2 *Every submodule and every quotient module of a semi-simple module are semi-simple.*

Proof. Let M be the direct sum of the M_i, $i \in I$, which are simple. If $M = N \oplus F$ where $F = \oplus_{j \in J} M_j$, $M = G \oplus F$ where $G = \oplus_{i \notin J} M_i$. $N \simeq M/F$ and $G \simeq M/F$ so N is isomorphic to G and must be semi-simple. Since $M/N = F$, every quotient module is semi-simple. ∎

Corollary 5.1.1 *If M is a direct sum of simple modules $(M_i)_{i \in I}$, every simple submodule of M is isomorphic to one of the M_i.*

Proposition 5.1.1 *The decomposition of a semi-simple module as a direct sum of simple modules is unique up to order and isomorphisms.*

Proof. Let $M = M_1 \oplus \ldots \oplus M_k = N_1 \oplus \ldots \oplus N_l$ and assume that $k \leq l$.
    We will prove by induction that

$$M = (N_1 \oplus \ldots \oplus N_j) \oplus (M_{j+1} \oplus \ldots \oplus M_k)$$

where $N_1 \simeq M_1, \ldots, N_j \simeq M_j$, $0 \leq j \leq k$.
    If we already know that

$$M = N_1 \oplus \ldots \oplus N_{j-1} \oplus M_j \oplus \ldots \oplus M_k,$$

M is the direct sum of

$$N_1 \oplus \ldots \oplus N_{j-1} \oplus M_{j+1} \oplus \ldots \oplus M_k$$

and some N_p, where $p \geq j$, by Lemma (5.1.1). Comparing the last two decompositions of M, we get $M_j = \oplus_{p \geq j} N_p$, so $\oplus N_p$ reduces to a single term which we will write as N_j and the proposition follows. ∎

Definition 5.1.1 *If N is a simple A-module, let λ be the class of A-modules isomorphic to N : the sum of the terms in $\oplus_{i \in I} M_i$ which are isomorphic to N are called the* isotypic component of type λ *of the semi-simple module M. If all M_i are isomorphic, M is said to be* isotypic.

5.1.4 Semi-simple rings.

We will first prove the

Lemma 5.1.3 *The following are equivalent :*

1. *A_S is semi-simple.*

2. *Every A-module M is semi-simple.*

Proof. Let $A_s^{(I)}$ be the direct sum of a family of modules indexed by I, all of which are equal to A_s. Each $x \in A_s^{(I)}$ can be written uniquely as

$$x = \sum_{i \in I} \xi^i e_i.$$

To this element we will associate $\sum_{i \in I} \xi^i a_i$, if a_i, $i \in I$ are a set of generators for the module M. A homomorphism u will be defined from $A_s^{(I)}$ to M such that its kernel is the submodule N of elements in $A_s^{(I)}$ for which $\sum \xi^i a_i = 0$; then $M \simeq A_s^{(I)}/N$. If A_s is semi-simple, $A_s^{(I)}$ is semi-simple, hence M is semi-simple (by Lemma (5.1.2)).

1 is a special case of 2. ∎

A ring for which 1 or 2 holds is called *semi-simple*.

Proposition 5.1.2 *If A is a semi-simple ring and if J is a left ideal of A, there exists an idempotent element $e \in J$ such that $J = Ae = Je$.*

Proof. Let J' be the supplement to J in A, then J' is semi-simple by Lemma (5.1.1). $1 = e + e'$ for some $e \in J$, $e' \in J'$, so $\forall x \in J$, $x = xe + xe'$, $xe \in J$, $xe' \in J'$ and $x = xe$ and $xe' = 0$ follow. (In particular, $ee' = 0$). From this we can deduce $e^2 = e$, $J = Je$ and $Ae \subseteq J = Je \subseteq Ae$ implies $J = Ae$. ∎

Proposition 5.1.3 *Let A be a semi-simple ring, then the isotypic components of A can be identified with the minimal two-sided ideals of A. Every two-sided ideal of A is the direct composition of minimal two-sided ideals.*

Proof. $M = A_S$, M_λ being the isotypic component of type λ, M_λ is a two-sided ideal.

If δ_b is the right multiplication by b in A, δ_b is an endomorphism of A_s and $\delta_b(M_\lambda)$ is isomorphic to a quotient of M_λ, hence it is part of M_λ and M_λ is a two-sided ideal.

If a two-sided ideal N contains M_i, which is of type λ, $M_i = Ae$ where e is idempotent and N must contain all M_j of type λ.

Indeed, M_j being of type λ must be isomorphic to M_i by an isomorphism φ :

$$\varphi(x) = \varphi(xe) = x\varphi(e) = xb, \quad b = \varphi(e), \quad x \in M_i.$$

$M_j = \{xb, x \in M_i\} \subseteq N$.

Every two-sided ideal is a submodule of A_S, hence it is the sum of some simple submodules such as M_i; every two-sided ideal is therefore the sum of isotypic components and it is minimal if it is itself isotypic.

One easily verifies that the composition is direct : if $A_s = \oplus_{\lambda_i} M_{\lambda_i}$ where the M_{λ_i} are the isotypic components of type λ_i,

$$x = x_{\lambda_1} + x_{\lambda_2} + \ldots + x_{\lambda_k}, \quad x_{\lambda_i} \in M_{\lambda_i},$$

$x_{\lambda_i} x_{\lambda_2} \in M_{\lambda_i}$, $x_{\lambda_1} x_{\lambda_2} \in M_{\lambda_2}$ so $x_{\lambda_1} x_{\lambda_2} = 0$. ∎

Proposition 5.1.4 *Every simple module on a semi-simple ring A is isomorphic to a minimal left ideal of A.*

Proof. Indeed, every simple A-module is isomorphic to a quotient of A_S cf. section 1.2, hence to a simple submodule of A, since A_S is semi-simple.   ∎

5.1.5 Simple rings.

Proposition 5.1.5 *Let A be semi-simple, then the following are equivalent :*

1. *All simple A-modules are isomorphic.*

2. *The semi-simple A-module A_S is isotypic.*

3. *A and (0) are the only two-sided ideals of A*

Such a ring is called simple.

Proof. Clearly 1 implies 2 and 2 and 3 are equivalent by Proposition (5.1.1). 1 follows from 2 by Proposition (5.1.4) since every simple A-module is isomorphic to a minimal left ideal of A, i.e. to a simple submodule of A_s, so there exists only one type.   ∎

5.1.6 The radical.

Definition 5.1.2 *A maximal submodule of M is a submodule which differs from M and is not contained in any other submodule than M. The* radical *of M is the intersection of all maximal submodules of M.*

If A is a ring, the radical of A, written $\mathcal{R}(A)$, is the radical of A_S, i.e. the intersection of all maximal left ideals in A.

Lemma 5.1.4 *The radical of A is a two-sided ideal, either zero or proper.*

Proof. $\alpha \in \mathcal{R}(A)$ means that α belongs to all maximal left ideals of A, so it lies in the annihilator of each element $x \neq 0$ of each simple A-module; α is therefore an element of the annihilator of each simple left A-module. But the annihilator of an A-module is a two-sided ideal, so the radical is the intersection of two-sided ideals and therefore itself a two-sided ideal.   ∎

Proposition 5.1.6 *Every finite-dimensional algebra A over $\mathbf{K}$ whose radical is zero, is semi-simple.*

Proof. Let A_s be a finite-dimensional space over $\mathbf{K}$.

Consider the set $\mathcal{E}$ of submodules which are the intersection of a finite number of maximal submodules of A_s. The elements of $\mathcal{E}$ are also vector spaces; let R be an element of maximal dimension and let N be a maximal submodule of A_s. Since $N \cap R \subseteq R$ by the defining property for R, $N \cap R = R$, so $R \subseteq N$ for all maximal N, which can only hold if $R = 0$ and therefore $\mathcal{R}(A) = 0$.

Hence there exists a finite family M_i, $1 \le i \le p$ of maximal submodules of A_s whose intersection is not zero. Let $\varphi_i : x \in A_s \to \varphi_i(x) \in A_s/M_i$ be the canonical homomorphism, then $\prod_i(\varphi_i)$ is an injective homomorphism from A_S to $\prod_i(A_S/M_i)$ where

$$\prod_{i=1}^{p} \varphi_i : x \to (\varphi_1(x), \ldots, \varphi_p(x)) \in \prod_{i=1}^{p}(A_S/M_i),$$

since $\prod_i(\varphi_i)(x) = \prod_i \varphi_i(x')$ implies that $x - x' \in M_i$, $i = 1, 2, \ldots, p$ and $x = x'$.

M_i being maximal, A_S/M_i is simple, hence $\prod_{i=1}^{p}(A_S/M_i)$ is semi-simple and $\prod_i(\varphi_i)$ being an injective homomorphism, A_s is semi-simple. ∎

Corollary 5.1.2 *A finite-dimensional algebra over the field $\mathbf{K}$ is a simple ring if and only if it has no proper two-sided ideals.*

Proof. This follows from the previous results and from the definition of simple rings. ∎

This corollary justifies the terminology of Chapter 3, where a Clifford algebra was defined to be simple if its two-sided ideals were trivial.

5.1.7 Some results on the idempotent elements of semi-simple rings.

1. We have proved that if A is a semi-simple ring and if J is one of its left ideals, there exists an idempotent element $e \in J$ such that

$$J = Ae = Je, \quad x = xe, \quad x \in J.$$

   Note that the same left ideal can be determined by several different idempotent elements. A non-trivial example of this fact will be given in the Clifford algebra (Chapter 8, 1.2).

2. From the proof of Proposition (5.1.2), we know that if A is the direct sum of two left ideals J and J', there exist idempotent elements e and e' ($J' = Ae' = J'e'$) such that

$$e + e' = 1, \quad e^2 = e, \quad (e')^2 = e'$$
$$ee' = e'e = 0$$

   and this result can be extended to any direct sum :

$$e_1 + e_2 + \ldots + e_k = 1, \quad (e_i)^2 = e_i,$$
$$e_i e_j = e_j e_i = 0, \quad i, j = 1, 2, \ldots, k, \quad i \neq j$$

with a similar proof. Such a set $(e_1, e_2, \ldots, e_k)$ is called an *orthogonal system of idempotent elements*.

This system allows to express any element $x \in A$ uniquely in the form

$$x = \oplus_i x_i, \quad x_i \in Ae_i.$$

3. If J is a left ideal of A and e is an idempotent element of it, J is a direct sum of minimal left ideals $J_1, J_2, \ldots, J_l$. If we define

$$e = e_1 + e_2 + \ldots + e_l, \quad e_i \in J_i,$$

$e_1 = e_1 e$, by 1, since $x = xe$ for $x \in J$,

$$e_1 = (e_1)^2 + e_1 e_2 + \ldots + e_1 e_l,$$

$J_1, J_2, \ldots, J_l$ being left ideals, the uniqueness of the decomposition implies that

$$(e_1)^2 = e_1, \quad e_1 e_2 = \ldots = e_1 e_l = 0.$$

The same holds for $e_2, \ldots, e_l$ so that

$$e = e_1 + e_2 + \ldots + e_l, \quad (e_i)^2 = e_i, \quad e_i e_j = e_j e_i = 0, \quad (i \neq j),$$

e_i being an idempotent element of J_i.

4. A is now clearly seen to be the direct sum of two-sided ideals $J_1, J_2, \ldots, J_k$ if and only if $1 = e_1 + e_2 + \ldots + e_k$ and the idempotent elements determining the J_i are in the center of A. For Clifford algebras, we have encountered this situation in Chapter 3, 3.2.

Definition 5.1.3 *An idempotent element e is called* primitive *if no decomposition $e = e' + e''$ exists, where $(e')^2 = e'$, $(e'')^2 = e''$ and $e'e'' = e''e' = 0$.*

Proposition 5.1.7 *If e is primitive, Ae is a minimal ideal. Conversely, if J is a minimal ideal, every idempotent element of J is primitive.*

Proof. If J is not minimal, and $J = Ae$, ideals $J_1, J_2, \ldots, J_l$ can be found as before in 3. Then for $e' = e_1$ and $e'' = e_2 + \ldots + e_l$, $(e')^2 = e'$, $(e'')^2 = e''$ and $e'e'' = e''e' = 0$.

If J is minimal and if $J = Ae$ where $e = e' + e''$ as in the definition, then $\forall x \in A$, $x = xe' + xe''$, the sum being direct (since $xe' = ye''$ implies $xe'e'' = ye'' = 0$). So Ae' and Ae'' are non-trivial left ideals in J, which is a contradiction. ∎

Proposition 5.1.8 $J = Ae$ *is a minimal left ideal (or, equivalently, e is primitive), if and only if eAe is a (possibly non-commutative) field.*

We will need the

Lemma 5.1.5 *Let A be a ring with unit, $A \neq 0$ whose left ideals are trivial. Then A is a field, and vice versa.*

Proof. Let A be a field and J one of its left ideals, $J \neq 0$. There exists an $a \in J$ such that $a \neq 0$; then $x = (xa^{-1})a \in J$, $\forall x \in A$.

Conversely, let $x \in A$, $x \neq 0$, then the left ideal Ax contains x and must therefore be equal to A. Then $x' \in A$ can be found for which $x'x = 1$, $x' \neq 0$. Replacing x by x', an element x'' is found such that $x''x' = 1$. Now $x'' = x''x'x = x$ and $x'x = xx' = 1$, proving that A is a field. ∎

Proof of the proposition. If $J = Ae$ is a minimal left ideal, let J_1 be a non-zero left ideal of eAe (which is a subring with unit).

$$AJ_1 \subseteq AeJe \subseteq Je = J.$$

AJ_1 is a left ideal of A, contained in J, so $J = AJ_1$.

$$eAe = eJ = eAJ_1 = eAeJ_1 \subseteq J_1,$$

from which $eAe = J_1$ follows, J_1 being a subset of eAe by assumption. The only left ideals of eAe are therefore eAe and (0), so by the lemma, eAe is a field.

Conversely, if J is not minimal,

$$J = J_1 \oplus J_2$$

yields a decomposition

$$e = e_1 + e_2,$$

$e_1 = e_1e = ee_1$, $e_2 = e_2e = ee_2$ by 3 above; $e_1 = ee_1e$, $e_2 = ee_2e$, but $e_1e_2 = 0$ proves that eAe cannot be a field. ∎

Proposition 5.1.9 *Let J and J' be minimal left ideals : $J = Ae$, $J' = Ae'$ where e and e' are primitive. Then these left ideals are isomorphic if and only if $J' = Ja'$ for some nonzero $a' \in J'$.*

Proof. If $J' = Ja'$, $x \to xa'$ clearly is an isomorphism from J to J'. Conversely, if φ is an isomorphism, $\varphi(xe) = x\varphi(e)$ for all x in A.

But $xe = x$ if $x \in J$ and hence $\varphi(x) = x\varphi(e)$. Setting $a' = \varphi(e)$, $J' = Ja'$ follows. ∎

Remarks. Ja' is a left ideal of A, $Ja' \subseteq J'$. $Ja' = J'$ by the minimality of J', so J is isomorphic to J' is and only if $JJ' = J'$.

From $\varphi(x) = x\varphi(e)$, it follows that $\varphi(e) = a' = ea'$ and $a' = ea'e'$; every isomorphism from J to J' therefore corresponds to

$$y \rightarrow yex_oe', \quad x_o \in A, \quad ex_oe' \neq 0$$

since $a'e' = x_0e'$, $x_0 \in A$.

Note that 'isomorphism' in this context means A-isomorphism.

There also exists a $b \in J$ such that $J = J'b$, hence $J = Ja'b$. $J = Ae$ contains an invertible element z, eAe being a field, hence $z = za'b$, $a'b = 1$, $b = a'^{-1}$ or $J' = a'^{-1}Ja'$ so that J' is the image of J under an inner automorphism.

If e' and e are two primitive idempotents determining isomorphic ideals, a $u \in C(Q)^*$ can be found such that $e' = ueu^{-1}$.

Proposition 5.1.10 *If A is an algebra over $\mathbf{K} = \mathbf{C}$ and if e is a primitive idempotent, the set of exe, $x \in A$, is $\mathbf{C}e$.*

Proof. If f is a homomorphism from the A-module V to the A-module W and if both V and W are simple, either f is zero, or it is an isomorphism (this is called the *Schur Lemma*). Indeed, only two cases can occur : $\ker f = V$, $\operatorname{im} f = 0$ or $\ker f = 0$, $\operatorname{im} f = W$. If $V = W$, u is a homomorphism of the A-module V and λ is an eigenvalue of u, $u - \lambda \operatorname{Id} = u_1$ is again a homomorphism of V but cannot be injective, hence $u_1 = 0$ and

$$u = \lambda \operatorname{Id}.$$

Applying a remark above to the right multiplication by exe, $e = e'$ and $J = J'$, the proposition follows. ∎

5.2 Spin representations.

Using the notations of Chapter 3, 3.3, we have obtained for non-degenerate Q, if $\dim E = n$, that

- If $n = 2r$, $C(Q)$ is a simple algebra.

- If $n = 2r + 1$, $C(Q)$ is either simple or semi-simple.

The most interesting case for applications occurs when $\mathbf{K} = \mathbf{C}$. The reader may consult Theorem (3.3.3) where this situation is summarized.

5.2.1 The spinors for $n = 2r$ and $\mathbf{K} = \mathbf{C}$.

All representations considered here will be of finite dimension, i.e. they involve finite-dimensional vector spaces only.

By 1.5, all irreducible representations of $C(Q)$ are equivalent, and $C(Q)$ being simple, they are all faithful.

The representation of $C(Q)$ in a (2^r)-dimensional space S, constructed in Proposition (3.3.1), is irreducible, since $\rho(C(Q)) = \text{End } S$. Hence all irreducible representations of $C(Q)$ are of degree 2^r.

The elements of S are called *spinors* (their dependence on $C(Q)$ being implicit). We generally use the 'standard' representation ρ given in Chapter 3 :

$$E = F \oplus F'$$

being a Witt decomposition, F and F' are maximal totally isotropic subspaces, $(x_1, x_2, \ldots, x_r, y_1, y_2, \ldots, y_r)$ is a 'special' Witt basis, where

$$B(x_i, y_i) = \tfrac{1}{2}\delta_{ij}, \quad B(x_i, x_i) = B(y_i, y_i) = 0, \quad x_i y_j + y_j x_i = \delta_{ij}.$$

Setting $f = y_1 y_2 \ldots y_r$, S is the minimal left ideal of $C(Q)$ generated by the

$$\{x_{i_1} x_{i_2} \ldots x_{i_h} f\}, \quad 1 \le i_1 < i_2 < \ldots < i_h \le r,$$

and by $x_\emptyset f = f$.

The representation ρ is given by

$$\rho(u)vf = uvf, \quad u \in C(Q).$$

Remark. Observe that $(f)^2 = 0$, so f is not an idempotent; nevertheless, if we put :

$$x_1 = \frac{e_1 + e_n}{2}, x_2 = \frac{e_2 + e_{n-1}}{2}, \ldots, x_r = \frac{e_r + e_{r+1}}{2},$$
$$y_1 = \frac{e_1 - e_n}{2}, y_2 = \frac{e_2 - e_{n-1}}{2}, \ldots, y_r = \frac{e_r - e_{r+1}}{2}$$

and

$$f' = x_1 x_2 \ldots x_r,$$

$e = f'f$ is a primitive idempotent and corresponds to the minimal left ideal under consideration.

ρ induces a representation on $C^+(Q)$, denoted by ρ^+, and representations ρ of the group G, ρ^+ of G^+, ρ_0 of G_0, ρ_0^+ of G_0^+ and, in general, of all spin groups.

Semi-spinors. $C^+(Q)$ is semi-simple, but not simple, by the results of Chapter 3, 3. ρ^+ *is the direct sum of two non-equivalent irreducible representations.*

This follows from Proposition (3.3.1, part 2), $C^+(Q)$ being the direct composition of two simple ideals, each of which is isomorphic to the endomorphism algebra of a (2^{r-1})-dimensional complex vector space, in agreement with Proposition (5.1.3).

The representation spaces are $C^+(Q)f$ and $C^-(Q)f$. The elements of the representation spaces of irreducible representations of $C^+(Q)$ are called *semi-spinors*. As a special case, $S^+ = C^+(Q)f$ and $S^- = C^-(Q)f$ are semi-spinor spaces.

Spin representations of G, G^+ and G_0^+. A representation of a group is defined to be a homomorphism from that group into the isomorphism group of a vector space.

For the groups under consideration here, the group operation is induced by the multiplication in the algebra, so the representations of $C(Q)$ induce representations of the groups related to it.

The spin representation of $G = \Gamma$ induced by ρ is simple.

Indeed, since every non-isotropic vector of E is an element of G and since Q is non-degenerate, E is generated by $E \cap G$ and $C(Q)$ is, as an algebra, generated by elements of G, proving the statement.

The spin representation of $G^+ = \Gamma^+$ induced by ρ is the direct sum of two non-equivalent irreducible representations.

As an algebra, $C^+(Q)$ is generated by products of even numbers of non-isotropic elements of E, hence $C^+(Q)$ is, as an algebra, generated by G^+, the representations are therefore irreducible. If the representations were equivalent as group representations, this would extend to $C^+(Q)$ and contradict their non-equivalence as algebra representations.

Similar results hold for ρ_0 and ρ_0^+, since one can assume that for the elements $e_1, e_2, \ldots, e_n$ of an orthonormal basis, $Q(e_i) = 1$, so that $C(Q)$ is generated, as an algebra, by elements of G_0.

Important remark.   If $\mathbf{K} = \mathbf{R}$ and if Q is of maximal index, all these results remain valid, except for those on G_0 and G_0^+.

5.2.2   The spinors for $n = 2r + 1$ and $\mathbf{K} = \mathbf{C}$.

Knowing that $C^+(Q)$ is simple, we choose a (2^r)-dimensional space S from the Witt decomposition

$$E = F \oplus F' \oplus (e_n), \quad Q(e_n) \neq 0.$$

As before, the (x_i, y_i) are a Witt basis for $F \oplus F'$. The space

$$S = \{x_{i_1} x_{i_2} \ldots x_{i_h} f, 1 \leq i_1 < i_2 < \ldots < i_h \leq r\}, \quad f = y_1 y_2 \ldots y_r$$

is (2^r)-dimensional and linearly isomorphic to $C^+(Q)f$. Let i be the isomorphism from $C^+(Q)$ to $C(Q_1)$ (cf. Chapter 3, 3.2), Q_1 being (up to a non-zero factor) the metric induced by Q on $F \oplus F'$, then we define the representation ρ^+ by

$$\rho^+(u)(vf) = i(u)vf, \quad u \in C^+(Q), \quad vf \in S$$

S being considered as an ideal of $C(Q_1)$, i.e. $C(Q_1)f$.

The induced representations of $\Gamma^+ = G^+$ and G_0^+ can be proved simple as in the even case.

There are exactly two different extensions of the spin representation of $C^+(Q)$ in S to an irreducible representation of $C(Q)$ in S.

The center Z of $C(Q)$ is generated by $e_0 = 1$ and by an odd element z for which $z^2 \in \mathbf{K}$ (z is, up to a non-zero factor, $e_N = e_1 e_2 \ldots e_n$).

We can assume that $z^2 = 1$.

If $u \in C(Q)$, $u = u_1 + u_2 z$ uniquely, with $u_1, u_2 \in C^+(Q)$ and $u_2 \to u_2 z$ is injective since $z^2 = 1$.

The linear mappings $\varphi : u \to u_1 + u_2$ and $\varphi' : u_1 - u_2$ are homomorphisms from $C(Q)$ to $C^+(Q)$ and the mappings

$$\rho = \rho^+ \circ \varphi, \quad \rho' = \rho^+ \circ \varphi'$$

are therefore irreducible representations of $C(Q)$ extending ρ^+.

We will now prove their uniqueness up to equivalence.

If ρ_1 is a representation of $C(Q)$ extending ρ^+ and if z is as before, we define

$$\sigma = \rho_1(z).$$

σ^2 is the identity on S since $z^2 = 1$ and σ commutes with all operators in $\rho^+(C^+(Q))$, z being a central element of $C(Q)$.

σ^2 being the identity, S is the direct sum of the space S_1 of elements w such that $\sigma(w) = w$ and of the space S_2 of elements such w' such that $\sigma(w') = -w'$. By the commutation property, these spaces are invariant under $\rho^+(C^+(Q))$. ρ^+ being simple, one of the S_1, S_2 must be S and the other, 0. Hence $\sigma = \pm \mathrm{Id}$ and ρ_1 coincides with one of the representations given before, since

$$\rho_1(u_1 + u_2 z) = \rho^+(u_1) \pm \rho^+(u_2).$$

These two representations are non-equivalent, since if

$$\rho'(u) = \psi \circ \rho(u) \circ \psi^{-1},$$

$\rho'(z) = \psi \circ \rho(z) \circ \psi^{-1}$ or $-\mathrm{Id} = \psi \circ \mathrm{Id} \circ \psi^{-1} = \mathrm{Id}$, which is impossible. These representations induce representations of G and Γ; they are called the *spin representations* of $C(Q)$, G and Γ.

5.2.3   The spinors in the real case.

We will use the table in Chapter 3, 2.1 and the classical fact that the matrix algebras $M_n(\mathbf{R})$, $M_n(\mathbf{K})$ and $M_n(\mathbf{H})$ are simple.

Under these conditions, one sees that it is sufficient to consider the values of $p - q \,(\mathrm{mod}\, 8)$, independently of $p + q$.

Simple computations lead to the conclusion that, S being an irreducible representation space of $C(Q)$, if $p - q = 0, 1, 2 \,(\mathrm{mod}\, 8)$, $\dim_{\mathbf{R}}(S) = 2^{[n/2]}$ where $[\cdot]$ stands for the integer part, whereas if $p - q = 3, 4, 5, 6, 7 \,(\mathrm{mod}\, 8)$, $\dim_{\mathbf{R}}(S) = 2^{[n/2]+1}$.

Take for example the Minkowski space $E_{1,3}$ with orthogonal basis e_0, e_1, e_2, e_3, $(e_0)^2 = 1$, $(e_i)^2 = -1$, $i = 1, 2, 3$.

$p - q = 6 \,(\mathrm{mod}\, 8)$ so $\dim_{\mathbf{R}}(S) = 8$ (note that for $E_{3,1}$, $\dim_{\mathbf{R}}(S) = 4$).

The elements $e = (1 + e_0)/2$, $e' = (1 + e_0 e_3)/2$, $e'' = (1 + e_1 e_2 e_3)/2$ are primitive idempotents : it can be easily verified that $C(Q)e$, $C(Q)e'$ and $C(Q)e''$ are eight-dimensional or, equivalently, that eue, $e'ue'$ and $e''ue''$ belong to a field, $\forall u \in C(Q)$.

The most natural spin representation is given by the left multiplication. The algebra being simple, all three left ideals found are minimal and $C(Q)$-isomorphic.

Now consider the real space $E_{2,1}$ with a pseudo-metric of signature $(2, 1)$ and an orthogonal basis (e_1, e_2, e_3) :

$$(e_1)^2 = 1, \quad (e_2)^2 = 1, \quad (e_3)^2 = -1.$$

The elements $e_I = (1 + e_1)(1 + e_2 e_3)/4$, $e_{II} = (1 + e_2)(1 + e_1 e_3)/4$, $e_{III} = (1 - e_1)(1 - e_2 e_3)$ and $e_{IV} = (1 - e_2)(1 - e_1 e_3)$ are easily seen to be an orthogonal system of primitive idempotents (cf. 1.7). If $A, B = I, II, III$ or IV,

$$(e_A)^2 = e_A, \quad \sum_A e_A = 1, \quad e_A e_B + e_B e_A = 0.$$

Then S is two-dimensional and

$$C(Q) = \oplus_A C(Q) e_A.$$

Since $p - q = 1$, the algebra $C(2, 1)$ is the direct composition of two simple four-dimensional ideals. By Chapter 3, 3.2, idempotent elements u, v can be chosen

$$u = \frac{1 + e_1 e_2 e_3}{2}, \quad v = \frac{1 - e_1 e_2 e_3}{2}$$

such that $C(Q) = C(Q)u \oplus C(Q)v$; $u = e_I + e_{IV}$, $v = e_{II} + e_{III}$, (u and v are not primitive). The results on semi-simple algebras are thus confirmed.

Important remark. In the Minkowski space, we found three idempotent elements e, e' and e'' from which three families of idempotent elements can be deduced if we apply inner automorphisms to them,

$$e \to ueu^{-1}, \quad u \in C(3, 1)^*.$$

It is important to distinguish the inner automorphisms for which $u \in G$ or some other Clifford or spin group. Indeed, there cannot be a $u \in G$ such that $e' = ueu^{-1}$, and no such relation can be found between any pair of e, e', e''.

The situation is quite different for the idempotent elements e_A in $C(2, 1)$. The following definitions are therefore in order.

5.2.4   Algebraic and geometric equivalence.

Definition 5.2.1

1. *Two idempotent elements such that for some $\gamma \in G$ (the Clifford group),*

$$e' = \gamma e \gamma^{-1}$$

are called geometrically equivalent.

2. *Two idempotent elements such that for some $u \in C(Q)^*$,*

$$e' = ueu^{-1}$$

are called algebraically equivalent.

By the remark following Proposition (5.1.9) we know that two primitive idempotents are algebraically equivalent if the Clifford algebra is simple.

Proposition 5.2.1 *Let $n = 2r$ and let Q be of maximal index r, then the primitive idempotents of the form $f'f$ (using the notations of 2.1) are geometrically equivalent.*

Proof. An element of the Clifford group can be found such that it sends a Witt basis (x_i, y_i) to any other given Witt basis (x_i', y_i'). ∎

The great importance of the idempotent elements $f'f$ (i.e. of the isotropic r-vectors f) for the study of spinor structures is illustrated by the previous proposition : as will be shown in Chapter 8, an orthogonal system of 2^r such primitive idempotents can easily be obtained, leading to an expression for the Clifford algebra as a direct sum of 2^r spinor spaces.

Under the assumptions of Proposition (5.2.1), it is easily seen that

$$x_1 y_1, x_2 x_1 y_1 y_2, \ldots, x_k x_{k-1} \ldots x_1 y_1 y_2 \ldots y_k, \quad k < r,$$

are idempotent, but (as the reader can verify) they are not primitive, by the criterion given in Proposition (5.1.8).

Similarly, the idempotent elements $(1+x_i+y_i)/2 = (1+e_i)/2$, $Q(e_i) = 1$, generally are not primitive. The decomposition

$$1 = \frac{1 + e_1}{2} + \frac{1 - e_1}{2}, \quad (1 + e_1)(1 - e_1) = 0$$

shows that $C(Q)$ is the direct sum of two geometrically equivalent (2^{2r-1})-dimensional left ideals, where $2^{2r-1} > 2^r$ unless $r = 1$.

Starting from $C(1,1)$, primitive idempotents

$$(\lambda_1 + x_1) y_1, \quad x_1(\lambda_1 + y_1), \quad (\lambda_1 \in \mathbf{R}^*)$$

can be constructed, eAe being $\mathbf{R}$ in this case.

Taking tensor products, two primitive idempotents in $C(1,1)$ are seen to have a primitive idempotent product in $C(2,2)$. Repeating this, the primitive idempotents

$$(\lambda_1 + x_1)y_1(\lambda_2 + x_2)y_2 \ldots (\lambda_r + x_r)y_r, \quad x_1(\lambda_1 + y_1)x_2(\lambda_2 + y_2) \ldots x_r(\lambda_r + y_r)$$

are obtained in $C(r,r)$, here $\lambda_i \in \mathbf{R}$. In particular, this can be applied to the complex even-dimensional case.

Setting $\lambda_1 = \lambda_2 = \ldots = \lambda_r = 0$, the idempotent element

$$x_1 y_1 x_2 y_2 \ldots x_r y_r$$

considered in 2.1 is reobtained; its geometric meaning is particularly simple.

Some of these idempotent elements are not geometrically equivalent; for instance, taking $n = 2$, $r = 1$, $(\lambda_1 + x_1)y_1$ and $x_1 y_1$ are not geometrically equivalent, although they must be algebraically equivalent. Indeed, if $\gamma \in G$ was such that

$$\gamma(\lambda_1 + x_1)y_1 \gamma^{-1} = x_1 y_1,$$

setting $\gamma x_1 \gamma^{-1} = ax_1 + by_1$, $\gamma y_1 \gamma^{-1} = a'x_1 + b'y_1 \neq 0$,

$$(\lambda_1 + ax_1 + by_1)(a'x_1 + b'y_1) = x_1 y_1,$$

$$\lambda_1(a'x_1 + b'y_1) + ab'x_1 y_1 + a'by_1 x_1 = x_1 y_1$$

which, if $\lambda_1 \neq 0$, is impossible.

5.2.5 Some properties of the primitive idempotents in a Clifford algebra.

In this subsection we consider a faithful spin representation ρ in a space S. S will be a vector space over $\mathbf{K}$ such that the image of the Clifford algebra $C(Q)$ under ρ is the set of linear $\mathbf{K}$-endomorphisms of S. This is the case for the real Clifford algebras with signature (p,q) which are isomorphic to matrix algebras over $\mathbf{R}$, $\mathbf{C}$ or $\mathbf{H}$, when $p - q \neq 1 \,(\mathrm{mod}\ 4)$. By the 'periodicity' properties,

$$\begin{aligned} \mathbf{K} &= \mathbf{R}, \quad \text{if } p - q = 0, 1, 2 \,(\mathrm{mod}\ 8) \\ \mathbf{K} &= \mathbf{C}, \quad \text{if } p - q = 3, 7 \,(\mathrm{mod}\ 8) \\ \mathbf{K} &= \mathbf{H}, \quad \text{if } p - q = 4, 5, 6 \,(\mathrm{mod}\ 8). \end{aligned}$$

Proposition 5.2.2 *If ρ is a faithful spin representation on S, if $\rho(C(Q)) \simeq \mathrm{End}\, S$ and if e is a primitive idempotent determining the minimal left ideal $J = C(Q)e$, the kernel of $\rho(e)$ is a hyperplane $H \subseteq S$ and $\rho(J)H = 0$.*

Proof. $e^2 = e$ leads to $(\rho(e))^2 = \rho(e)$ so that the minimal polynomial of $\rho(e)$ is $X^2 - X = X(X - 1)$. By a classical result of linear algebra, S can be decomposed as

$$S = \ker(\rho(e) - \mathrm{Id}) \oplus \ker \rho(e).$$

Setting $H = \ker \rho(e)$, $H' = \ker(\rho(e) - \text{Id})$, if $ve \in J$,

$$\rho(ve)H = \rho(v)\rho(e)H = 0$$

hence $\rho(J)H = 0$.

To prove that H is a hyperplane in S, set

$$s = h + h', \quad h \in H, \quad h' \in H'.$$

If $v \in C(Q)$ satisfies $\rho(v)h = 0$, $\forall h \in V$,

$$\rho(v)s = \rho(v)h', \quad \forall s \in S,$$
$$\rho(ve)s = \rho(v)\rho(e)s = \rho(v)h'.$$

ρ being faithful, $ve = v$ and therefore $v \in J$; J is the set of $u \in C(Q)$ such that $\rho(u)H = 0$.

But the set of $u \in C(Q)$ such that $\rho(u)$ maps a subspace $H_1 \subseteq S$ to 0 is a non-zero left ideal J_1 and $H_1 \subseteq H$ implies $J_1 \supseteq J$. J being minimal, H must be a hyperplane. ∎

For example, if we take S and the representation as in 2.1, the idempotent element $e = f'f$ is seen to annihilate the hyperplane determined by

$$1 \cdot f, x_i f, \ldots, x_{i_1} x_{i_2} \ldots x_{i_k} f, \quad 1 \leq i_1 < i_2 < \ldots < i_k \leq r, \quad k < r.$$

Lemma 5.2.1 *Under the assumptions of the previous proposition, if J' is a minimal right ideal, $\rho(J')$ maps S onto a subspace of dimension 1 over $\mathbf{K}$.*

Proof. Let $J' = e'C(Q)$, $(e')^2 = e'$.

$$S = \ker(\rho(e') - \text{Id}) \oplus \ker \rho(e') = D' \oplus D$$

Since $\rho(e')(S) = D'$, $\rho(e'v)S \subseteq D'$ and $\rho(J')(S) = D$. Conversely, if $\rho(v')(S) \subseteq D'$, $\rho(e')$ being the identity on D', $\rho(e'v') = \rho(v')$ and $v' = e'v'$. Hence J' is the set of $u \in C(Q)$ such that $\rho(u)(S) = D'$.

The set of $u \in C(Q)$ for which $\rho(u)$ maps S into a subspace $D'_1 \subset S$ is a right ideal J'_1, and $D'_1 \subseteq D'$ implies $J'_1 \subseteq J'$; D' being non-zero, it must be one-dimensional over $\mathbf{K}$. ∎

Proposition 5.2.3 *The intersection of a minimal left ideal and a minimal right ideal of $C(Q)$ is a subspace of dimension 1 over $\mathbf{K}$.*

Proof. If J and J' are such ideals, $J \cap J'$ is the set of $v \in C(Q)$ such that

$$\rho(v)H = 0, \quad \rho(v)S = D'.$$

If $x = h + h'$, $h \in H$, $h' \in H'$, and if $v \in J \cap J'$,

$$\rho(v)x = ay_0 = \rho(v)h', \quad a \in \mathbf{K}$$

y_0 being a non-zero element of D'.

If x_0 is a non-zero element of H',

$$\rho(v)x_0 = \lambda y_0, \quad \lambda \in \mathbf{K}.$$

$v \to \lambda y_0$ is an injective linear mapping, since $\rho(v')x_0 = \lambda y_0$ implies $\rho(v')h' = \rho(v)h'$ and $\rho(v) = \rho(v')$, so $v = v'$.

This mapping is an isomorphism and $\dim(J \cap J') = 1$. ∎

Immediate consequences. If e and e' are primitive idempotents determining J and J', $e'C(Q)e$ is a subspace of $C(Q)$ of dimension 1 over $\mathbf{K}$.

An anti-automorphism β exists, providing a bijection between left and right ideals, hence $\beta(e)C(Q)e$ has dimension 1 over $\mathbf{K}$.

Every primitive idempotent therefore determines a subspace of $C(Q)$ of dimension 1 over $\mathbf{K}$.

If $\beta(e)e \neq 0$, the element determines the subspace.

Note that for an idempotent $f'f$, $\beta(e)e = 0$ and the one-dimensional subspace contains f.

Corollary 5.2.1 *For real Clifford algebras of signature* (p, q),

$$eC(Q)e \text{ is a field isomorphic to } \mathbf{R} \text{ if } p - q = 0, 1, 2 \,(\mathrm{mod}\,8),$$
$$eC(Q)e \text{ is a field isomorphic to } \mathbf{C} \text{ if } p - q = 3, 7 \,(\mathrm{mod}\,8),$$
$$eC(Q)e \text{ is a field isomorphic to } \mathbf{R} \text{ if } p - q = 4, 5, 6 \,(\mathrm{mod}\,8).$$

Proof. Immediate. ∎

5.3 Selected references.

- N. Bourbaki, *Algèbre, Chapitre 8, Modules et anneaux semi-simples*, Hermann, Paris, 1938.

- C. Chevalley, *The algebraic theory of spinors*, Columbia University Press, 1954.

- A. Crumeyrolle, *Algèbres de Clifford et spineurs*, Publ. dépt. Math., Université de Toulouse, 1974.

- Curtis, Reiner, *Representation theory*, Interscience, 1966.

- P. Lounesto, *On primitive idempotents of Clifford algebras*, Reports Math. Inst. Tek. Hog., Helsinki, 1977.

Chapter 6

FUNDAMENTAL LIE ALGEBRAS AND LIE GROUPS IN THE CLIFFORD ALGEBRAS.

In this chapter $\mathbf{K} = \mathbf{R}$, Q is non-degenerate of arbitrary signature, $\dim E = n$, $C(Q)^*$ is, as before, the multiplicative group of all invertible elements of $C(Q)$, $\mathbf{R}^n$ has the usual topology and $C(Q)$ has the topology of $\mathbf{R}^{n'}$, $n' = 2^n$.

6.1 The exponential function in the Clifford algebras.

6.1.1 Definition of the exponential function.

Given $X \in C(Q)$, let $\theta(X) = \theta_X : Y \in C(Q) \to XY \in C(Q)$.

Lemma 6.1.1 *$\theta : X \to \theta_X$ is a homeomorphism from $C(Q)$ to a subspace of the endomorphisms of $C(Q)$. This homeomorphism and its inverse are analytic.*

Proof. $X \to \theta_X$ is injective since $XY = X'Y$, $\forall Y \in C(Q)$, implies $X = X'$. Given a basis of $C(Q)$, the matrix of θ_X obviously consists of components depending analytically on the components of X.

Since $\theta_X(1) = X$, if the sequence θ_{X_n} has a limit, so does $\theta_{X_n}(1) = X_n$ and θ has a continuous inverse.

Note that θ is linear, hence $d\theta = \theta$. ∎

Definition 6.1.1 *Define for $X \in C(Q)$,*

$$\exp(\theta_X) = \lim_{n \to +\infty} \sum_{k=0}^{n} \frac{(\theta_X)^k}{k!} = \lim_{n \to +\infty} \theta(\sum_{k=0}^{n} \frac{X^k}{k!}). \tag{1}$$

θ_X being an endomorphism, the first limit must exist; by the lemma it follows that $\sum_{k=0}^{n} X^k/k!$ converges $\forall X \in C(Q)$. Therefore we define

$$\exp X = \lim_{n \to +\infty} \sum_{k=0}^{n} \frac{X^k}{k!}. \tag{2}$$

Also,
$$\exp(\theta(X)) = \theta(\exp(X)). \tag{3}$$

Elementary properties of the exponential function.

- $X \to \exp X$ *is continuous.* Indeed, it can be obtained from continuous mappings :
$$X \to \theta(X) \to \exp(\theta(X)) = \theta(\exp X) \to \exp X$$

- *If* $XY = YX$,
$$\exp X \exp Y = \exp(X + Y). \tag{4}$$

  From $\theta(X)\theta(Y) = \theta(Y)\theta(X)$ and a known result on the exponential of endomorphisms,
  $$\exp(\theta_X)\exp(\theta_Y) = \exp(\theta_X + \theta_Y).$$
  By (3), $\theta(\exp X)\theta(\exp Y) = \exp(\theta(X + Y))$ and $\theta(\exp X \exp Y) = \theta(\exp(X + Y))$ leads to $\exp X \exp Y = \exp(X + Y)$.

- $\exp X$ *is invertible and* $(\exp(X))^{-1} = \exp(-X)$. Indeed, $\exp 0 = 1$ and $\exp(X)\exp(-X) = \exp(0) = 1$.

6.1.2   The Lie algebra of $C(Q)^*$.

The mappings
$$(X, Y) \to XY, \quad X \to X^{-1}$$

being analytic (X^{-1} can be found, given a basis, by using Cramer's solution to a system of linear equations), $C(Q)^*$ is a Lie group.

Obviously $t \to \exp(tX)$ is an analytic homomorphism from the additive group $\mathbf{R}$ to $C(Q)^*$ and its derivative at $t = 0$ is X.

$\mathcal{L}(C(Q)^*)$ being the Lie algebra of $C(Q)^*$, X is an element of this Lie algebra and hence $C(Q) \subseteq \mathcal{L}(C(Q)^*)$. But the dimensions force
$$\mathcal{L}(C(Q)^*) = C(Q). \tag{5}$$

$\theta(C(Q))$ is an associative sub-algebra of $\mathrm{End}(C(Q))$, hence it is a Lie sub-algebra of $\mathrm{End}(C(Q))$ with the usual bracket :
$$[\theta_X, \theta_Y] = \theta_X \theta_Y - \theta_Y \theta_X = \theta(XY - YX)$$
$$[\theta_X, \theta_Y] = [d\theta(X), d\theta(Y)] = d\theta([X, Y]) = \theta([X, Y]),$$

which motivates the notation
$$[X, Y] = XY - YX.$$

Proposition 6.1.1 $C(Q)$ is the Lie algebra of $C(Q)^*$, the bracket $[X,Y]$ being $XY - YX$.

For $w \in C(Q)$, we define

$$\varphi_u(w) = uwu^{-1}, \quad u \in C(Q)^*,$$

φ_u is an inner automorphism of $C(Q)$. Also, for $X, Y \in C(Q)$,

$$(\operatorname{ad} X)(Y) = [X, Y]. \tag{6}$$

Proposition 6.1.2 $\varphi : u \to \varphi_u$ is a linear representation of $C(Q)^*$ and

$$\varphi(\exp tX) = \exp(t \operatorname{ad} X). \tag{7}$$

φ is called the adjoint representation of $C(Q)^*$.
Proof. Define

$$\varphi(\exp tX)(w) = Y(t) = \exp(tX)w \exp(-tX), \quad t \in \mathbf{R},$$

then

$$
\begin{aligned}
Y(t+h) - Y(t) &= \exp(hX)Y(t)\exp(-hX) - Y(t) \\
&= (1 + hX + \tfrac{1}{2}h^2 X^2 + \ldots)Y(t)(1 - hX + \tfrac{1}{2}h^2 X^2 - \ldots) - Y(t) \\
&= h(XY(t) - Y(t)X) + \ldots
\end{aligned}
$$

and

$$(\frac{dY}{dt})_{t=0} = (\operatorname{ad} X)(w).$$

For $t = 0$ the derivative of $t \to \varphi(\exp tX)$ is $\operatorname{ad} X$. Noticing that for $t = 0$, the analytic homomorphisms $\varphi(\exp tX)$ and $\exp(t \operatorname{ad} X)$ have equal derivatives then proves the proposition. ∎

φ is often written Ad,

$$\operatorname{Ad}_{\exp tX} = \exp(t \operatorname{ad} X).$$

6.1.3 The Lie algebra of the Clifford group.

First consider the Clifford group G.

Proposition 6.1.3 G is a closed Lie subgroup of $C(Q)^*$.

Proof. If all elements of the sequence u_n are in G, $\varphi(u_n)$ preserves E and so does the limit u if u_n tends to u.

G being closed in $C(Q)^*$, it has the structure of a topological Lie subgroup of $C(Q)^*$, by a classical result, then $X \in \mathcal{L}(G)$ if and only if $\exp tX \in G$.   ∎

The following remark will prove useful : $\varphi(\exp tX)$ maps E into itself for all $t \in \mathbf{R}$, so $\mathrm{ad}\, X$ also maps E into itself, and the converse.

Let $x, y \in E$, then $\forall z \in E$,

$$
\begin{aligned}
xyz - zxy &= 2xB(y,z) - (xz + zx)y \\
&= 2B(y,z)x - 2B(x,z)y
\end{aligned}
$$

so that

$$\mathrm{ad}(xy)(z) = 2B(y,z)x - 2B(x,z)y \in E$$

and $xy \in \mathcal{L}(G)$.

If E is odd-dimensional, the center Z of $C(Q)$ is generated by 1 and by an odd element e_N. For any $u \in Z$, $\mathrm{ad}\, u = 0$ and hence $Z \subseteq \mathcal{L}(G)$.

Let $\mathcal{C}$ be the space which is linearly generated by the products (xy), $x, y \in E$ and by $e_N \in Z$ if n is odd. A basis for $\mathcal{C}$ is given by the $e_i e_j$, $i < j$, by $e_0 = 1$ and, if n is odd, by e_N.

So $\dim \mathcal{C} = n(n-1)/2 + 1$ if n is even, $n(n-1)/2 + 2$ if n is odd.

The image of G under φ is $O(Q)$ or $SO(Q)$ depending on the parity of n and we know that the Lie algebra of $O(Q)$ or $SO(Q)$ is $(n(n-1)/2)$-dimensional.

If n is odd : $\ker \varphi = G \cap Z$ and since $SO(Q) \simeq G/(G \cap Z)$,

$$\dim \mathcal{L}(SO(Q)) = \dim \mathcal{L}(G) - \dim \mathcal{L}(G \cap Z)$$

and $\dim \mathcal{L}(G) = n(n-1)/2 + 2$ leads to $\mathcal{L}(G) = \mathcal{C}$.

If n is even, a similar reasoning leads to $\dim \mathcal{L}(G) = n(n-1)/2 + 1$.

If Γ is the twisted Clifford group, the situation is the same regardless of the parity of n since the kernel of the representation is in both cases $\mathbf{R}^*$ and $p(\Gamma) = O(Q)$.

Hence :

Proposition 6.1.4 $\mathcal{L}(G)$ *is linearly generated by the elements* (xy), $x, y \in E$ *and by* $e_N \in Z$ *if n is odd.* $\mathcal{L}(\Gamma)$ *is linearly generated by the elements* (xy), $x, y \in E$, *regardless of the parity of n.*

If we consider the special Clifford group $G^+ = \Gamma^+$, $\exp tX$ is the limit of a sequence in $C^+(Q)$ if $X \in C^+(Q)$, and therefore it lies in $C^+(Q)$. Conversely, if $\exp tX \in C^+(Q)$, its derivative at $t = 0$ must lie in $C^+(Q)$, so that

$$\mathcal{L}(\Gamma^+) = \mathcal{L}(\Gamma) \cap C^+(Q)$$

and

Proposition 6.1.5 $\mathcal{L}(G^+) = \mathcal{L}(\Gamma^+)$ *is linearly generated by the elements* (xy), $x, y \in E$. *Its dimension is* $n(n-1)/2 + 1$.

6.1.4 The Lie algebra of $\mathrm{Spin}(Q)$.

Recall that $\mathrm{Pin}^\alpha(Q)$ was defined as the subgroup of $g \in \Gamma$ such that $\tilde{N}(g) = \pm 1$ and $\mathrm{Spin}^\alpha(Q)$ as its subgroup $\mathrm{Pin}^\alpha(Q) \cap C^+(Q)$. When n is even $\mathrm{Pin}^\alpha(Q) = \mathrm{Pin}(Q)$, when n is odd, $\mathrm{Spin}(Q) = \mathrm{Spin}^\alpha(Q)$.

These groups obviously are topological Lie subgroups of Clifford groups.

Let us determine the Lie algebra of $\mathrm{Spin}(Q)$.

Proposition 6.1.6 *The Lie algebra of* $\mathrm{Spin}(Q)$ *is the set of* $X \in \mathcal{L}(G^+) = \mathcal{L}(\Gamma^+)$ *such that* $\beta(X) + X = 0$.

Proof. $X \in \mathcal{L}(\mathrm{Spin}(Q))$ means that $\exp tX \in \mathrm{Spin}(Q)$, which can be expressed by $\beta(\exp tX) \exp(tX) = 1$. Taking the derivative at $t = 0$,

$$\beta(X) + X = 0 \tag{8}$$

and $\mathcal{L}(\mathrm{Spin}(Q))$ is $(n(n-1)/2)$-dimensional. ∎

This result can be explained by noticing that $\mathrm{Spin}(Q)$ is a covering of $\mathrm{SO}(Q)$ so that the Lie algebras of $\mathrm{Spin}(Q)$ and $\mathrm{SO}(Q)$ are isomorphic.

Note also that $\mathcal{L}(\mathrm{Spin}(Q))$ is the subspace of C^+ generated by the products (xy), $x \perp y$, since $\beta(xy) + xy = 0$ if and only if $xy + yx = 2B(x, y) = 0$. A basis of $\mathcal{L}(\mathrm{Spin}(Q))$ is given by $\{e_i e_j, i < j\}$ if $(e_1, e_2, \ldots, e_n)$ is an orthogonal basis of E.

Using classical results from the theory of Lie groups and algebras, the isomorphism from the Lie algebra of the orthogonal group to the Lie algebra of the spin group can be found explicitly.

If $X \in \mathcal{L}(\mathrm{Spin}(Q))$, $X = a^{ij} e_i e_j$, $1 \le i, j \le n$, $a^{ij} = -a^{ji}$,

$$\exp(tX) e_k \exp(-tX) = A_k^j(t) e_j, \quad \forall t \in \mathbf{R},$$

(A_k^j) being the matrix of an orthogonal transformation. Taking the derivative of both members at $t = 0$, we obtain :

$$X e_k - e_k X = b_k^j e_j, \quad b_k^j \in \mathcal{L}(\mathrm{O}(Q)).$$

Replacing X by $a^{ij} e_i e_j$, an easy computation shows that if we define

$$b^{ij} = b_k^i g^{kj}, \quad b^{ij} = -b^{ji},$$

(g^{kj}) being the inverse matrix of $(g_{kj}) = (B(e_k, e_j))$,

$$b^{ij} = 4a^{ij},$$

and the following proposition is obtained :

Proposition 6.1.7 *The natural isomorphism between the Lie algebra of the orthogonal group and the Lie algebra of the spin group is given by*

$$(\mu^{ij}) \to \tfrac{1}{4} \mu^{ij} e_i e_j, \quad \mu^{ij} = -\mu^{ji}.$$

6.2   Connectedness.

Proposition 6.2.1 *If Q is a definite quadratic form and if $\dim E > 1$, $\text{Spin}(Q) = G_0^+$ is connected.*

Proof. We have seen that $\varphi(\text{Spin}(Q)) = SO(Q)$ and that $\ker \varphi$, restricted to $\text{Spin}(Q)$, is $\{1, -1\}$ (cf. Chapter 4).

Note that (-1) lies in the connected component of 1 in $\text{Spin}(Q)$, since if $x, y \in E$ such that $B(x, y) = 0$ and $Q(x) = Q(y) = 1$ (or $Q(x) = Q(y) = -1$),

$$(xy)^2 = -Q(x)Q(y) = -1.$$

Taking the series expansion of the exponential function and noticing that $(xy)^2 = -1$, we obtain :

$$\exp(txy) = \cos t + (xy)\sin t, \quad \exp(\pi xy) = -1, \tag{1}$$

so that a continuous path from 1 to (-1) can be found in $\text{Spin}(Q)$.

Knowing that $\text{Spin}(Q)$ is a covering of $SO(Q)$ there exists a connected neighborhood $\mathcal{V}'$ of 1 in $SO(Q)$ which, by a classical result, is homeomorphic to a connected neighborhood $\mathcal{V}$ of 1 in $\text{Spin}(Q)$.

$$\forall x \in \text{Spin}(Q), \quad x' = \varphi(x) = x_1' x_2' \ldots x_k',$$

and such a factorization is always possible in terms of $x_i' \in \mathcal{V}'$ since $SO(Q)$, being connected, is generated by the neighborhood $\mathcal{V}'$ of the identity. Then

$$x' = \varphi(x_1)\varphi(x_2)\ldots\varphi(x_k), \quad x_i \in \mathcal{V}, \quad i = 1, 2, \ldots, k.$$

Now $\varphi(x) = \varphi(x_1 x_2 \ldots x_k)$ and by the results on the kernel of φ,

$$x = \epsilon x_1 x_2 \ldots x_k, \quad \epsilon = \pm 1.$$

The connectedness of $\mathcal{V}$ is equivalent to its arcwise connectedness and implies the existence of continuous paths from 1 to $x_1, x_2, \ldots, x_k$, hence also from 1 to x, since there exists a continuous path from 1 to -1 which can be used if $\epsilon = -1$.

Then $\text{Spin}(Q)$ is arcwise connected, hence connected.   ∎

Proposition 6.2.2 *If Q is an indefinite quadratic form,*

$G_0^+ \subset \text{Spin}(Q)$, G_0^+ has index 2 in $\text{Spin}(Q)$.
G_0^+ is connected if $\dim E > 2$.
G_0^+ consists of two connected components if $\dim E = 2$.

Proof. Every element of G^+ is a product $\lambda x_1 x_2 \ldots x_k$, $k = 2h$, $\lambda \in \mathbf{R}^*$, x_i non-isotropic, $i = 1, 2, \ldots, k$, $N(x_i) = \pm 1$.

Since $x_1 x_2 = x_2(x_2^{-1} x_1 x_2) = x_2 y_1$ with $N(y_1) = N(x_1)$, we may assume that the x_i for which $N(x_i) = -1$ are written first. If $g \in G_0^+$, $N(g) = 1$, so every element of $\mathrm{Spin}(Q)$ is the product of elements with spin norm (-1) and an element of G_0^+.

We can assume $\lambda = 1$ in $\mathrm{Spin}(Q)$.

On the other hand, if two of the x_i are linearly dependent, permutations of the same type as before will lead to a factor ± 1 and we can assume that in the sequence $x_1, x_2, \ldots, x_k$ the x_i are pairwise linearly independent, the x_i for which $N(x_i) = -1$ leading.

Obviously G_0^+ is normal in $\mathrm{Spin}(Q)$ and if two elements g, g' of $\mathrm{Spin}(Q)$ are equivalent modulo G_0^+, $N(g) = \pm N(g')$. Hence there are two quotient classes, $\mathrm{Spin}(Q)/G_0^+ = \{1, -1\}$.

Now let (x, y) be a pair of elements for which $N(x) = N(y)$ and let P be the plane (x, y). Either P is isotropic (but not totally, since $N(x) = \pm 1$), or non-isotropic of index 0 or 1. There are three cases to consider :

1. P is isotropic.

   Let $z \in P \cap P^\perp$, $N(z) = 0$. $P = (x) \oplus (z)$, $B(x, z) = 0$, $Q(x) = Q(y) = \epsilon$, $\epsilon = \pm 1$. If $y = \alpha x + \beta z$, $Q(\alpha x + \beta z) = \epsilon$ if and only if $\alpha^2 = 1$, so

$$y = \pm x + \lambda z, \quad \lambda \in \mathbf{R}.$$

   Letting t vary from 0 to λ, a continuous path from x to y or from $-x$ to y is obtained, yielding a continuous path from xy to $\pm x^2 = \pm 1$.

2. P is non-isotropic of index 0.

   An orthonormal basis (x, z) can be found in P, $y = \lambda x + \mu z$ and $Q(y) = \epsilon$ if and only if $\lambda^2 + \mu^2 = 1$. Setting

$$y = x \cos t + z \sin t$$

   a continuous path from xy to $x^2 = \pm 1$ is obtained.

3. P is non-isotropic of index 1.

   A Witt decomposition gives

$$P = (z) \oplus (z')$$

   where z, z' are isotropic and $B(z, z') = 1$.

$$y = \alpha_0 z + \beta_0 z', \quad Q(y) = 2\alpha_0\beta_0, \quad y = \alpha_0 z + \epsilon \frac{z'}{2\alpha_0}, \quad \alpha_0 \neq 0.$$

Similarly, $x = \alpha_1 z + \beta_1 z'$, $x = \alpha_1 z + \epsilon z'/(2\alpha_1)$, $\alpha_1 \neq 0$.

If $\alpha_0 \alpha_1 > 0$, a continuous path from x to y is obtained, hence also from xy to $x^2 = \epsilon$.

If $\alpha_0 \alpha_1 < 0$, a continuous path from $-x$ to y is obtained, hence also from xy to $-x^2 = -\epsilon$.

In all three cases a continuous path from an arbitrary element of G_0^+ to 1 or -1 is found.

Now distinguish two cases :

- If $\dim E = 2$.

  Q having a non-zero index, its index is one and 3 applies. $\alpha_0 \alpha_1 > 0$ and $\alpha_0 \alpha_1 < 0$ being mutually exclusive, it is impossible to reach both 1 and -1 by a continuous path starting from xy. G_0^+ has therefore exactly two connected components.

- If $\dim E > 2$.

  Using the reduced diagonal form of Q one sees that there always exist orthogonal x, y such that
  $$N(x) = N(y) = \epsilon = \pm 1.$$

  The reasoning used in the definite case can therefore be applied to the plane (x, y) and offers a continuous path from 1 to -1. Hence G_0^+ is connected.

  ∎

Corollary 6.2.1 If Q is indefinite, $\varphi(G_0^+)$ is connected regardless of the dimension.

Proof. $\varphi(G_0^+)$ is generated by products

$$\varphi(x_1 y_1)\varphi(x_2 y_2)\ldots\varphi(x_k y_k), \quad N(x_i) = N(y_i)$$

and ± 1, which is mapped to the identity by φ, is reached by a continuous path starting from $x_i y_i$.   ∎

Consequence. For $n > 2$, G_0^+, which is of index 2 in $\mathrm{Spin}(Q)$, must be the connected component of the identity in $\mathrm{Spin}(Q)$.

By the isomorphism theorems on the groups

$$\frac{\varphi(\mathrm{Spin}(Q))}{\varphi(G_0^+)} \simeq \frac{\mathrm{Spin}(Q)}{G_0^+},$$

$\varphi(G_0^+)$ is the connected component of $\varphi(\mathrm{Spin}(Q)) = \mathrm{SO}(Q)$.

Using the representation p defined in Chapter 4, 2 (and writing $\text{Pin}(Q)$ for $\text{Pin}^\alpha(Q)$) :

$$\frac{p(\text{Pin}(Q))}{p(\text{Spin}(Q))} = \frac{O(Q)}{SO(Q)} \simeq \frac{\text{Pin}(Q)}{\text{Spin}(Q)} \qquad (2)$$

knowing that $SO(Q)$ has index 2 in $O(Q)$, $O(Q)$ *has four connected components if* Q *is indefinite, regardless of the dimension*—a well-known result we have rapidly reobtained.

By (2), $\text{Pin}(Q)$ also has four connected components.

The connected component G_0^+ corresponds to the $g \in C^+(Q)$ such that $N(g) = 1$, $\text{Spin}(Q)$ to the $g \in C^+(Q)$ such that $N(g) = \pm 1$ and $\text{Pin}(Q)$ to the $g \in C(Q)$ such that $N(g) = \pm 1$.

The elements of $\varphi(G_0^+)$ are products of an even number of symmetries relative to hyperplanes which are orthogonal to non-isotropic, linearly independent vectors $x_1, x_2, \ldots, x_{2h}$, where $2k \leq 2h$ of the x_i satisfy $N(x_i) = -1$.

An element of $SO(Q)$ may consist of an even or odd number of elements x_i such that $N(x_i) = -1$.

Similarly, an element of $O(Q)$ can be factored in an even or odd number of symmetries defined by $x_1, x_2, \ldots, x_l$.

Summarizing,

$$
\begin{aligned}
&\text{if } g \in \varphi(G_0^+), \quad g = x_1 x_2 \ldots x_{2k} x_{2k+1} \ldots x_{2h}, \quad N(x_i) = -1, \quad i \leq 2k, \\
&\text{if } g \in SO(Q), \quad g = x_1 x_2 \ldots x_l x_{l+1} \ldots x_{2h}, \quad N(x_i) = -1, \quad i \leq l, \\
&\text{if } g \in O(Q), \quad g = x_1 x_2 \ldots x_l,
\end{aligned}
$$

where all x_i must satisfy $N(x_i) = \pm 1$.

In the signature $(1, 3)$, for example, $\varphi(G_0^+)$ is the restricted Lorentz group, $SO(Q)$ is the unimodular Lorentz group and $O(Q)$ is the 'complete' Lorentz group.

6.3 Selected references.

- C. Chevalley, *Theory of Lie groups*, Princeton University Press, 1946.

- S. Helgason, *Differential geometry and Symmetric spaces*, Academic Press, 1962.

Chapter 7

THE MATRIX APPROACH TO SPINORS IN THREE AND FOUR-DIMENSIONAL SPACES.

We will survey the traditional methods widely used by theoretical physicists and show how they fit in the modern theory.

In order to understand the reasons underlying these theories, we should consider the work of E. Cartan and Dirac. To linearize the Klein-Gordon operator,

$$(\Box - m^2)\psi = 0, \quad \Box = -g^{\alpha\beta}\frac{\partial^2}{\partial x^\alpha \partial y^\beta}, \tag{1}$$

m being a real number (the mass) and ψ a 'wave function', Dirac introduced the spinors. To reduce (1) to an equation of the first degree, he rewrote it as

$$(\gamma^\alpha \frac{\partial}{\partial x^\alpha} - m)(\gamma^\beta \frac{\partial}{\partial x^\beta} + m) = 0,$$

obtaining the sufficient conditions :

$$\gamma^\alpha \frac{\partial}{\partial x^\alpha} \pm m = 0, \tag{2}$$

the *Dirac equation*, where

$$\gamma^\alpha\gamma^\beta + \gamma^\beta\gamma^\alpha = -2g^{\alpha\beta}, \tag{3}$$

which, up to small differences, are our defining relations for a Clifford algebra. Dirac considered the γ^α as matrices.

We will recall the local isomorphisms of $SU(2, \mathbf{C})$ and $SO(3, \mathbf{R})$ and those of $SL(2, \mathbf{C})$ and the restricted Lorentz group, used by E. Cartan, who discovered the spinors in his study of the representations of the Lie algebra of the orthogonal group, but did not give them a special name.

7.1 The complex Clifford algebra in $\mathbb{C}^3$.

Let $E = \mathbb{C}^3$ and let (e_1, e_2, e_3) be its canonical basis. For $x = x^i e_i$, let $Q(x) = \sum_{i=1}^{3}(x^i)^2$.

$C(Q)$ has a basis 1, e_1, e_2, e_3, e_{12}, e_{13}, e_{23}, e_{123}, where e_{12} stands for $e_1 e_2$, e_{23} for $e_2 e_3$ etc. and the rules

$$(e_i)^2 = 1, \quad i = 1, 2, 3$$
$$e_i e_j + e_j e_i = 0, \quad i \neq j$$

apply.

$C^+(Q)$ has an irreducible representation in a space of dimension $2^1 = 2$ on $\mathbb{C}$. $C^+(Q)$ is isomorphic to $\text{End}(\mathbb{C}^2)$, as can be verified if we define the matrices

$$\sigma_x = \begin{pmatrix} 0 & i \\ i & 0 \end{pmatrix}, \quad \sigma_y = \begin{pmatrix} 0 & 1 \\ -1 & 0 \end{pmatrix}, \quad \sigma_z = \begin{pmatrix} i & 0 \\ 0 & -i \end{pmatrix}, \quad \sigma_0 = \begin{pmatrix} 1 & 0 \\ 0 & 1 \end{pmatrix} \quad (1)$$

which form a basis of $\text{End}(\mathbb{C}^2)$ and are (up to a factor $(-i)$) the so-called *Pauli matrices*. They satisfy the equations :

$$\begin{aligned} \sigma_x \sigma_y &= -\sigma_z & \text{(cycl.)} \\ \sigma_x^2 &= -\sigma_0 & \text{(cycl.)} \\ \sigma_x \sigma_y + \sigma_y \sigma_x &= 0 & \text{(cycl.)} \end{aligned} \quad (2)$$

If we associate σ_x to e_{12}, σ_y to e_{23}, σ_z to e_{31} and σ_0 to 1, an isomorphism ρ^+ from $C^+(Q)$ to $\text{End}(\mathbb{C}^2)$ is defined and ρ^+ can be called the spin representation of $C^+(Q)$, the elements of $\mathbb{C}^2$ being called 'spinors'.

By Chapter 5, 2.2, this representation can be extended to a representation $\rho = \rho^+ \circ \varphi$ of $C(Q)$ in $\mathbb{C}^2$. We define

$$u = u_1 + i e_{123} u_2, \quad u_1, u_2 \in C^+(Q), \quad (i e_{123})^2 = 1.$$

If $u = e_1$, $u_2 = i e_{23}$ and $u_1 = 0$, $\rho(e_1) = i\sigma_y$ and similarly $\rho(e_2) = i\sigma_z$, $\rho(e_3) = i\sigma_x$.

The image F of E under ρ is a three-dimensional subspace of $\text{End}(\mathbb{C}^2)$, given by the endomorphisms with matrix

$$x^1 \sigma_x + x^2 \sigma_y + x^3 \sigma_z, \quad x^i \in \mathbb{C}.$$

It is easily seen that F can be identified to the set of matrices with zero trace. The restriction of ρ to E is a linear isomorphism from E to F defining on F a 'metric' Q' :

$$Q'(H) = Q(h) \quad \text{if} \quad H = \rho(h).$$

Let us determine the image of the group G^+ under ρ^+.

If $u \in G^+$, $\forall h \in E$, $h' = uhu^{-1} \in E$ and therefore $\rho(h') = \rho(u)\rho(h)\rho(u^{-1}) \in F$, or with obvious notations,

$$H' = U \circ H \circ U^{-1} \in F, \quad U \in \text{GL}(\mathbb{C}^2).$$

Conversely, if $H' = UHU^{-1}$ where $H \in F$ and $U \in \mathrm{GL}(\mathbf{C}^2)$, the trace of H' is zero ($\mathrm{tr}\, H' = \mathrm{tr}\, H = 0$), so $H' \in F$.

The restrictions of ρ to $C^+(Q)$ and E being bijective, we deduce from

$$H' = U \circ H \circ U^{-1}, \quad \text{if} \quad H = \rho(h), \quad U = \rho(u), \quad u \in C^+(Q)$$

and $H' = \rho(h')$, $h' \in E$:

$$h' = uhu^{-1} \in E \quad \text{and} \quad u \in G^+.$$

Therefore ρ^+ is an isomorphism from G^+ to $\mathrm{GL}(\mathbf{C}^2)$.

In the older publications on this subject, the representation $\varphi : u \to \varphi_u$ is described using UHU^{-1}, written $\varphi_U(H)$, so that

$$\varphi(\mathrm{GL}(\mathbf{C}^2)) = \mathrm{SO}(3, \mathbf{C})$$

by the results on the Clifford group.

Now we can prove that $\varphi(U) \in \mathrm{SO}(3, \mathbf{R})$ if and only if U is unitary up to a complex factor.

Indeed, a 'real' vector in F can be identified with a matrix H such that $\bar{H} = -H^T$ (for such a vector, x^1, x^2 and x^3 are real).

$\varphi(U) \in \mathrm{SO}(3, \mathbf{R})$ if and only if

$$\overline{UHU^{-1}} = -(UHU^{-1})^T = (U^{-1})^T \bar{H} U^T,$$

whenever $\bar{H} = -H^T$. $\bar{H} = (U^T \bar{U}) \bar{H} ((\bar{U}^{-1})(U^{-1})^T)$ for all H obtained from real x^1, x^2, x^3. $U^T \bar{U} \in Z \cap G^+$ implies $\bar{U}^T U = \lambda I$ and $U/\sqrt{\lambda}$ is unitary.

The kernel of the restriction of φ to $\mathrm{SU}(2, \mathbf{C})$ is a subset of the set of invertible elements of the center $\mathrm{End}(\mathbf{C}^2)$, i.e. $\{\sigma_0, -\sigma_0\}$. We have proved :

$$\mathrm{SU}(2, \mathbf{C}) \text{ is a double covering of } \mathrm{SO}(3, \mathbf{R}).$$

Remarks.

1. $\mathrm{SU}(2, \mathbf{C})$ is the set of endomorphisms with matrix

$$\alpha \sigma_x + \beta \sigma_y + \gamma \sigma_z + \delta \sigma_0$$

where α, β, γ and δ are real and $\alpha^2 + \beta^2 + \gamma^2 + \delta^2 = 1$ because of (2).

$\mathrm{SU}(2, \mathbf{C})$ is isomorphic to the multiplicative group of unit quaternions and is therefore a double covering of $\mathrm{SO}(3, \mathbf{R})$.

We have reobtained the result which, historically, led to the first definition of the quaternions.

2. If U is a real matrix and, as an element of SU(2, **C**), determines a rotation of angle α, the matrix of $\varphi(U)$ can be written, relying on

$$H' = UHU^{-1},$$

as

$$\begin{pmatrix} \cos 2\alpha & 0 & -\sin 2\alpha \\ 0 & 1 & 0 \\ \sin 2\alpha & 0 & \cos 2\alpha \end{pmatrix}.$$

If OX, OY, OZ is a positive orthogonal reference frame and if we identify $\mathbf{C}^2$ with the plane OXZ, a rotation of angle 2α and axis OY corresponds to U. So when $h \in \mathbf{C}^2 \subset \mathbf{C}^3$ 'turns' over 2α around OY, h, considered as a spinor, 'turns' over α around OY. This is no paradox : to a rotation of angle α around OY there corresponds, under φ, a rotation of angle 2α around the same axis.

7.2 Minkowski Clifford algebras and Dirac matrices.

7.2.1 Minkowski space and the Lorentz group.

We start by a general definition holding for all dimensions.

A real n-dimensional vector space with a non-degenerate indefinite normal hyperbolic quadratic form will be called a *Minkowski space* $E_\mathbf{R}$ or $E_{n-1,1}$. There exist orthonormal bases $e_0, e_1, e_2, \ldots, e_{n-1} = \{e_0, e_i\}$ for which

$$Q(x) = -(x^0)^2 + \sum_{i=1}^{n-1}(x^i)^2, \quad \forall x \in E;$$

a latin index such as i will vary from 1 to $n - 1$; a greek index will vary from 0 to n.

x^0 is called the *time coordinate*, the x^i are called *space coordinates*. Using the Einstein convention, we write

$$Q(x) = -g_{\alpha\beta}x^\alpha x^\beta$$

where $g_{ij} = -\delta_{ij}$, $g_{0i} = g_{i0} = 0$ and $g_{00} = 1$ in the basis $\{e_0, e_i\}$; $g^{\alpha\beta}$ stands for the inverse matrix of $g_{\alpha\beta}$.

Remark. We could also have chosen the signature $(1, n - 1)$ without affecting the essential results of this chapter.

An isometry given by

$$X^\alpha = A^\alpha_\lambda x^\lambda,$$

satisfies

$$g_{\alpha\beta} = A^\lambda_\alpha A^\mu_\beta g_{\lambda\mu}, \tag{1}$$

or in matrix form :

$$g = A^T g A,$$

which is also equivalent to

$$g^{\alpha\beta} = A^\alpha_\lambda A^\beta_\mu g^{\lambda\mu}. \tag{2}$$

From (1) and (2) one easily deduces that

$$\det A = \pm 1, \quad (A^0_0)^2 = 1 + \sum_{i=0}^{n-1}(A^i_0)^2 = 1 + \sum_{i=1}^{n-1}(A^0_i)^2, \quad |A^0_0| \geq 1.$$

Fundamental subgroups of the Lorentz group. The set of x such that $Q(x) < 0$ defines the 'interior' of the isotropic cone.

Definition 7.2.1 *The isometry group of the space $E_{n-1,1}$ is called the Lorentz group L_n. Its subgroup of elements which preserve every interior sheet of the isotropic cone is called the orthochronic subgroup.*

The orthochronic subgroup is characterized by $A^0_0 \geq 1$.
Indeed, $X^0 = A^0_0 x^0 + A^0_i x^i$ leads, by the Schwarz inequality, to

$$(A^0_i x^i)^2 \leq \sum_{i=1}^{n-1}(A^0_i)^2 \sum_{i=1}^{n-1}(x^i)^2$$

where $\sum_{i=1}^{n-1}(A^0_i)^2 \leq (A^0_0)^2$ and $\sum_{i=1}^{n-1}(x^i)^2 \leq (x^0)^2$, so that

$$(A^0_i x^i)^2 \leq (A^0_0 x^0)^2,$$

and if $A^0_0 \leq -1$, X^0 and x^0 have opposite sign, but if $A^0_0 \geq 1$, they have the same sign.

The connected component of the identity in the Lorentz group L_n (i.e. $O(n-1,1)$) is the group L^0_n which was written $\varphi(G^+_0)$ in Chapter 6. It consists of matrices A whose determinant is 1 and contains all elements of $SO(n-1)$ for which $A^0_0 = 1$; therefore this subgroup is characterized by

$$\det A = 1, \quad A^0_0 \geq 1.$$

These conditions lead to the existence of exactly four connected components (confirming the results of Chapter 6), the other three being obtained by

$$\det A = 1, \quad A^0_0 \leq 1,$$
$$\det A = -1, \quad A^0_0 \geq 1,$$
$$\det A = -1, \quad A^0_0 \leq 1,$$

the elements of which can be factored, for $n = 4$, as

$$g = \gamma g_0, \quad g_0 \in L^0_n,$$

where γ is either the identity, the space symmetry

$$\begin{cases} X^i = -x^i, & i = 1,2,3 \\ X^0 = x^0 \end{cases},$$

the time symmetry

$$\begin{cases} X^i = x^i, & i = 1,2,3 \\ X^0 = -x^0 \end{cases},$$

or the spacetime symmetry

$$\begin{cases} X^i = -x^i, & i = 1,2,3 \\ X^0 = -x^0 \end{cases}.$$

A similar factorization holds for all even n.

The case where n is odd is left as an exercise for the reader.

The orthochronic group is generated by L_n^0 and the space symmetry; it has two connected components.

7.2.2   Four-dimensional minkowskian Clifford algebras.

Let $C'(Q)$ be the complexified Clifford algebra $C(Q) = C(3,1)$. $C'(Q)$ is isomorphic to the endomorphism algebra of a four-dimensional complex vector space, which will itself be isomorphic to $E_{\mathbf{C}}$.

The Dirac matrices.   Let $F_{\mathbf{C}}$ be the complex vector space generated in $\text{End}(\mathbf{C})$ by the four matrices γ_α, $\alpha = 0,1,2,3$,

$$\gamma_0 = i \begin{pmatrix} \sigma_0 & 0 \\ 0 & -\sigma_0 \end{pmatrix}, \gamma_1 = \begin{pmatrix} 0 & \sigma_x \\ -\sigma_x & 0 \end{pmatrix}, \gamma_2 = \begin{pmatrix} 0 & \sigma_y \\ -\sigma_y & 0 \end{pmatrix}, \gamma_3 = \begin{pmatrix} 0 & \sigma_z \\ -\sigma_z & 0 \end{pmatrix}. \tag{3}$$

Lemma 7.2.1 *The algebra generated by the four Dirac matrices is* $\text{End}(E_{\mathbf{C}})$.

A direct and elementary proof for this lemma runs as follows :
Proof. The generated algebra contains :

$$1, \gamma_0, \gamma_1, \gamma_2, \gamma_3, \gamma_\lambda \gamma_\mu, \gamma_\lambda \gamma_\mu \gamma_\nu, \gamma_0 \gamma_1 \gamma_2 \gamma_3$$

where $0 \le \lambda < \mu \le 3$, which are 16 linearly independent matrices, since, except for 1, their trace is zero.

Indeed, $(\gamma_0)^2 = -1$, $(\gamma_i)^2 = 1$, $i = 1,2,3$, $\gamma_\alpha \gamma_\beta + \gamma_\beta \gamma_\alpha = 0$ if $\alpha \ne \beta$.

$$\text{tr}(\gamma_\lambda \gamma_\mu) = \text{tr}(\gamma_\mu \gamma_\lambda) = -\text{tr}(\gamma_\mu \gamma_\lambda),$$

so $\text{tr}(\gamma_\lambda \gamma_\mu) = 0$. If $\sigma \ne \lambda, \mu, \nu$,

$$\text{tr}(\gamma_\lambda \gamma_\mu \gamma_\nu) = \epsilon \, \text{tr}(\gamma_\lambda \gamma_\mu \gamma_\nu \gamma_\sigma \gamma_\sigma) = \epsilon \, \text{tr}(\gamma_\sigma \gamma_\lambda \gamma_\mu \gamma_\nu \gamma_\sigma)$$

where $\epsilon = \pm 1$ and we have relied on the invariance of the trace under cyclic permutation. Given the properties of the matrices in (3), this implies

$$-\epsilon\, \mathrm{tr}(\gamma_\lambda \gamma_\mu \gamma_\nu \gamma_\sigma \gamma_\sigma) = -\,\mathrm{tr}(\gamma_\lambda \gamma_\mu \gamma_\nu).$$

Finally,

$$\mathrm{tr}(\gamma_0 \gamma_1 \gamma_2 \gamma_3) = \mathrm{tr}(\gamma_0(\gamma_1 \gamma_2 \gamma_3)) = \mathrm{tr}(\gamma_1 \gamma_2 \gamma_3 \gamma_0) = -\,\mathrm{tr}(\gamma_0 \gamma_1 \gamma_2 \gamma_3).$$

Now if

$$F = \lambda + \mu^0 \gamma_0 + \mu^1 \gamma_1 + \ldots + \mu^{0123} \gamma_0 \gamma_1 \gamma_2 \gamma_3 = 0,$$

$\mathrm{tr}(F) = 4\lambda = 0$ and, similarly, $\mathrm{tr}(\gamma_0 F) = \mathrm{tr}(\mu^0 \gamma_0 \gamma_0)$, $\gamma_0^2 = -1$, so $\mu^0 = 0$ etc.

If we set $\rho(e_\alpha) = \gamma_\alpha$, ρ is seen to extend to an algebra homomorphism sending a basis of $C(Q)$ to a basis of $\mathrm{End}(E_\mathbf{C})$, hence ρ must be a spin representation. ∎

Remark. A faster, but less elementary proof of Lemma (7.2.1) can be given as follows : define an endomorphism u from $E_\mathbf{C}$ to $F_\mathbf{C}$ by $u(e_\alpha) = \gamma_\alpha$. By the multiplication table of the γ_α,

$$(u(\xi))^2 = Q(\xi)$$

for all $\xi \in E_\mathbf{C}$, so by the universal property of Clifford algebras, there exists a homomorphism $\bar{u}$ from $C'(Q)$ into the algebra generated by the Dirac matrices, $\bar{u}$ clearly is surjective and $C'(Q)$ being simple, $\bar{u}$ is injective, so the algebra generated by the (γ_α) is sixteen-dimensional and must coincide with $\mathrm{End}(E_\mathbf{C})$.

Restriction of the field to R. Let us consider the Clifford algebra $C(Q)$ of the Minkowski space $E_{3,1}$ and the real vector space F generated by the real linear combinations of the Dirac matrices. $C(Q)$ is known to be a simple algebra. Using the universal property, an R-endomorphism from E to F can be defined by the method of the previous remark. $C(Q)$ is therefore isomorphic to the real algebra generated by the four Dirac matrices, which, in turn, is isomorphic to $\mathrm{End}(E)$.

The 'vector' and the 'spin' structure of F. The isomorphism between E and F gives F a Minkowski space structure. $C(Q)$ is isomorphic to $\mathrm{End}(E)$, hence to $\mathrm{End}(F)$. The representations of $C(Q)$ can be constructed using $\mathrm{End}(F)$, F is then a spinor space.

If $\Lambda \in G$, $\Lambda \in \mathrm{End}(F)$ and Λ^{-1} exists, $\Lambda F \Lambda^{-1} \subseteq F$. Setting $\varphi_\Lambda(x) = \Lambda x \Lambda^{-1}$, φ_Λ is an isometry and for $x = \gamma_\alpha$, $\Lambda \gamma_\alpha \Lambda^{-1} = A_\alpha^\lambda \gamma_\lambda$.

(A_α^λ) is the matrix of a Lorentz isometry.

In this way, an element of F can either be considered as a vector in a Minkowski space on which the Lorentz group acts, or as a spinor on which the group G, isomorphic to the Clifford group of $E_{3,1}$, acts. The associated operations are related by

$$\Lambda \gamma_\alpha \Lambda^{-1} = A_\alpha^\beta \gamma_\alpha. \tag{4}$$

This concept is found in many publications by theoretical physicists; we see that it is closely linked to the four-dimensional case ($2^n = n^2$ if $n = 4$).

7.2.3 The homomorphism from SL(2, C) to the restricted Lorentz group and to SO(3, C). The complex vectorial formalism.

Let H be an arbitrary hermitian matrix which can be identified to an element of $\mathbf{R}^4$ (also written H) :

$$H = \begin{pmatrix} x^0 + ix^3 & x^1 - ix^2 \\ x^1 + ix^2 & x^0 - ix^3 \end{pmatrix}, \quad (x^0, x^1, x^2, x^3) \in \mathbf{R}^4.$$

If $A \in \mathrm{SL}(2, \mathbf{C})$, we set

$$H' = AHA^* = \varphi_A(H),$$

A^* being the inverse of H^T.

$$\det H = \det H' = (x^0)^2 - \sum_{k=1}^{3}(x^k)^2 = (x^{0'})^2 - \sum_{k=1}^{3}(x^{k'})^2.$$

This choice of coordinates shows the signature $(1, 3)$ of the metric. φ maps A to an element φ_A or $\varphi(A)$ of $\mathrm{O}(1, 3)$, somewhat abusively denoted by L_4.

One easily shows that $\mathrm{SL}(2, \mathbf{C})$ is arcwise connected so that $\varphi(A)$ belongs to the restricted Lorentz group L_4^0.

It is obvious that $\varphi(A) = \varphi(A')$ implies $A' = \pm A$.

The Lie algebra of $\mathrm{SL}(2, \mathbf{C})$ is six-dimensional over $\mathbf{R}$ and consists of the 2×2 complex matrices with zero trace. A real basis for it is

$$X = \begin{pmatrix} 0 & 0 \\ i & 0 \end{pmatrix}, Y = \begin{pmatrix} 0 & i \\ 0 & 0 \end{pmatrix}, Z = \begin{pmatrix} 1 & 0 \\ 0 & -1 \end{pmatrix}, iX, iY, iZ.$$

Note that

$$[X, Z] = 2X, \quad [Y, Z] = -2Y, \quad [X, Y] = Z. \tag{5}$$

From the following one-parameter subgroups of $\mathrm{SL}(2, \mathbf{C})$:

$$\alpha_1(t) = \begin{pmatrix} \cos t/2 & i \sin t/2 \\ i \sin t/2 & \cos t/2 \end{pmatrix}, \qquad \alpha_2(t) = \begin{pmatrix} \cos t/2 & \sin t/2 \\ -\sin t/2 & \cos t/2 \end{pmatrix},$$

$$\alpha_3(t) = \begin{pmatrix} \exp(it/2) & 0 \\ 0 & \exp(-it/2) \end{pmatrix}, \quad \beta_1(t) = \begin{pmatrix} \mathrm{ch}\, t/2 & -\mathrm{sh}\, t/2 \\ -\mathrm{sh}\, t/2 & \mathrm{ch}\, t/2 \end{pmatrix},$$

$$\beta_2(t) = \begin{pmatrix} \mathrm{ch}\, t/2 & i\, \mathrm{sh}\, t/2 \\ -i\, \mathrm{sh}\, t/2 & \mathrm{ch}\, t/2 \end{pmatrix}, \qquad \beta_3(t) = \begin{pmatrix} \exp(-t/2) & 0 \\ 0 & \exp(t/2) \end{pmatrix}$$

we can construct elements in the Lie algebra of $\mathrm{SL}(2, \mathbf{C})$:

$$a_1 = \tfrac{1}{2}\sigma_x, \quad a_2 = \tfrac{1}{2}\sigma_y, \quad a_3 = \tfrac{1}{2}\sigma_z, \quad b_1 = ia_1, \quad b_2 = ia_2, \quad b_3 = ia_3.$$

If $\tilde{\varphi}$ is the tangent map to φ, we have at the identity,

$$\tilde{\varphi}(a_1) : H \to (\alpha_1(t)H\alpha_1^*(t))|_{t=0} = a_1 H - H a_1^T.$$

But $H = x^0\sigma_0 - ix^1\sigma_x - ix^2\sigma_y - ix^3\sigma_z$ so that

$$
\begin{aligned}
\tilde{\varphi}(a_1) &= i(x^2\sigma_z - x^3\sigma_y),\\
\tilde{\varphi}(a_2) &= i(x^3\sigma_x - x^1\sigma_z),\\
\tilde{\varphi}(a_3) &= i(x^1\sigma_y - x^2\sigma_x),\\
\tilde{\varphi}(b_k) &= i\tilde{\varphi}(a_k), \quad k = 1, 2, 3.
\end{aligned}
\tag{6}
$$

It can easily be proved that the image of the Lie algebra of $SL(2,\mathbf{C})$ obtained in this way is six-dimensional over $\mathbf{R}$. It is therefore the Lie algebra of $O(1,3)$ which, by the exponential function, generates the connected component of the identity in $O(1,3)$, i.e. L_4^0.

   Therefore $SL(2,\mathbf{C})$ is a connected double covering of the restricted Lorentz group.
   For $x^0 = 0$, $\varphi(A) \in SO(3,\mathbf{R})$ (by the connectedness of $SO(3,\mathbf{R})$).
   By $\tilde{\varphi}$ the Lie algebra of $SO(3,\mathbf{R})$ is obtained and as $\tilde{\varphi}(b_k) = i\tilde{\varphi}(a_k)$, $k = 1, 2, 3$, we see that the image of the Lie algebra of $SL(2,\mathbf{C})$ under $\tilde{\varphi}$ can be interpreted as the Lie algebra of $SO(3,\mathbf{C})$; $SO(3,\mathbf{C})$ being connected, $SL(2,\mathbf{C})$ *is a covering of* $SO(3,\mathbf{C})$.
   The Lie algebras of the groups $SL(2,\mathbf{C})$, $SO(3,\mathbf{C})$, $Spin(1,3)$ and $SO(1,3)$ are isomorphic.
   Using a Lorentz basis (e_0, e_1, e_2, e_3) satisfying $(e_0)^2 = 1$, $(e_k)^2 = -1$, $k = 1, 2, 3$ the basis

$$e_1e_2, e_2e_3, e_3e_1, e_0e_1, e_0e_2, e_0e_3$$

of $\mathcal{L}(L_4^0)$ is obtained. One easily verifies that

$$[e_3e_1/2, e_2e_3/2] = -e_1e_2/2, \quad [e_0e_1/2, e_0e_3/2] = e_3e_1/2$$

and all cyclic permutations of these equations.

   Setting $J(e_2e_3) = e_0e_1$ etc. an operator J is obtained for which $J^2 = -\,\mathrm{Id}$, giving a complex structure to $\mathcal{L}(L_4^0)$.

   The isomorphism between the Lie algebra of $SO(3,\mathbf{C})$ (given a real structure) and that of the Lorentz group is at the origin of the 'complex vectorial formalism' of relativity theory (cf. ultra Chapter 8, 3).

7.3   Selected references.

- M. Cahen, R. Debever, L. Defrise, *A complex vectorial formalism in general Relativity*, Journal of Math. and Mech., 16 - 7, 1967.

- E. Cartan, *Leçons sur la théorie des spineurs*, Hermann, Paris, 1938.

- D. Kastler, *Introduction a l'électrodynamique quantique*, Dunod, Paris, 1961.

Chapter 8

THE SPINORS IN MAXIMAL INDEX AND EVEN DIMENSION.

8.1 Pure spinors and associated m.t.i.s.

8.1.1

Q is assumed non-degenerate of maximal index r, $\dim E = n = 2r$ over $\mathbf{K}$.

In Chapter 5, 2.1, an analogous situation was considered for $\mathbf{K} = \mathbf{C}$; we will use the same notations.

$f = y_1 y_2 \ldots y_r$ will be called an *isotropic r-vector*. The spinor space S is $C(Q)f$ and the spin representation ρ is just the multiplication from the left.

Lemma 8.1.1 $fC(Q)$ *is a minimal right ideal.*

Proof. This follows at once from the existence of the anti-automorphism β, $C(Q)f$ being a minimal left ideal (cf. Chapter 5). ∎

Proposition 8.1.1 f *and* f_1 *being isotropic r-vectors,* $C(Q)f \cap f_1C(Q)$ *is a one-dimensional subspace (over $\mathbf{K}$) of* $C(Q)$.

Proof. This is a special case of Proposition (5.2.2), since $C(Q)f$ and $f_1C(Q)$ are minimal ideals, the first a left ideal, the second a right ideal of $C(Q)$.

This result being very important, we also give a direct proof : first let us determine the intersection of $C(Q)f$ and $fC(Q)$. If $uf = fv$, $u, v \in C(Q)$, we may assume that u and v are linear combinations of $x_1, x_2, \ldots, x_r$. Taking a term of maximal degree in u, $u_0 = \alpha^{i_1 \ldots i_k} x_{i_1} x_{i_2} \ldots x_{i_k}$ (where there is no summation), and multiplying from the left by the complement of the sequence $x_{i_1}, x_{i_2}, \ldots x_{i_k}$ in $x_1, x_2, \ldots, x_r$, we obtain :

$$\alpha x_1 \ldots x_r f = u' fv, \quad \alpha \in \mathbf{K}^*,$$

u' does not contain all x_i if u_0 is not scalar.

Since $\alpha f x_1 \ldots x_r f = \alpha k f \neq 0$, $k \in \mathbf{K}^*$, and $fu'fv = 0$, a contradiction is obtained.

So $uf = (\lambda + u_1)f$, $\lambda \in \mathbf{K}^*$.

If $u_1 = 0$, $uf = \lambda f$ and the proof is complete; otherwise $u_1 f = f(v - \lambda)$ and we are back in the previous case for u_1, u_1 contains a non-scalar term of minimal degree and a contradiction is reached.

In general, if $uf = f_1 v$ there exists, by the Witt theorem, an element γ of the Clifford group such that $f_1 = \gamma^{-1}f\gamma$, and therefore $\gamma^{-1}uf = f\gamma^{-1}v$,

$$\gamma^{-1}uf = \lambda f, \quad uf = \lambda\gamma f;$$

we note however that λ is not unique, if $f_1 = \gamma_1 f\gamma_1^{-1}$, $\gamma^{-1}\gamma_1 f = f\gamma^{-1}\gamma_1 = \sigma f$, $\sigma \in \mathbf{K}$, $\gamma_1 f = \gamma\sigma f$ and the same subspace is determined. ∎

Definition 8.1.1 *A non-zero element of the one-dimensional subspace defined by the intersection of the minimal ideals $C(Q)f$ and $f_1 C(Q)$ is called a pure spinor.*

F' and F_1 being m.t.i.s. determined by f and f_1, we will say that the pure spinors determined by f and f_1 are also determined by F and F_1, since by Proposition $(2.2.5)$, if $y_1', y_2', \ldots, y_r'$ form another basis of F', $y_1'y_2'\ldots y_r'$ and $y_1 y_2 \ldots y_r$ are collinear.

For example, $C(Q)f \cap fC(Q)$ contains f, and F, associated to $(x_1 x_2 \ldots x_r)$ determines, with F', the pure spinor $x_1 x_2 \ldots x_r f = f'f$, which is idempotent.

To simplify the notations we will assume that F', associated to $y_1 y_2 \ldots y_r$, is fixed and we will say that any other m.t.i.s. 'determines an equivalence class of collinear pure spinors'.

Lemma 8.1.2 *F_1 being a m.t.i.s., there exists a $g \in G$ (the Clifford group) such that $gF'g^{-1} = F_1$; then gf is a pure spinor determined by F_1.*

Proof. The existence of g follows from the Witt theorem. If φ is the projection of G on $O(Q)$, setting $\sigma = \varphi(g)$,

$$gF'g^{-1} = \varphi_g(F') = F_1,$$

gfg^{-1} is a product f_1 of elements of a basis for F_1, and

$$gf = f_1 g \in C(Q)f \cap f_1 C(Q).$$

∎

Proposition 8.1.2 *Given a spin representation in $C(Q)f$, every m.t.i.s. is characterized by an equivalence class of pure spinors.*

Proof. Every pure spinor lies in the intersection of two ideals $C(Q)f$ and $f_1 C(Q)$. If $F_1 = gF'g^{-1}$, $uf = \lambda gf$ for all pure spinors determined by F_1, according to Lemma $(8.1.2)$. ∎

Conversely, to each pure spinor there corresponds a unique m.t.i.s. F_1, as is proved by the following lemma :

Lemma 8.1.3 *If F_1 is associated to gf :*

1. $F_1 = \{x \in E, xgf = 0\}$, *i.e. F_1 is the intersection of E and the annihilator of gf, which is an element of the representation module.*

2. *If $uf \in S$ is such that $xuf = 0$, $\forall x \in F_1$, then $uf = \lambda gf$, $\lambda \in \mathbf{K}^*$ and uf characterizes F_1.*

Proof. Indeed, if $x \in F_1$, $xgf = xf_1g$ and as $xf_1 = 0$, $xgf = 0$.

Conversely, if $xgf = 0$, let $E = F_1 \oplus F_1'$ be a Witt decomposition, then $x = x_1 + x_1'$, $(x_1 + x_1')gf = 0$, $(x_1 + x_1')f_1g = 0$, $(x_1 + x_1')f_1 = 0$ and, as $x_1f_1 = 0$, $x_1'f_1 = 0$ so that $x_1' = 0$ and $x \in F_1$.

For 2, $xuf = 0$ if and only if $\gamma x \gamma^{-1}(\gamma uf) = 0$, $\gamma \in G$ and the property of $x \in F_1$ we are considering will also hold for $\gamma F_1 \gamma^{-1}$. Hence we can assume that $F_1 = F$.

$C(Q)f$ is linearly isomorphic to $\wedge(F)$ and uf is, by this isomorphism, associated to an element of degree r in $\wedge(F)$ since $xuf = 0$, therefore

$$uf = \lambda x_1 x_2 \ldots x_r f$$

and F being associated to $x_1 x_2 \ldots x_r f$, uf is in the class of pure spinors determined by F. ∎

Important remark. $uf = \lambda gf$ for all pure spinors determined by $F_1 = gF'g^{-1}$, but one should note that g is by no means uniquely determined, even modulo a scalar factor : there can exist $\gamma \in G$ such that $\gamma f \gamma^{-1} = kf$, $k \in \mathbf{K}^*$ and F_1 is associated to $g\gamma f$.

Corollary 8.1.1 *If $(u_1, u_2, \ldots, u_r)$ determine a m.t.i.s. F_1', if $(z_1, z_2, \ldots, z_r)$ determine a m.t.i.s. F_2 and if $\lambda z_1 z_2 \ldots z_r = u_1 u_2 \ldots u_r$, $F_1' = F_2$.*

Proof. Indeed, $f_1 = u_1 u_2 \ldots u_r$, $f_r = \lambda z_1 z_2 \ldots z_r$, setting

$$f_1 = \gamma f \gamma^{-1},$$

$f_2 = \lambda \gamma f \gamma^{-1}$ and γf is a pure spinor determined by F_1, and also by F_2. ∎

Finally, we conclude that having chosen a spinor space $C(Q)f$, a class of pure spinors can be bijectively associated to each m.t.i.s.—or equivalently, that a m.t.i.s. is characterized by a class of pure spinors.

Remark. A pure spinor is a semi-spinor, since we can assume it is gf, $g \in G$ where g is a product of non-isotropic vectors in E; g has a well-determined parity. If g is changed to $g\gamma$ where $\gamma f \gamma^{-1} = kf$, $k \in \mathbf{K}^*$, then $\gamma f = kf\gamma = \lambda f$, $\lambda \in \mathbf{K}^*$, γ is even and therefore $g\gamma$ and g have the same parity.

Definition 8.1.2 *A m.t.i.s. will be called* even *or* odd *according to the parity of its associated pure spinor.*

Obviously two m.t.i.s. are of the same parity if and only if one of them can be mapped to the other by an element of $SO(Q)$.

8.1.2   The subgroup of $O(Q)$ inducing the identity on a m.t.i.s.

Let $E = F \oplus F'$ be a Witt decomposition and let (x_i, y_i) be the corresponding Witt basis (cf. Chapter 5, 2.1). If the isometry σ fixes all points of F, it gives rise to another Witt decomposition :

$$E = F \oplus F_1', \quad F_1' = \sigma(F').$$

Now set $\sigma(x_i) = x_i$, $\sigma(y_i) = a_i^j x_j + b_i^k y_k$, $i = 1, 2, \ldots, r$. The isometry property leads to $B(a_i^l x_l + b_i^k x_k, x_j) = \delta_{ij}$ so $b_i^j = \delta_i^j$, and $B(a_i^l x_l + \delta_i^k x_k, a_j^l x_l + \delta_j^k y_l) = 0$ implies that

$$a_i^j + a_j^i = 0.$$

We have obtained :

Lemma 8.1.4 $\sigma \in O(Q)$ *induces the identity on* F *if and only if* $\sigma(x_i) = x_i$ *and* $\sigma(y_i) = a_i^j x_j + y_i$ *where* $a_i^j + a_j^i = 0$.

Note that F' is not uniquely determined by F and the requirements that $E = F \oplus F'$ be a Witt decomposition and that any F_1' be obtained by some σ.

Remark concerning the idempotent elements.   We have noted that $f'f = x_1 x_2 \ldots x_r y_1 y_2 \ldots y_r$ is a primitive idempotent determining the minimal left ideal $C(Q)f$,

$$f = y_1 y_2 \ldots y_r.$$

$y_1 y_2 \ldots y_r x_1 x_2 \ldots x_r$ determines $C(Q)f'$, $f' = x_1 x_2 \ldots x_r$. Replacing $y_1 y_2 \ldots y_r$ by $\sigma(y_1)\sigma(y_2) \ldots \sigma(y_r)$ we obtain a new idempotent element which also determines $C(Q)f'$.

This provides us with the example announced in Chapter 5, 1.7 part 1. These two idempotents are essentially different (we would call them trivially different if they only differed by a sign), yet they are geometrically equivalent[1].

The exponential function in $\wedge^2(F)$.   The definition of the exponential function on this space does not raise any convergence problems since the powers of $X = \sum_{i<j\leq r} a^{ij}(x_i \wedge x_j)$ vanish beyond some finite value of the exponent.

Setting $X = \sum_{\alpha=1}^{N} X_\alpha$,

$$\begin{aligned} \exp X &= 1 + \textstyle\sum_\alpha (X_\alpha) + \sum_{\alpha_1 < \alpha_2} X_{\alpha_1} X_{\alpha_2} + \sum_{\alpha_1 < \alpha_2 < \alpha_3} X_{\alpha_1} X_{\alpha_2} X_{\alpha_3} + \cdots \\ &= \textstyle\prod_\alpha (1 + X_\alpha) \end{aligned}$$

since $(x_i \wedge x_j)^2 = 0$. Therefore,

$$\exp(X + X') = \exp(X)\exp(X'), \quad X, X' \in \wedge^2(F),$$

and if $X = x \wedge y$, $\exp(X) = 1 + X$ and $\exp(-X) = (\exp(X))^{-1} = 1 - X$.

---

[1]cf. ultra, Chapter 20.

Proposition 8.1.3 *(Using the notations of Chapter 4.) If $u \in \wedge^2(F)$, $\exp(u) \in G_0^+$ and $\varphi(\exp u)$ fixes the elements of F. Conversely, every element of $O(Q)$ which fixes all elements of F is in $\varphi(G_0^+)$ and can be written as $\varphi(\exp u)$, $u \in \wedge^2(F)$.*

Proof. Obviously the elements of the form $\exp u$, $u \in \wedge^2(F)$, generate a group. Each element of $\wedge^2(F)$ being the sum of decomposable elements, it is sufficient to show that if $x_1, x_2 \in F$, $\exp(x_1 x_2) \in G$ to prove that $\exp u \in G$.

But from $x_i y + y x_i = 2B(x_i, y)$, $i = 1, 2$, $y \in F'$ we deduce that

$$\exp(u)y(\exp(u))^{-1} = (1+u)y(1-u),$$

where $u = x_1 x_2$, or $\exp(u)y(\exp(u))^{-1} = y + 2B(x_2, y)x_1 - 2B(x_1, y)x_2 \in E$.

Obviously $\exp(u) \in G^+$. $\beta(x_1 x_2) = -x_1 x_2$, so if u is decomposable,

$$\beta(\exp(u)) = \exp(-u) = (\exp(u))^{-1},$$

hence $\exp(u) \in G_0^+$. This result is easily extended to arbitrary $u \in \wedge^2(F)$.

Finally, since u commutes with all elements of F, $\varphi(\exp u)$ fixes each element of F.

Conversely : we need only establish that $\varphi(\exp(\wedge^2(F)))$ is the set of elements in $O(Q)$ which fix all points of F.

If $u = x_1 x_2$, $\exp(u)y_j(\exp(u))^{-1} = y_j + 2\delta_{2,j}x_1 - 2\delta_{1,j}x_2$.

If $u = a^{12}x_1 x_2$, $\exp(u)y_j(\exp(u))^{-1} = y_j + 2a^{12}(\delta_{2,j}x_1 - \delta_{1,j}x_2)$.

Now if $u = a^{i_1 i_2} x_{i_1} x_{i_2} = \sum X_\alpha$, $a^{i_1 i_2} = -a^{i_2 i_1}$,

$$\exp(u) = \prod_\alpha (1 + X_\alpha).$$

Repeating the previous computations yields

$$
\begin{aligned}
(\exp(u))y_j(\exp(u))^{-1} &= y_j + 2a^{i_1 i_2}(\delta_{i_2,j}x_{i_1} - \delta_{i_1,j}x_{i_2}) \\
&= y_j + 4a^{j i_1} x_{i_1},
\end{aligned}
$$

which is the general element described in Lemma (8.1.4). ∎

8.1.3 The general form of pure spinors.

This general form can be deduced from the following two lemmas and from Proposition (8.1.4).

Lemma 8.1.5 *If V, V', V'' are m.t.i.s. such that $V \cap V' = V \cap V''$, there exists a $\sigma \in \varphi(G_0^+)$ fixing all elements of V and mapping V' to V''.*

Proof. Under an isometry τ the problem is not changed, σ is replaced by $\sigma' = \tau \circ \sigma \circ \tau^{-1}$ and if $\sigma \in \varphi(G_0^+)$, so will σ'; on the other hand if $\tau(V) = V_1$, $\sigma'(V_1) = V_1$ and σ' fixes all elements of $V_1 : \tau(V) \cap \tau(V') = \tau(V) \cap \tau(V'')$.

We may assume that $V = F$.

Let (x_i, y_i) be a Witt basis corresponding to the decomposition $E = F \oplus F'$ and let $F \cap V' = F \cap V'' = (x_1, x_2, \ldots, x_h)$.

$F_1 = (x_{h+1}, \ldots, x_r, y_1, y_h)$ is a m.t.i.s., and $F_1 \cap V' = 0$. Indeed, if $z \in F_1 \cap V'$, it is orthogonal to all elements of F_1 and V', so to

$$(x_1, x_2, \ldots, x_h, x_{h+1}, \ldots, x_r, y_1, \ldots, y_h);$$

but the space which is orthogonal to this $(r + h)$-dimensional space and contains $x_{h+1}, \ldots, x_r$ is $(x_{h+1}, \ldots, x_r)$. Therefore $z \in (x_{h+1}, \ldots, x_r)$ but $z \in V'$ and $F \cap V' = (x_1, \ldots, x_h)$ then imply that z is zero.

Hence $E = F_1 \oplus V' = F_1 \oplus V''$ and, applying another isometry, we must only prove that if

$$E = F \oplus V' = F \oplus V'',$$

there exists a $\sigma \in p(\exp(\wedge^2(F)))$ mapping V' to V''.

A Witt basis for the decomposition $F \oplus V''$ can be found : taking x_1' in F, a $y_1' \in V''$ is associated, requiring that $B(x_1', y_1') = 1$. (x_1', y_1') is non-isotropic and E can be written as $(x_1', y_1') \oplus H_1 \oplus H_1'$ where $\dim H_1 = \dim H_1' = r - 1$, H_1 and H_1' are totally isotropic, etc.

Let u be an isometry sending the Witt basis $\{x_i, y_j\}$ of $F \oplus F'$ to the Witt basis $\{x_i', y_j'\}$ of $F \oplus V' : x_i' = u(x_i) = A_i^k x_k$, $y_i' = u(y_i) = B_i^k y_k + C_i^k x_k$.

Now necessarily $A^T B = \mathrm{Id}$ and u is the product of two isometries :

$$x_i \to A_i^k x_k = X_i \to X_i,$$
$$y_i \to B_i^k y_k = Y_i \to Y_i + C_l^k (A^{-1})_j^l X_k,$$

the first one, preserving F and F' separately, can be interpreted as a simple change of Witt bases; this basis change completed, an isometry σ is obtained; it fixes every point of F and sends F' to V'. The same holds for F and V'', proving the lemma. ∎

Lemma 8.1.6 *Let $(x_1, x_2, \ldots, x_h)$ be a free system in the m.t.i.s. F and let M be the space generated by it; let $M' = F' \cap M^\perp$. Then $V = M + M'$ is a m.t.i.s. and $x_1 x_2 \ldots x_h f$ is a representative spinor of V (if the sequence is empty, f is obtained).*

Proof. We may assume that each $x_1, x_2, \ldots, x_h$ belongs to a Witt basis $\{x_i, y_j\}$ adapted to the decomposition $E = F \oplus F'$. Clearly V is a m.t.i.s. since the orthogonal space to M in E contains F and the $r - h$ vectors $y_{h+1}, \ldots, y_h$,

$$M^\perp = (x_1, x_2, \ldots, x_r, y_{h+1}, \ldots, y_r),$$

$M^\perp \cap F' = (y_{h+1}, \ldots, y_{h+r})$ and V is the space of the $(x_1, \ldots, x_h, y_{h+1}, \ldots, y_r)$.

We have proved that if $uf \in S$ is such that $xuf = 0$, $\forall x \in V$, then uf is a pure spinor representing V (Lemma (8.1.3)), now

$$(\sum_{i=1}^{h} \alpha^i x_i + \sum_{i=h+1}^{r} \beta^i y_i)x_1 \ldots x_h y_1 \ldots y_r = 0.$$

∎

Proposition 8.1.4

1. uf is a pure spinor if and only if $u = \lambda \exp(v)x_1 x_2 \ldots x_h$ where $x_1, x_2, \ldots, x_h$ are linearly independent in F, $\lambda \in K^*$ and $v \in \wedge^2(F)$.

2. If uf represents F_1, $(x_1, x_2, \ldots, x_h)$ is a basis of $F \cap F_1$.

Proof. By Lemma (8.1.6), if $x_1, x_2, \ldots, x_h$ are linearly independent, $x_1 x_2 \ldots x_h f$ represents V such that $V \cap F = (x_1, x_2, \ldots, x_h)$; applying $\exp v$, $(\exp v)V(\exp v)^{-1} = F_1$, $F \cap F_1 = (x_1, x_2, \ldots, x_h)$ and 2 is proved; furthermore $\exp(v)x_1 x_2 \ldots x_h f$ is a spinor representing F_1.

Conversely, if F_1 is a m.t.i.s. and if $(x_1, x_2, \ldots, x_h)$ is a basis of $F_1 \cap F$, let us define V as in Lemma (8.1.6), then $V \cap F = F_1 \cap F$ so by Lemma (8.1.5) there exists a $v \in \wedge^2(F)$ such that $\varphi(\exp v)$ maps V onto F_1 and $\exp(v)x_1 x_2 \ldots x_h$ is a pure spinor defining F_1.   ∎

Corollary 8.1.2 If the dimension n is 4 or 6, all semi-spinors are pure.

Proof. In four dimensions, a pure spinor can be written as $\lambda(1 + ax_1 x_2)f$,

$$\lambda(1 + ax_1 x_2)(bx_1 + cx_2)(b'x_1 + c'x_2)f = \lambda(bc' - b'c)x_1 x_2 f,$$
$$\lambda(1 + ax_1 x_2)(bx_1 + cx_2)f = \lambda(bx_1 + cx_2)f$$

and a semi-spinor $(\alpha x_1 + \beta x_2)f$, $(\alpha' + \beta' x_1 x_2)f$ is of one of these forms.
The proof in the six-dimensional case is similar.   ∎

Corollary 8.1.3 If V and V' are two m.t.i.s. and if $h = \dim(V \cap V')$, V and V' are of the same parity if and only if $h = r \pmod 2$.

Proof. By the Witt theorem there exists a $\sigma \in O(Q)$ mapping V to F. Note that V and V' have the same parity if and only if $\sigma(V)$ and $\sigma(V')$ have the same parity (if $\tau(V) = V'$, $\sigma\tau\sigma^{-1}$ maps $\sigma(V)$ to $\sigma(V')$, and $\tau \in SO(Q)$ if and only if $\sigma\tau\sigma^{-1} \in SO(Q)$). Therefore we can assume that $V = F$.

A representative spinor for F is $x_1 x_2 \ldots x_r f$ and a representative spinor for V' is $u' = \lambda \exp(v)x_1 \ldots x_h f$; $(x_1, x_2, \ldots, x_h)$ being a basis of $V' \cap F$, the result is proved.   ∎

Corollary 8.1.4 *Every $(r-1)$-dimensional totally isotropic subspace (t.i.s.) of E is contained in exactly one even m.t.i.s. and in exactly one odd m.t.i.s.*

Proof. We reduce the problem to a subspace U of F with basis $x_1, x_2, \ldots, x_{r-1}$. If V is an m.t.i.s. containing U, of the same parity as F, $\dim(V \cap F) = r \pmod 2$ and $\dim(V \cap F) \geq r-1$, so $V = F$, otherwise the same reasoning leads to $V \cap F = U$.

If s is a pure spinor representing V, $s = \lambda x_1 \ldots x_{r-1} f$, since at least one of the $x_1, \ldots, x_{r-1}$ must occur in $\exp v$, h being equal to $r-1$: $(\exp v) x_1 \ldots x_{r-1} = x_1 \ldots x_{r-1}$. The space V is uniquely determined by s.

Conversely, $x_1 \ldots x_{r-1} f$ is a pure spinor and represents a m.t.i.s. of different parity, also containing U.   ∎

8.1.4 Decomposition of the Clifford algebra as a direct sum of minimal left ideals.

Lemma 8.1.7 *The isotropic r-vectors f and f' determine the same minimal left ideal if and only if they are collinear.*

Proof. By the Witt theorem, a $\delta \in \mathrm{Pin}(Q)$ can be found, such that $\delta f \delta^{-1} = f'$; if $C(Q)f = C(Q)f'$, there exists a $s \in C(Q)$ such that $f\delta^{-1} = sf$, which, by Proposition (8.1.1), implies that $f\delta^{-1} = \lambda_1 f$, $\lambda_1 \in \mathbf{K}^*$. Using $\beta : \delta f \to N(\delta)\lambda_1 f$, $f' = \lambda f$, $\lambda \in \mathbf{K}^*$; the converse is obvious.

If f and f' are not collinear, they determine different left ideals, whose intersection must be 0 by minimality.   ∎

Consider the Witt decomposition and its adapted basis $\{x_i, y_j\}$, $i, j = 1, \ldots, r$. A new adapted Witt basis is obtained if x_k and $y_{k'}$ are exchanged. If $k \neq k'$, the isotropic r-vectors $y_1 \ldots x_k \ldots y_r$ and $y_1 \ldots x_{k'} \ldots y_r$ are not collinear, and the same holds for any pair of isotropic r-vectors obtained by replacing a sequence of p different elements $y_{k_1} y_{k_2} \ldots y_{k_p}$ by $x_{k_1} x_{k_2} \ldots x_{k_p}$ $(1 < p \leq r, \ k_1 < k_2 < \ldots < k_p)$. In this way, a system of 2^r isotropic r-vectors determining 2^r minimal left ideals is found; the intersection of any two of these ideals is 0. The Clifford algebra is the direct sum of these 2^r ideals, so we have :

Proposition 8.1.5 *The Clifford algebra $C(Q)$, Q being neutral, is the direct sum of minimal left ideals determined by 2^r isotropic r-vectors.*

8.2 Invariant bilinear forms on the spinor space.

8.2.1 Definition of an invariant bilinear form.

Let $uf, vf \in C(Q)f = S$, β being the main anti-automorphism; if $x_1, x_2, \ldots, x_h \in F$,

$$\beta(x_1 x_2 \ldots x_h) = (-1)^{h(h-1)/2} x_1 x_2 \ldots x_h$$

and an analogous result holds for $\beta(y_1 y_2 \ldots y_h)$, $y_i \in F'$.

Consider $\beta(uf)vf = \beta(f)\beta(u)vf = \epsilon f \beta(u)vf$ where $\epsilon = (-1)^{r(r-1)/2}$.

$$\beta(uf)vf \in C(Q)f \cap fC(Q),$$

so it is a pure spinor and by Proposition (8.1.1), we can write :

$$\beta(uf)vf = \mathcal{B}(uf, vf)f, \tag{1}$$

$\mathcal{B}$ being a bilinear form on $S \times S$.

The following properties are immediately obtained :

$$\begin{aligned}
\mathcal{B}(guf, gvf) &= N(g)\mathcal{B}(uf, vf), \quad \text{if } g \in G, \quad N(g) = \beta(g)g \in \mathbf{K}^*, & (2)\\
\mathcal{B}(xuf, xvf) &= Q(x)\mathcal{B}(uf, vf), \quad \text{if } x \in E, & (3)\\
\mathcal{B}(xuf, vf) &= \mathcal{B}(uf, xvf), \quad \text{if } x \in E. & (4)
\end{aligned}$$

(2) means that $\mathcal{B}$ is an invariant bilinear form for the spin representation of G_0 (since on G_0, $N(g) = 1$).

Proposition 8.2.1 *The bilinear form $\mathcal{B}$ is non-degenerate and*

$$\mathcal{B}(vf, uf) = \epsilon \mathcal{B}(uf, vf). \tag{5}$$

Proof. If $uf = \alpha^{i_1 i_2 \ldots i_h} x_{i_1} x_{i_2} \ldots x_{i_h} f \neq 0$, a vf can be found such that $uv = x_1 x_2 \ldots x_r$, if we take a term of minimal degree in u and multiply u by $k x_{j_1} x_{j_2} \ldots x_{j_h}$, $(j_1, j_2, \ldots, j_h)$ being the complement relative to $(1, 2, \ldots, r)$ of the sequence occurring in the minimal term, and $k \in \mathbf{K}^*$ suitably chosen.

$\forall uf \in S$ such that $uf \neq 0$, $f\beta(u) \neq 0$ and there exists a $vf \in S$ such that $\beta(u)v = x_1 x_2 \ldots x_r$; then $\mathcal{B}(uf, vf)f = f x_1 x_2 \ldots x_r f = \epsilon f$, implying that $\mathcal{B}$ is non-degenerate.

Finally, $\mathcal{B}(vf, uf)f = \beta(f)\beta(v)uf = \epsilon \beta(f)\beta(u)vf$ and

$$\beta(\mathcal{B}(vf, uf)f) = \epsilon \mathcal{B}(vf, uf)f = \beta(f)\beta(u)vf.$$

∎

Proposition 8.2.2 *Let V and V' be two m.t.i.s. in E and let uf and $u'f$ be their representative spinors, then $V \cap V' \neq 0$ if and only if $\mathcal{B}(uf, u'f) = 0$.*

Proof. By (2) we may assume that $V = F$. Then

$$uf = x_1 x_2 \ldots x_r f, \quad u'f = \lambda \exp(v) x_1 x_2 \ldots x_h f$$

relying on Proposition (8.1.4), and

$$\mathcal{B}(uf, u'f) = \beta(uf)u'f = kf x_1 \ldots x_r \exp(v) x_1 \ldots x_h f$$

must vanish since $\exp(v) = \prod(1 + a^{ij}x_i x_j)$, except if $x_1 \ldots x_h f = f$, in which case $B(uf, u'f) \neq 0$.

But if $(x_1, x_2, \ldots, x_h)$ is a basis of $V' \cap F$, then $B(uf, u'f) \neq 0$ if $V' \cap F = 0$ and $B(uf, u'f) = 0$ if $V' \cap F \neq 0$.   ∎

Remark. If r is odd, B vanishes on $S^+ \times S^+$ and on $S^- \times S^-$. If r is even, B vanishes on $S^+ \times S^-$ and on $S^- \times S^+$. $B(uf, vf) = 0$ if and only if the homogeneous component of degree r in $\beta(u)v$ vanishes.

Finding all invariant bilinear forms for the representation of G_0^+. α being the main automorphism in $C(Q)$, $\tilde{\beta} = \beta \circ \alpha$ is an antiautomorphism whose restriction to E is $-$ Id.

$$\tilde{\beta}(f) = (-1)^{r(r+1)/2} f.$$

Defining $\tilde{B}(uf, vf)f = \tilde{\beta}(uf)vf$, $\tilde{B}$ is a bilinear form on $S \times S$ and analogous formulas to (2), (3), (4) are obtained :

$$\tilde{B}(guf, gvf) = \epsilon(g)N(g)\tilde{B}(uf, vf), \tag{6}$$
$$\tilde{B}(xuf, xvf) = -Q(x)\tilde{B}(uf, vf), \quad x \in E, \tag{7}$$
$$\tilde{B}(xuf, vf) = -\tilde{B}(uf, xvf), \quad x \in E \tag{8}$$

where $\epsilon(g) = 1$ if $g \in G^+$ and $\epsilon(g) = -1$ if $g \in G^-$. (6) means that $\tilde{B}$ is invariant for the spin representation of G_0^+. Furthermore,

$$\tilde{B}(vf, uf) = \tilde{\epsilon}\tilde{B}(uf, vf), \quad \tilde{\epsilon} = (-1)^{r(r+1)/2}. \tag{9}$$

Proposition 8.2.3 *All invariant bilinear forms for the spin representation ρ_0^+ of G_0^+ are linear combinations of B and $\tilde{B}$.*

Proof. Let B' be such a bilinear form. S^* being the dual of the spinor space S, there exists a dual representation ω to ρ_0^+, $\omega(g) = (\rho_0^+)^T(g^{-1})$ by definition.

Define Φ by $\Phi_s : t \to B(s, t)$ and Φ' by $\Phi_s' : t \to B'(s, t)$ (here $uf = s$, $vf = t$). By (2) :

$$B(\rho_0^+(g)s, t) = B(s, \rho_0^+(g^{-1}t)),$$

i.e.

$$\langle \Phi \circ \rho_0^+(g)s, t \rangle = \langle \Phi(s), \rho_0^+(g^{-1})t \rangle = \langle \rho_0^{+T}(g^{-1}) \circ \Phi(s), t \rangle,$$

or $\omega(g) \circ \Phi = \Phi \circ \rho_0^+(g)$ and, similarly, $\omega(g) \circ \Phi' = \Phi' \circ \rho_0^+(g)$.

Φ being bijective (since B is non-degenerate), this leads to

$$\Phi^{-1}\Phi' \circ \rho_0^+(g) = \rho_0^+(g) \circ (\Phi^{-1}\Phi').$$

$\psi = \Phi^{-1}\Phi'$ commutes with all $\rho_0^+(g)$, hence with all $\rho^+(u)$ for which u is generated in $C(Q)$ by the elements of G_0^+. But all $u \in C^+(Q)$ can be generated in this way (cf. Chapter 5, 2). Now $C^+(Q)$ is isomorphic to $\text{End}(S^+) \times \text{End}(S^-)$, hence ψ commutes with all endomorphisms whose matrix has the form $\begin{pmatrix} A & 0 \\ 0 & B \end{pmatrix}$, $A, B \in M_{r'}(\mathbf{K})$, $r' = 2^{r-1}$.

It is easy to prove that the matrix of ψ must be $\begin{pmatrix} \lambda I_{r'} & 0 \\ 0 & \mu I_{r'} \end{pmatrix}$, so that $\psi = \lambda\alpha_1 + \mu\alpha_2$, $\alpha_1, \alpha_2 \in \text{End}(S)$ and $\Phi' = \Phi \circ \psi = \lambda\Phi_1 + \mu\Phi_2$, $\Phi_1, \Phi_2 \in \text{Hom}(S, S^*)$. $\mathcal{B}' = \lambda\mathcal{B}_1 + \mu\mathcal{B}_2$, $\mathcal{B}_1$ and $\mathcal{B}_2$ being bilinear forms. $\mathcal{B}'$ belongs to a two-dimensional space containing $\mathcal{B}$ and $\tilde{\mathcal{B}}$. But $\mathcal{B}$ and $\tilde{\mathcal{B}}$ are linearly independent since if

$$\alpha_1 \mathcal{B}(s,t) + \beta_1 \tilde{\mathcal{B}}(s,t) = 0, \quad \forall s, t \in S,$$

taking $(s,t) \in S^+ \times S$ such that the degree of st is r, $\mathcal{B}(s,t) \neq 0$, and $\tilde{\mathcal{B}} = \mathcal{B}$ holding on $S^+ \times S$, $\alpha_1 + \beta_1 = 0$. Similarly, if $(s,t) \in S^- \times S$, $\alpha_1 - \beta_1 = 0$ follows. Finally, $\mathcal{B}' = k_1 \mathcal{B} + k_2 \tilde{\mathcal{B}}$, $k_1, k_2 \in \mathbf{K}$.   ∎

8.3   The tensor product of a spin representation with itself or with its dual.

We still assume that $\dim E = 2r$ and that Q is non-degenerate of maximal index r.

8.3.1

Definition 8.3.1 ρ being a spin representation of the Clifford group G in the space S, we set :

$$(\rho \otimes \rho)_g(s \otimes t) = \rho(g)(s) \otimes \rho(g)(t), \quad \forall s, t \in S, \quad \forall g \in G.$$

This is the classical definition of the tensor product of a representation ρ with itself. In what follows, we will take $S = C(Q)f$, the Clifford algebra $C(Q)$ being simple, and we use the notations of Chapter 5, 2.1.

Proposition 8.3.1 $S \otimes S$ can be identified with $C(Q)$ and the representation $\rho \otimes \rho$ can be identified with the representation χ given by

$$\chi_g(w) = N(g)gwg^{-1}, \quad g \in G, \quad w \in C(Q). \tag{1}$$

Proof. Let $\eta : (uf, vf) \in S \times S \rightarrow uf\beta(v) \in C(Q)$, β being the main antiautomorphism, then η is bilinear and by a classical result (Chapter 2, 1.7), there exists a linear mapping $\psi : S \otimes S \rightarrow C(Q)$ such that

$$\psi(uf \otimes vf) = uf\beta(v).$$

We prove that $\psi(S \otimes S) = C(Q)$. $C(Q)$ being simple, it is sufficient to verify that $\psi(S \otimes S)$ is a two-sided ideal, $S \otimes S$ and $C(Q)$ having the same dimension.

If $w \in C(Q)$, $wuf\beta(v) \in \psi(S \otimes S)$ and

$$uf\beta(v)w = uf\beta(\beta(w)v) \in \psi(S \otimes S),$$

so $\psi(S \otimes S) = C(Q)$ and ψ is a linear isomorphism. Identifying the two spaces using ψ, $guf \otimes gvf$ is identified with $guf\beta(gv) = guf\beta(v)\beta(g)$. But $\beta(g) = g^{-1}N(g)$ (Chapter 4, 1.2), so $guf\beta(gv) = N(g)guf\beta(v)g^{-1} = N(g)g(uf \otimes vf)g^{-1}$, and since $C(Q)$ is linearly generated by the $uf \otimes vf$, the proposition is proved.   ∎

Proposition 8.3.2 ω being the dual representation of ρ (i.e. $\omega(g) = \rho^T(g^{-1})$, $g \in G$), $S \otimes \beta(S)$ can be identified with $C(Q)$ and the representation $\rho \otimes \omega$ of G can be identified with the representation Ad for which

$$(\text{Ad}\, g)w = gwg^{-1}. \tag{2}$$

Proof. Defining as in 2.1 the bilinear form $\mathcal{B}$ by

$$\mathcal{B}(uf, vf)f = \beta(uf)vf, \quad (\epsilon f = \beta(f)),$$

$(\beta(uf), vf) \to \mathcal{B}(uf, vf)$ is a non-degenerate bilinear form from $fC(Q) \otimes C(Q)f$ to $\mathbf{K}$, allowing us to linearly identify $fC(Q)$ with the dual of $C(Q)f$:

$$\langle fu, vf \rangle = \epsilon\mathcal{B}(\beta(u)f, vf).$$

Considering $\rho^T(g^{-1})$,

$$\begin{aligned}
\langle fu, g^{-1}vf \rangle &= \epsilon\mathcal{B}(\beta(u)f, g^{-1}vf) = \epsilon N(g)\mathcal{B}(g\beta(u)f, vf) \\
&= \epsilon\mathcal{B}(\beta(ug^{-1})f, vf) = \langle fug^{-1}, vf \rangle,
\end{aligned}$$

$\rho^T(g^{-1})$ is seen to be the multiplication by g^{-1} from the right, in $fC(Q)$,

$$(\rho \otimes \omega)_{(g)}(uf \otimes vf) = guf \otimes fvg^{-1}.$$

Replacing η by $(uf, vf) \to ufv$ in the proof of Proposition (8.3.1), $guf \otimes fvg^{-1}$ and $g(ufv)g^{-1}$ can be identified in the same way.   ∎

Corollary 8.3.1 The representation $\rho \times \rho$ defines a representation $(\rho \times \rho)_h$ of G in C_h/C_{h-1}, C_h being the subspace of $C(Q)$ generated by the products of at most h elements in E (and $C_0 = \mathbf{K}$). The same holds for $\rho \times \omega$.

Proof. If the x_i are arbitrary elements of E, $i = 1, 2, \ldots, h$, $gx_1x_2\ldots x_hg^{-1} = gx_1g^{-1}gx_2g^{-1}\ldots gx_hg^{-1} = y_1y_2\ldots y_h$. Recall that $\wedge(E)$ and $C(Q)$ can be linearly identified by an isomorphism $\bar{\lambda}$ such that if the e_i form a basis of E,

$$\bar{\lambda}(e_{i_1} \wedge e_{i_2} \wedge \ldots \wedge e_{i_h}) = e_{i_1}e_{i_2}\ldots e_{i_h}.$$

This leads to

$$C_h = \oplus_{0 \leq h' \leq h} \wedge^{h'}(E)$$

and C_h/C_{h-1} can be identified with $\wedge^h(E)$.

The corollary then follows at once if we note that a change of basis does not modify C_h/C_{h-1}.   ∎

8.3.2   The representations θ_h of $O(Q)$.

If for each $w \in C(Q)$ we construct gwg^{-1}, the same result is obtained if g is replaced by kg, $k \in \mathbf{K}^*$, so in fact gwg^{-1} only depends on $\varphi(g) \in O(Q)$ where $\varphi(g)(x) = gxg^{-1}$, $x \in E$. Defining

$$\theta_{\varphi(g)}(w) = gwg^{-1}$$

a representation of $O(Q)$ into $C(Q)$, or into $\wedge(E)$, after a linear identification, is obtained. $\theta_{\varphi(g)}$ preserves C_h/C_{h-1} (which is linearly identified with $\wedge^h(E)$) yielding a representation θ_h of $O(Q)$ in that space.

Proposition 8.3.3 *The representation θ_h of $O(Q)$ is equivalent to the representation $\mathcal{Z}_h$ of $O(Q)$ in $\wedge^h(E)$ given by*

$$\mathcal{Z}_h(\sigma)(x_1 \wedge x_2 \wedge \ldots \wedge x_h) = \sigma x_1 \wedge \sigma x_2 \wedge \ldots \wedge \sigma x_h, \quad \sigma \in O(Q).$$

Proof. Set $x_i' = \sigma(x_i) = gx_ig^{-1}$ where $\varphi(g) = \sigma$. The double bar denoting the equivalence class modulo C_{h-1},

$$\theta_h(\sigma)(\overline{\overline{x_1 x_2 \ldots x_h}}) = \overline{\overline{x_1' x_2' \ldots x_h'}}.$$

By the linear isomorphism $\bar{\lambda}$,

$$\begin{aligned}
\bar{\lambda}(x_1 \wedge x_2 \wedge \ldots \wedge x_h) &= \bar{\lambda}(A_1^{k_1} e_{k_1} \wedge A_2^{k_2} e_{k_2} \wedge \ldots \wedge A_h^{k_h} e_{k_h}) \\
&= A_1^{k_1} A_2^{k_2} \ldots A_h^{k_h} \bar{\lambda}(e_{k_1} \wedge e_{k_2} \wedge \ldots \wedge e_{k_h})
\end{aligned}$$

the (e_i) belonging to an orthonormal basis of E. If $(k_1, k_2, \ldots, k_h)$ is a sequence of different elements,

$$\bar{\lambda}(e_{k_1} \wedge e_{k_2} \wedge \ldots \wedge e_{k_h}) = e_{k_1} e_{k_2} \ldots e_{k_h},$$

but if some elements occur more than once :

$$\bar{\lambda}(e_{k_1} \wedge e_{k_2} \wedge \ldots \wedge e_{k_h}) = 0,$$

$e_{k_1} e_{k_2} \ldots e_{k_h}$ being a product of less than h elements of E in $C(Q)$. For each system $x_1, x_2, \ldots, x_h \in E$,

$$\bar{\lambda}(x_1 \wedge x_2 \wedge \ldots \wedge x_h) = x_1 x_2 \ldots x_h \quad (\text{mod } C_{h-1}).$$

This implies

$$\theta_h(\sigma) \circ \bar{\lambda} = \bar{\lambda} \circ \mathcal{Z}_h(\sigma),$$

proving the proposition. ∎

Corollary 8.3.2 *θ is equivalent to the direct sum of the representations*

$$\mathcal{Z}_0, \mathcal{Z}_1, \ldots, \mathcal{Z}_h.$$

Proof. Obvious, since $\theta_{\varphi(g)}$ preserves $\wedge^h(E)$. ∎

In fact, Proposition (8.3.2) does not require the hypothesis that $\dim E$ be even and that Q be of maximal index. We will give an example related to the *complex vectorial formalism* introduced in Chapter 7, 2.3.

Take g in the connected component of the identity in $\text{Spin}(1,3)$, written as $\text{Spin}_0(1,3)$ and the space C_2/C_1 which can be identified with the Lie algebra of this group.

$$\theta_2(g)w = gwg^{-1},$$

w being a **R**-linear combination of

$$e_1e_2, e_2e_3, e_3e_1, e_0e_1, e_0e_2, e_0e_3$$

or a **C**-linear combination of

$$e_2e_3 + ie_0e_1, e_3e_1 + ie_0e_2, e_1e_2 + ie_0e_3$$

by the existence of a complex structure J :

$$J(e_2e_3) = e_0e_1 \quad (\text{cycl.}).$$

The action of g gives rise to a new **C**-linear basis

$$e_2'e_3' + ie_0'e_1', e_3'e_1' + ie_0'e_2', e_1'e_2' + ie_0'e_3',$$

(e_0', e_1', e_2', e_3') being a new Lorentz basis.

Taking exterior products a pseudo-metric on $\wedge^2(E)$ given by

$$(\wedge^2 B)(e_r \wedge e_s, e_i \wedge e_j) = \tfrac{1}{2}(B_{sj}B_{ir} - B_{rj}B_{is})$$

is obtained from the pseudo-metric with components B_{ij}, and the bases such as $e_2 \wedge e_3 + ie_0 \wedge e_1$ etc. (linearly identified with $e_2e_3 + ie_0e_1$ etc.) are orthonormal for a 'metric' with isometry group $O(3,\mathbf{C})$; by the connectedness only $SO(3,\mathbf{C})$ is obtained as the image of $\text{Spin}_0(1,3)$ and

$$SO(3,\mathbf{C}) = \text{Spin}_0(1,3)/\mathbf{Z}_2 = L_4^0.$$

$\text{Spin}_0(1,3)$ is just G_0^+, whose index in $\text{Spin}(1,3)$ is 2, so that

$$SO(3,\mathbf{C}) = SO(1,3)/\mathbf{Z}_2.$$

8.3.3   The $\wedge^h(E)$-valued bilinear forms $\mathcal{B}_h$.

Let uf and vf be arbitrary elements of the spinor space S.

We have seen (in the proof of Proposition (8.3.1)) that $uf \otimes vf$ could be identified with $uf\beta(v)$ and $\wedge^h(E)$ with C_h/C_{h-1}. Now we define

$$uf\beta(v) = \sum_{h=0}^{n} \mathcal{B}_h(uf, vf) \tag{3}$$

$\mathcal{B}_h$ being the homogeneous component of degree h in $\wedge(E)$.

Proposition 8.3.4 $\mathcal{B}_h$ is a bilinear mapping from $S \times S$ to $\wedge^h(E)$ and $\forall g \in G$,

$$\mathcal{B}_h(guf, gvf) = N(g)\mathcal{Z}_h(\varphi(g))\mathcal{B}_h(uf, vf) \qquad (4)$$

Proof. Only the formula must be verified :

$$
\begin{aligned}
guf\beta(gv) &= \epsilon gu\beta(gvf) &&= guf\beta(v)\beta(g) \\
&= N(g)guf\beta(v)g^{-1} &&= N(g)\textstyle\sum_h g\mathcal{B}_h(uf, vf)g^{-1}
\end{aligned}
$$

where $\epsilon f = \beta(f)$. ∎

Proposition 8.3.5 If $h = r\,(\mathrm{mod}\,2)$, $\mathcal{B}_h(S^+ \times S^-) = \mathcal{B}_h(S^- \times S^+) = 0$. If $h = r+1\,(\mathrm{mod}\,2)$, $\mathcal{B}_h(S^+ \times S^+) = \mathcal{B}_h(S^- \times S^-) = 0$.

Proof. Obvious. ∎

It is very easy to verify that

$$\mathcal{B}_h(uf, vf) = \epsilon\epsilon_1 \mathcal{B}_h(vf, uf) \qquad (5)$$

where $\epsilon_1 = (-1)^{h(h-1)/2}$.

For $h = n$ we reobtain, possibly up to a change of sign, the fundamental bilinear form $\mathcal{B}$ of section 2. This is easily seen if uf and vf are of the form $x_{i_1} x_{i_2} \dots x_{i_h} f$.

8.4   Selected references.

- E. Cartan, *Leçons sur la théorie des spineurs*, Hermann, Paris, 1938.

- C. Chevalley, *The algebraic theory of spinors*, Columbia University Press, 1954.

Chapter 9

THE SPINORS IN MAXIMAL INDEX AND ODD DIMENSION.

In this chapter, Q is assumed non-degenerate and of maximal index r, $\dim E = n = 2r + 1$.

Such a situation was already considered for $\mathbf{K} = \mathbf{C}$ in Chapter 5, 2.2.

By the results of Chapter 3, sections 3.2 and 3.3, $C(Q)$ is isomorphic to the Clifford algebra $C^+(\hat{Q}_1)$ of a $(2r+2)$-dimensional space $(\hat{E}_1, \hat{Q}_1)$ and $C^+(Q)$ is isomorphic to the Clifford algebra of a $(2r)$-dimensional space (E_1, Q_1).

These isomorphisms are not unique. $C^+(Q)$ is simple, $C(Q)$ is semi-simple, being the direct composition of two simple ideals.

Two possibilities arise for the study of these Clifford algebras : either E is considered as a subspace of $\hat{E}_1$, or E_1 is considered as a subspace of E; the former is suited to the study of m.t.i.s. of E, the latter is better adapted for the construction of the irreducible representations of $C^+(Q)$ and of the standard spinors.

9.1 Standard spinors.

9.1.1

Let F and F' be two m.t.i.s. in E and let $z_0 \in E$ be non-isotropic, such that

$$E = F \oplus F' \oplus (z_0), \quad Q(z_0) = a \in \mathbf{K}^*,$$

is a Witt decomposition, $E_1 = F \oplus F'$ being orthogonal to z_0.

Just as in the proof of Theorem (3.3.2), we can introduce the 'metric' Q_1 on E_1 with :

$$Q_1(y) = -aQ(y), \quad \forall y \in E_1,$$

then Q_1 is, up to a non-zero factor, the induced metric. (If $\mathbf{K} = \mathbf{C}$, the factor can be assumed equal to 1.)

The isomorphism $j : C(Q_1) \to C^+(Q)$ is the natural extension of $y \in E_1 \to z_0 y \in C^+(Q)$.

The standard spin representation of $C(Q_1)$ is in the space

$$S = \{x_{i_1} x_{i_2} \dots x_{i_h} f, f = y_1 y_2 \dots y_r\},$$

$\{x_i, y_j\}$ being a Witt basis for $(F \oplus F', Q)$; note that S is a subset of $C(Q_1)$. By the isomorphism j, S corresponds to a set of elements $u^+ f = (u_1^+ + z_0 u_2^-)f$ if r is even, $u^+ f z_0 = (u_1^+ + z_0 u_2^-)f z_0$ if r is odd.

We can restrict ourselves to the first form for any parity of r and choose for the irreducible representation of $C^+(Q)$ the space of

$$u^+ f = (u_1^+ + z_0 u_2^-)f, \quad u^+, u_1^+ \in C^+(Q), \quad u_2^- \in C^-(Q),$$

on which $C^+(Q)$ acts by left multiplication.

9.1.2   A fundamental invariant 2-form for G_0^+.

If $(u^+ f, v^+ f)$ is a pair of standard spinors, we set as before :

$$\beta(u^+ f)(v^+ f) = \beta((u_1^+ + z_0 u_2^-)f)(v_1^+ + z_0 v_2^-)f.$$

If $\beta(f) = \epsilon f$,

$$\beta(u^+ f)(v^+ f) = \epsilon f \beta(u_1^+) v_1^+ f + a \epsilon f \beta(u_2^-) v_2^- f + \epsilon f \beta(u_1^+) z_0 v_2^- f + \epsilon f \beta(u_2^-) z_0 v_1^+ f.$$

If r is odd, the first two terms vanish, if r is even, the last two terms vanish.

Definition 9.1.1 *The fundamental 2-form $\mathcal{B}$ is defined by*

$$\begin{aligned} \mathcal{B}(u^+ f, v^+ f)f &= \beta(u^+ f)(v^+ f) && \text{if } r \text{ is even,} \\ \mathcal{B}(u^+ f, v^+ f)z_0 f &= \beta(u^+ f)(v^+ f) && \text{if } r \text{ is odd.} \end{aligned}$$

Obviously $\mathcal{B}$ is non-degenerate and invariant under G_0^+. If $r = 0, 1 \,(\mathrm{mod}\, 4)$, $\mathcal{B}$ is symmetric; if $r = 2, 3 \,(\mathrm{mod}\, 4)$, $\mathcal{B}$ is antisymmetric. $\tilde{\mathcal{B}}$ is defined analogously. Both $\mathcal{B}$ and $\tilde{\mathcal{B}}$ can be extended to $C(Q)f$.

9.2   M.t.i.s. and pure spinors.

Given z_0, $Q(z_0) = a \neq 0$, the m.t.i.s. of $(z_0)^\perp = E_1$ are the same for the quadratic form Q as for Q_1. Hence, in E_1, the results of Chapter 8 hold. But we want to study the m.t.i.s. of E rather than E_1.

Consider (E, Q) as a subspace of the $(2r + 2)$-dimensional space $(\hat{E}_1, \hat{Q}_1)$. $\hat{E}_1 = E \oplus (z_0')$ the sum being orthogonal and $\hat{Q}_1(z_0') = -a$, $\hat{Q}_1(x) = Q(x)$ if $x \in E$. $\hat{Q}_1$ has rank $2r + 2$, $\hat{F}_1 = F \oplus (x_0)$ and $\hat{F}_1' = F' \oplus (y_0)$ are m.t.i.s. of $\hat{E}_1$, (x_0) being determined by $z_0 + z_0'$ and (y_0) by $z_0 - z_0'$. We may assume that $(x_1, \ldots, x_r, x_0, y_1, \ldots, y_r, y_0)$ is a Witt basis of $\hat{E}_1$ adapted to the decomposition $\hat{F}_1 \oplus \hat{F}'_1$.

$C(Q)$ can be identified with the sub-algebra of $C(\hat{Q}_1)$ generated by E. $C^+(Q)$ and $C(Q)$ are semi-simple and isomorphic, so the spinor spaces are (2^r)-dimensional.

The spinor space S for $C^+(Q)$ is determined by the $x_{i_1} \ldots x_{i_h} f$, the even spinor space $\hat{S}_1^+$ for $C^+(Q_1)$ is determined by the basis consisting of the $x_{i_1} \ldots x_{i_h} x_0 f y_0$, h being odd, and the $x_{i_1} \ldots x_{i_h} f y_0$, h even.

The representations in the irreducible spaces S and $\hat{S}_1^+$ are equivalent; one can for instance associate $x_{i_1} \ldots x_{i_h} f y_0$ to $x_{i_1} \ldots x_{i_h} f$ if h is even and $x_{i_1} \ldots x_{i_h} x_0 f y_0$ to it if h is odd.

The representations of the Clifford groups G^+ and $\hat{G}_1^+$ of $C^+(Q)$ and $C^+(\hat{Q}_1)$ generate these algebras and are therefore also equivalent. Writing ρ^+ and $\hat{\rho}_1^+$ for them, i being the equivalence, $G^+ \subset \hat{G}_1^+$ leads to

$$\hat{\rho}_1^+(g) \circ i = i \circ \rho^+(g), \quad \forall g \in G^+.$$

The isomorphism between $C(Q)$ and $C^+(\hat{Q}_1)$ not being unique, if i' also has the previous property, viz. $\hat{\rho}_1^+(g) \circ i' = i' \circ \rho^+(g)$,

$$i' \circ \rho^+(g) \circ i'^{-1} = i \circ \rho^+(g) \circ i^{-1},$$

$\omega = i^{-1} \circ i'$ commutes with the left action of g and, G^+ generating the central algebra $C^+(Q)$, ω is a homothety.

Hence the equivalence between the representations in S and in $\hat{S}_1^+$ is unique up to a non-zero scalar factor.

9.2.1   The pure spinors in the odd-dimensional case (maximal index).

Let V be a m.t.i.s. of E, hence an r-dimensional subspace. It is also a m.t.i.s. of $\hat{E}_1$ for the quadratic form $\hat{Q}_1$ and by Corollary (8.1.4), it is contained in exactly one even m.t.i.s. $\hat{V}_1$ of $\hat{E}_1$.

Let $\hat{s}_1$ be a pure spinor representing $\hat{V}_1$ and let u be the spinor $s \in S$ for which $i(s) = \hat{s}_1$, then by the preceding results, u is determined up to a non-zero scalar factor.

Definition 9.2.1 *$i : S \to \hat{S}_1^+$ being the representation equivalence, a spinor of the space generated by $s \in S$ is called a pure spinor if $i(s)$ is a pure spinor.*

V is contained in a unique even m.t.i.s. of $\hat{E}_1$. Conversely, if $\hat{V}_1$ is an even m.t.i.s. of $\hat{E}_1$, $\hat{V}_1 \cap E$ is totally isotropic in E, hence necessarily r-dimensional (the only possible dimensions being r and $r+1$).

$\hat{V}_1 \cap E \subset \hat{V}_1$, so that :

Lemma 9.2.1 *There exists a bijection between the set of m.t.i.s. in E and the set of even m.t.i.s. in $\hat{E}_1$, every m.t.i.s. of S being represented by a pure spinor, unique up to a non-zero scalar factor.*

Proposition 9.2.1 *Let V be a m.t.i.s. of E whose representative pure spinor is s and let z be an invertible element of the center of $C(Q)$, $z \notin \mathbf{K}$.*

1. *V is the set of all $x \in E$ such that $\rho^+(zx)(s) = 0$.*

2. *If $s' \in S$ is such that $\rho^+(zx)(s') = 0$, $\forall x \in V$, then $s' = \lambda s$, $\lambda \in \mathbf{K}^*$.*

3. *If $g \in G^+$, $\rho^+(g)(s)$ represents $\varphi(g)(V)$.*

Proof. $\rho^+(zx)(s) = 0$ if and only if $i \circ \rho^+(zx)(s) = 0$, i.e.

$$(\rho_1^+(zx))(\hat{s}_1) = 0, \quad \hat{s}_1 = i(s),$$

or $(\rho_1^+(x))(\hat{s}_1) = 0$, z being invertible in the center; this implies by Lemma (8.1.3) that $x \in \hat{V}_1$, so $x \in V$ since x is also an element of E. This proves 1.

If $g \in G^+$,

$$i \circ \rho^+(g)(s) = \hat{\rho}_1^+(g)(\hat{s}_1)$$

so that $i \circ \rho^+(g)(s)$ represents $\varphi(g)(\hat{V}_1)$ by the results of Chapter 8. V being a subset of $\hat{V}_1$, $\varphi(g)V$ is contained in $\varphi(g)(\hat{V}_1)$, so $\rho^+(g)(s)$ represents $\varphi(g)V$, proving 3.

Taking the representation in $S = u^+ f$ (cf. 1.1), the stated property can be expressed by $zx(u^+ f) = 0$, $s' = u^+ f$, $\forall x \in V$; this property for V extends to F' (cf. the proof of Lemma (8.1.3)), the even m.t.i.s. containing F' is $F' \oplus (y_0) = \hat{F}_1'$.

$\rho^+(zx)(s') = 0$ if and only if

$$\hat{\rho}_1^+(x)(\hat{s}_1') = 0, \quad i(s') = \hat{s}_1'.$$

If this holds for all $x \in F'$, the representation $\hat{\rho}_1^+$ being being left multiplication in the space of the $x_{i_1} \dots x_{i_h} f y_0$, $\hat{\rho}_1^+(y_0)(\hat{s}_1') = 0$, y_0 anticommuting with $x_{i_1}, \dots, x_{i_h}$. Finally,

$$\hat{\rho}_1^+(x)(\hat{s}_1') = 0, \quad \forall x \in \hat{F}_1',$$

and invoking Lemma (9.2.3) used before, $\hat{s}_1' = \lambda \hat{s}_1$ so that $s' = \lambda s$, $\lambda \in \mathbf{K}^*$, proving 2. ∎

Chapter 10

THE HERMITIAN STRUCTURE ON THE SPACE OF COMPLEX SPINORS—CONJUGATIONS AND RELATED NOTIONS.

10.1 Some background and preliminary results.

10.1.1 Sesquilinear forms and hermitian forms.

K is the complex number field **C**. $\bar{\alpha}$ stands for the complex conjugate of α.

An n-dimensional complex vector space E can be given another complex vector space structure if we replace the product αx, $x \in E$, by $\bar{\alpha}x$, without changing anything else. This space will be written $\hat{E}$. Every basis of E is a basis of $\hat{E}$ and vice versa. If θ^i, $i = 1, 2, \ldots, r$ is a basis for the dual E^* of E, the $\hat{\theta}^i$ defined by $\hat{\theta}^i(x) = \bar{\theta}^i(x)$ form a basis of the dual $\hat{E}^*$.

A sesquilinear form B on $E \times E$ is a bilinear form on $\hat{E} \times E$, i.e.

$$B(\alpha x, y) = \bar{\alpha}B(x, y), \quad \forall x, y \in E, \quad \forall \alpha \in \mathbf{C}. \tag{1}$$

Nothing else changes and the theory of bilinear forms developed in Chapter 1 carries through. A sesquilinear form B such that $B(x, y) = \overline{B(y, x)}$ is called *hermitian sesquilinear* (or also *hermitian symmetric*).

The orthogonal space $V^\perp$ to the subspace V of E is defined as before, just as the concepts of isotropic subspace, totally isotropic subspace and non-isotropic subspace.

If B is non-degenerate and if V is non-isotropic, $E = V \oplus V^\perp$.

$B(x, x) = H(x)$ is the *hermitian form* associated to B. Conversely, B can be retrieved from H if we set

$$4B(x, y) = H(x + y) - H(x - y) + iH(x + iy) - iH(x - iy). \tag{2}$$

It can be proved that E, given a hermitian sesquilinear form, has bases consisting of pairwise orthogonal vectors; more precisely, vectors $x_1, x_2, \ldots, x_h \in E$ exist such that

$$B(x_i, x_j) = \lambda_{ij}\delta_{ij}, \quad 1 \leq i, j \leq p,$$

p being the rank of B, $\lambda_{ij} \in \mathbf{K}^*$. Since we are working over the complex field, a basis can be found for which

$$B(x,y) = \sum_{i=1}^{p} \epsilon^i \bar{\xi}^i \eta^i, \quad \epsilon^i = \pm 1, \tag{3}$$

ξ^i and η^i being the coordinates of x and y in this basis.

The index r of the hermitian sesquilinear form B is also the maximal dimension of the totally isotropic subspaces and $2r \leq n$. A Witt decomposition and a Witt theorem can be proved in this setting too : the reasoning is completely similar to the orthogonal case.

A finite-dimensional complex vector space provided with a positive definite hermitian sesquilinear form is called a hermitian space, i.e. $B(x,x) \geq 0$, $\forall x \in E$ and $B(x,x) = 0$ implies $x = 0$.

If the rank of B is n, B being non-degenerate, the space is called pseudo-hermitian. By (3),

$$B(x,y) = \epsilon(\sum_{i=1}^{r} \bar{\xi}^i \eta^i - \sum_{i=r+1}^{n} \bar{\xi}^i \eta^i), \quad \epsilon = \pm 1.$$

10.1.2 Conjugations in representation space.

The following definitions hold both for algebra representations and for group representations. Let us consider the representations of an associative algebra A in a complex vector space E. E is an A-module (cf. Chapter 5),

$$(a,x) \in A \times E \to ax \in E.$$

Definition 10.1.1 *A conjugation is a homomorphism J of the A-module E such that J is semi-linear, i.e. $J(\alpha x) = \bar{\alpha} J(x)$, and $J^2 = \pm \text{Id}$.*

J being an A-module homomorphism of E, $J(ax) = aJ(x)$ and $J(x_1 + x_2) = J(x_1) + J(x_2)$. Clearly the product of two conjugations is an A-linear isomorphism.

Recall that the quaternion skewfield $\mathbf{H}$ is a two-dimensional complex vector space. Now consider the cases $J^2 = \text{Id}$ and $J^2 = -\text{Id}$ separately.

1. $J^2 = -\text{Id}$.

   If E has the structure of a 'left' vector space over $\mathbf{H}$, it is a complex vector space of twice that dimension. Using the classical notations i, j, k for the standard basis of $\mathbf{H}$ over $\mathbf{R}$, set

   $$J(x) = jx, \quad x \in E$$
   $$J(\alpha x) = j\alpha x = \bar{\alpha} j x = \bar{\alpha} J(x).$$

   j defines a conjugation J such that $J^2 = -\text{Id}$.

Conversely, if an A-module homomorphism J of W exists and $J^2 = -\operatorname{Id}$, $J(\alpha x) = \bar{\alpha} J(x)$, then if $\mathcal{J}$ is the endomorphism defined by i,

$$\mathcal{J} J = -J \mathcal{J}$$

and J, $\mathcal{J}$ generate a quaternion algebra, W becomes a vector space over $\mathbf{H}$ in a natural way and hence :

Proposition 10.1.1 *A conjugation J such that $J^2 = -\operatorname{Id}$ corresponds to a quaternionic representation.*

2. $J^2 = \operatorname{Id}$.

Note that if E is real, $J^2 = \operatorname{Id}$ for $J = \pm 1$.

On $E_{\mathbf{C}} = \mathbf{C} \otimes_{\mathbf{R}} E$ we set $J(\alpha \otimes x) = \bar{\alpha} \otimes J(x)$ and $a(\alpha \otimes x) = \alpha \otimes (ax)$, $a \in A$. $E_{\mathbf{C}}$ then has a conjugation satisfying $J^2 = \operatorname{Id}$.

On $E_{\mathbf{C}}$, with its real structure, J can be considered as $\mathbf{R}$-linear; $E_{\mathbf{C}}$ as the direct sum of E_1 and E_2, the real proper subspaces of $E_{\mathbf{C}}$ associated to the eigenvalues 1 and -1 of J.

Conversely, if W is a complex representation space and J is a conjugation such that $J^2 = \operatorname{Id}$, $r(W)$ being the 'realified' space of W, the set of $x \in r(W)$ such that $J(x) = x$ or $J(x) = -x$ form real subspaces E_1 and E_2 whose direct sum is $r(W)$; $x \to ix$ exchanging these spaces, they are of equal dimension,

$$E_1 = (I + J)r(W), \quad E_2 = (I - J)r(W),$$

and E_1, E_2 are A-modules. Therefore :

Proposition 10.1.2 *To each representation into a complex space having a conjugation J such that $J^2 = \operatorname{Id}$, two equivalent real representations can be associated. $J^2 = \operatorname{Id}$ corresponds to a real representation.*

10.2   The fundamental hermitian sesquilinear form. Special cases.

10.2.1

We consider the complexification (E', Q') of a real vector space (E, Q), Q being non-degenerate, of signature $(p, n - p)$:

$$Q(x) = (x^1)^2 + \ldots + (x^p)^2 - (x^{p+1})^2 - \ldots - (x^n)^2,$$

B being the symmetric bilinear form associated to Q and, for instance, $p \leq n - p$, $\dim E = n = 2r$.

Let $E' = F \oplus F'$ be a Witt decomposition.

$$
\begin{aligned}
F &= (x_1 = (e_1 + e_n)/2, \ldots, x_p = (e_p + e_{n-p+1})/2, \\
&\quad x_{p+1} = (ie_{p+1} + e_{n-p})/2, \ldots, x_r = (ie_r + e_{n-r+1})/2), \\
F' &= (y_1 = (e_1 - e_n)/2, \ldots, y_p = (e_p - e_{n-p+1})/2, \\
&\quad y_{p+1} = (ie_{p+1} - e_{n-p})/2, \ldots, y_r = (ie_r - e_{n-r+1})/2),
\end{aligned}
$$

$$
B(x_i, y_i) = 1/2, \quad B(x_i, y_j) = 0, \quad i \neq j
$$

$$
x_i y_i + y_i x_i = 1, \quad i = 1, 2, \ldots, r,
$$

the (x_i, y_j) form a special 'real' Witt basis, adapted to the decomposition $F \oplus F'$ and associated to the ordered real basis $(e_1, \ldots, e_n)$, $(e_i)^2 = 1$, $i \leq p$, $(e_1)^2 = -1$, $i \geq p+1$. (The reader will have no difficulty to write down a special 'real' Witt basis if $p > n - p$.)

On the complexification E' of the real space E there exists a complex conjugation; taking a real basis in E', the complex conjugation is seen to be a linear mapping from E' to $\hat{E}'$ (cf. 1.1). From this one proves, along the same lines as in Chapter 3, 1.2 for the main automorphism α, that the complex conjugation is a semi-linear isomorphism of $C(Q')$ and that it commutes with α and β.

If f is an isotropic r-vector determining a m.t.i.s. (so it is unique up to a non-zero scalar factor), $\bar{f}$ determines the complex conjugate m.t.i.s. (generally, we will use the bar to denote the complex conjugation).

By Proposition (8.1.1), $f\bar{f}f = \lambda f$, $\lambda \in \mathbf{C}$.

If f is a m.t.i.s. defining F', then either

1. $\bar{f}f = 0$, f being collinear to $y_1 y_2 \ldots y_r$, this will happen if $p \neq 0$ and, in particular, if Q is neutral $(F' = \bar{F}')$.

2. $\bar{f}f \neq 0$, $f\bar{f}f = \lambda f$, $\lambda \neq 0$. This occurs if $p = 0$. $f\bar{f}f = \lambda f$ implies $\bar{f}f\bar{f} = \bar{\lambda}\bar{f}$, $(\bar{f}f)(\bar{f}f) = \lambda\bar{f}f$ and $(\bar{f}f)(\bar{f}f) = \bar{\lambda}\bar{f}f$ so λ is real.

In general, $f\bar{f}f \neq 0$ will mean that $F' \cap \bar{F}' = 0$, by Proposition (8.2.2), whereas $f\bar{f}f = 0$ will mean that $F' \cap \bar{F}' \neq 0$.

We will first study these two cases, which occur in the neutral case and in the elliptic case.

10.2.2

1. **The invariant hermitian sesquilinear form when** $\bar{f} = \exp(i\theta)f$, $\theta \in \mathbf{R}$. Let us consider the fundamental bilinear form $\mathcal{B}$ and $\mathcal{B}(\bar{u}\bar{f}, vf)$. We define

$$
\beta(\bar{u}\bar{f})vf = a\mathcal{H}(uf, vf)f \tag{1}
$$

where a is chosen such as to turn $\mathcal{H}$ into a hermitian sesquilinear form.

This condition means, explicitly,

$$\bar{a}\exp(i\theta) = \epsilon a,$$

$\epsilon = (-1)^{r(r-1)/2}$. Therefore it is sufficient to take some square root of $\epsilon\exp(i\theta)$ for a.

Obviously $\mathcal{H}$ is invariant under G_0 and non-degenerate.

Such considerations clearly hold for $\tilde{B}$ as well, and for any real linear combination of B and $\tilde{B}$, but the invariance then reduces to G_0^+.

2. **The invariant hermitian sesquilinear form when $\bar{f}f \neq 0$.** Note that the linear mapping

$$\bar{u}\bar{f} \in C(Q')\bar{f} \rightarrow \bar{u}\bar{f}f \in C(Q')f$$

is bijective, since $\bar{u}\bar{f}f = \overline{u'\bar{f}f}$ implies that $\bar{u}\bar{f}f\bar{f} = \overline{u'\bar{f}f}\bar{f}, \bar{u}\lambda\bar{f} = \bar{u}'\lambda\bar{f}$ or $\bar{u}\bar{f} = \bar{u}'\bar{f}$. Therefore this mapping is an equivalence between the spin representations in $C(Q')\bar{f}$ and in $C(Q')f$.

In this case we define

$$\beta(\bar{u}\bar{f}f)vf = \mathcal{H}(uv, vf)f \tag{2}$$

$\mathcal{H}$ has hermitian symmetry by an elementary computation and because λ is real.

$\mathcal{H}$ is non-degenerate and invariant under G_0; the same conclusions hold as in case 1.

10.2.3 The 'charge conjugation' and the 'Dirac adjoint'.

In the first case, $\bar{f} = \exp(i\theta)f$, consider

$$\mathcal{C}(uf) = \exp(i\alpha)\bar{u}\bar{f} = \exp(i(\alpha - \theta))\bar{u}f.$$

$\mathcal{C}$ is semi-linear and $\mathcal{C}^2 = \mathrm{Id}$. Furthermore, if $g \in G$, the real Clifford group,

$$\mathcal{C}(guf) = g\mathcal{C}(uf).$$

One easily verifies that

$$\mathcal{H}(\mathcal{C}(uf), vf) = \epsilon\mathcal{H}(\mathcal{C}(vf), uf). \tag{3}$$

In the second case,

$$f\bar{f}f = \lambda f, \quad \lambda \in \mathbf{R}^*.$$

Replacing f by σf, $\sigma \in \mathbf{C}^*$, we can assume that $\lambda = \pm 1$. Define

$$\mathcal{C}(uf) = \sqrt{\lambda}\exp(i\alpha)\bar{u}\bar{f}f.$$

(with $\sqrt{\lambda} = 1$ or i for example). $\mathcal{C}$ is a conjugation (it commutes with $g \in G$), $\mathcal{C}^2 = \text{Id}$ if $\lambda = 1$, $\mathcal{C}^2 = -\text{Id}$ if $\lambda = -1$. A routine computation shows that

$$\mathcal{H}(\mathcal{C}(uf), vf) = \epsilon \mathcal{H}(\mathcal{C}(vf), uf).$$

The semi-linear mapping $\mathcal{A}$ from $S = C(Q')f$ to $fC(Q')$ will be called the *Dirac adjoint*. $fC(Q')$ is sometimes called the space of co-spinors, due to the duality that can be expressed with $\mathcal{B}$. $\mathcal{A}$ satisfies

$$\mathcal{A} \circ \mathcal{C} = \mathcal{B}|_S. \tag{4}$$

Clearly in case 1,

$$\mathcal{A} : uf \to \mathcal{B}(\mathcal{C}(uf)) = \exp(i(\alpha - \theta))\mathcal{B}(\bar{u}f)$$

and in case 2 :

$$\mathcal{A} : uf \to \mathcal{B}(\mathcal{C}(uf)), \qquad \text{if } \lambda = 1$$
$$\mathcal{A} : uf \to -\mathcal{B}(\mathcal{C}(uf)), \qquad \text{if } \lambda = -1.$$

Letting $\mathcal{C}$ act naturally on $fC(Q')$, by transposition one obtains : $\mathcal{C} \circ \beta = \beta \circ \mathcal{C}$ and, finally,

$$\mathcal{A} \circ \mathcal{C} = \mathcal{C} \circ \mathcal{A}. \tag{5}$$

10.3 Hermitian sesquilinear forms and conjugations. The general case, $n = 2r$.

10.3.1 The conjugation pure spinor γf.

f being an arbitrary isotropic r-vector (under the hypotheses on the definiteness of $C(Q')$ stated in 2.1), and $\bar{f}$ its complex conjugate, let $f = x_{1*}x_{2*}\ldots x_{r*}$ be a decomposition of f as a product of r isotropic vectors. Then by the Witt theorem, there exists an element $\gamma \in \text{Pin}(Q')$ sending each x_α to $\bar{x}_{\alpha*}$, $\alpha = 1, \ldots, r$, $\bar{x}_{\alpha*} = \gamma x_{\alpha*} \gamma^{-1}$.

Then $\gamma f = \bar{f}\gamma$, γf belongs to the intersection of a minimal left ideal and a minimal right ideal, so it is a pure spinor determining $\bar{F}'$. Note that since we consider complex spaces, we may assume $\gamma \in G_0(Q')$, i.e. $N(\gamma) = 1$.

γ is not necessarily unique : other elements in $G_0(Q')$ can map f to $\bar{f}$. If $\gamma' \in G_0(Q')$ and

$$\bar{f} = \gamma' f \gamma'^{-1},$$

$f = gfg^{-1}$ where $g = \gamma'^{-1}\gamma$.

The following lemma, though elementary, is of great importance :

Lemma 10.3.1 *(Fundamental Lemma.) If g is an element of the Clifford group of an even-dimensional vector space E with neutral form Q, and for a given isotropic r-vector f,*

$$gfg^{-1} = \lambda f,$$

λ being a non-zero scalar, then there exists a non-zero scalar μ such that $gf = \mu f$, and vice versa. Furthermore, $\lambda = \mu^2 N(g)$.

Proof. $gf = \lambda fg$ is a pure spinor μf, and applying β,

$$
\begin{aligned}
fg^{-1}N(g) &= \mu f \\
\lambda f &= \mu N(g)gf \\
gf &= \mu f
\end{aligned}
$$

where $\lambda = \mu^2 N(g)$. Conversely, if $gf = \mu f$, applying β,

$$gfg^{-1} = \mu^2 N(g)f.$$

Relying on $f = gfg^{-1}$, $g \in G_0(Q') = G_0'$,

$$gf = \pm f$$

and g is an element of the subgroup H' of $g \in G_0'$ such that $gf = \pm f$. Finally, $\gamma' = \gamma\tau$, $\tau \in H'$, $\tau f = \pm f$.   ∎

We define $\rho = \bar{\gamma}\gamma$, the Fundamental Lemma implies that $\rho f = \epsilon' f$, $\epsilon' = \pm 1$.

Lemma 10.3.2 The scalar ϵ' such that $\rho f = \bar{\gamma}\gamma f = \epsilon' f$ where $\bar{f} = \gamma f \gamma^{-1}$ does not depend

1. on the choice of $\gamma \in G_0'$ such that $\bar{f} = \gamma f \gamma^{-1}$,

2. on the factorization of f as a product of r linearly independent isotropic vectors,

3. on the choice of f.

Hence ϵ' has an intrinsic geometric meaning.

Proof. Replacing γ by some $\gamma' = \gamma\tau$, $\tau f = \pm f$, i.e. $\tau f = \epsilon'' f$, $\bar{\gamma}\gamma$ is replaced by $\bar{\gamma}\bar{\tau}\gamma\tau$ and

$$\bar{\gamma}\bar{\tau}\gamma\tau f = \epsilon''\bar{\gamma}\bar{\tau}\gamma f = \epsilon''\bar{\gamma}\bar{\tau}\bar{f}\gamma = \bar{\gamma}\bar{f}\gamma = \bar{\gamma}\gamma f.$$

If $f = x'_{1*}x'_{2*}\ldots x'_{r*}$, a $\sigma \in H'$ can be found such that it maps $x_{\alpha*}$ to $x'_{\alpha*}$ and by an easy computation, γ must be replaced by

$$\bar{\sigma}\gamma\sigma^{-1} = \gamma', \quad \gamma' \in G_0', \quad \sigma f = \epsilon_1 f, \quad \epsilon_1 = \pm 1.$$

$$\gamma' f = \epsilon_1 \bar{\sigma}\gamma f = \epsilon_1 \bar{\sigma}\bar{f}\gamma = \bar{f}\gamma = \gamma f,$$

hence $\bar{\gamma}'\gamma' f = \bar{\gamma}\gamma f$.

Finally, if $f' = \sigma f \sigma^{-1}$, $\sigma \in G_0'$,

$$\bar{f}' = \bar{\sigma}\gamma\sigma^{-1}f'(\bar{\sigma}\gamma\sigma^{-1})^{-1},$$

we can take $\gamma' = \bar{\sigma}\gamma\sigma^{-1}$ modulo a factor in the group H' similar to the group H but constructed for f', and it easily follows that $\bar{\gamma}'\gamma' f' = \epsilon' f'$.

Nothing would change if we assumed that $\gamma \in \mathrm{Pin}(Q')$.   ∎

Remarks.

- One might wonder what relationship exists between ϵ' and the sign of λ (where $f\bar{f}f = \lambda f$, $\bar{f}f \neq 0$).

  Since $\bar{f}f$ belongs to the intersection of the ideals $C(Q')f$ and $\bar{f}C(Q')$, we can define (cf. Lemma (8.1.2))

  $$\bar{f}f = \alpha\gamma f, \quad \alpha \in \mathbf{C}^*,$$

  and an easy computation then shows that $\alpha\bar{\alpha} = \lambda\epsilon'$ so that λ and ϵ' have the same sign.

- If $\bar{f} = \exp(i\theta)f$, $\gamma f\gamma^{-1} = \exp(i\theta)f$ implies $\gamma f = \exp(i\theta/2)f$ and $\bar{\gamma}\gamma f = f$, $\epsilon' = 1$.

Relying on these remarks, some verifications can be carried out.

We will now study different questions concerning the m.t.i.s. and the isotropic r-vectors determining them, eventually giving the explicit geometric meaning of ϵ'.

In the following f determines the m.t.i.s. $\bar{F}'$.

Lemma 10.3.3 *If γf is a pure spinor determining $\bar{F}'$, $\gamma \in \mathrm{Pin}(Q')$, x is an element of $\bar{F}'$ if and only if $x\gamma f = 0$. If $\gamma \in \mathrm{Pin}(Q')$ and if $x\gamma f = 0$, $\forall x \in \bar{F}'$, γf is a pure spinor and determines $\bar{F}'$.*

Proof. This is a special case of Lemma (8.1.3). ∎

Lemma 10.3.4 *If B is the real symmetric bilinear form associated to Q and if $B(y, \bar{y}) = 0$ for all $y \in F'$, $B(\bar{y}, x) = 0$ for all $x \in F'$ and $\bar{F}' = F'$.*

Proof. Apply the hypothesis to $y + x$ and $y + ix$; $\bar{F}'$ is orthogonal to F' and F' being a m.t.i.s., $F' = \bar{F}'$ follows. ∎

Lemma 10.3.5 *If $B(y, \bar{y}) \neq 0$, for all non-zero $y \in F'$, a Witt basis*

$$(x_1, \ldots, x_r, y_1, \ldots, y_r)$$

can be found such that

$$\bar{y}_i = \delta x_i, \quad \delta = \pm 1, \quad i = 1, 2, \ldots, r, \quad F' \cap \bar{F}' = 0$$

and δ does not depend on i.

Proof. Choosing $y_1 \in F'$, we set $x_1 = \pm\bar{y}_1$; x_1 cannot be collinear with y_1. $H_1 = (x_1, y_1)$ is non-isotropic and $B(x_1, y_1) = 1/2$ can be assumed, $B(y_1, \bar{y}_1)$ being real; $y_2 \in F' \cap (H_1)^\perp$ and $x_2 = \pm\bar{y}_2$ are chosen such that $H_2 = (x_1, x_2, y_1.y_2)$ is non-isotropic and $B(x_i, y_j) = \delta_{ij}/2$, $i, j = 1, 2$, etc. Finally a Witt basis of the stated type is obtained, and $\bar{F}' = (x_1, x_2, \ldots, x_r)$, $F' = (y_1, y_2, \ldots, y_r)$.

$B(y, \bar{y})$ is real and has the same sign on all of F', so either $\delta = 1$ or $\delta = -1$ regardless of the index. ∎

Proposition 10.3.1 *If F' is a m.t.i.s. of the space $E_{\mathbf{C}} = E'$ and $\dim(F' \cap \bar{F}') = r - h$,*

1. *a Witt basis*

$$(x_1, x_2, \ldots, x_r, y_1, y_2, \ldots, y_r)$$

can be constructed, such that the isotropic vectors $y_1, y_2, \ldots, y_r$ generate F', $y_{h+1}, y_{h+2}, \ldots, y_r$ form a basis of $F' \cap \bar{F}'$ and $y_1 = \delta \bar{x}_1, \ldots, y_h = \delta \bar{x}_h$, δ being either 1 or -1,

2. *if γf is a pure spinor determining $\bar{F}'$, we can choose :*

$$\gamma = (x_1 + y_1)(x_2 + y_2)\ldots(x_h + y_h)$$

up to a scalar factor, and

$$\gamma f = x_1 x_2 \ldots x_h f, \qquad f = y_1 y_2 \ldots y_r$$

up to a scalar factor.

Proof. If $h = 0$, 1 is obvious since $F' = \bar{F}'$ and $\gamma = 1$ will do.

Assume that $h > 0$. By Lemma (10.3.4), there exists a $y_1 \in F'$ such that $B(y_1, \bar{y}_1) = \pm 1/2$. Let $H_1 = (y_1, \bar{y}_1)$, then H_1 is non-isotropic and so is $(H_1)^{\perp}$. $F'_1 = (H_1)^{\perp} \cap F'$ is $(r-1)$-dimensional. If $B(y, \bar{y}) = 0$ for all $y \in F'_1$, $\bar{F}'_1 = F'_1$ by Lemma (10.3.4), therefore a Witt basis can be constructed with $x_1 = \pm \bar{y}_1$ and $y_2, y_3, \ldots, y_r \in \bar{F}'_1 \cap F'_1 = \bar{F}' \cap F'$, proving 1.

If there exists a $y_2 \in F'_1$ such that $B(y_2, \bar{y}_2) = \pm 1/2$, we construct the non-isotropic space $H_2 = (y_1, y_2, \bar{y}_1, \bar{y}_2)$ and obtain the same situation, so we can proceed as in the proof of Lemma (10.3.5).

We still have to prove that δ does not depend on the index $i = 1, 2, \ldots, h$. Suppose that $y_h = \bar{x}_h$ and $y_{h-1} = -\bar{x}_{h-1}$. We can construct $z_1 = y_h + \lambda y_{h-1}$ such that $\lambda \bar{\lambda} = 1$, $\bar{z}_1 = x_h - \bar{\lambda} x_{h-1}$ implying that $B(z_1, \bar{z}_1) = 0$. To z_1 we can add

$$z_2 = y_h - \lambda y_{h-1}.$$

The $(r - h + 2)$-dimensional space $F'_2 = (z_1, z_2, y_{h+1}, \ldots, y_r)$ satisfies the conditions of Lemma (10.3.4), but F'_2 is the intersection of F' with the non-isotropic subspace $K = G^{\perp}$, where $G = (x_1, \ldots, x_{h-2}, y_1, \ldots, y_{h-2})$.

Lemma (10.3.4) can be applied to the non-degenerate restriction of B to K. Then $F'_2 = \bar{F}'_2$, so $F' \cap \bar{F}'$ contains F'_2, which is impossible, the dimension of $F' \cap \bar{F}'$ being $r - h$.

The proof of 2 uses Lemma (10.3.3), verifying that $\forall x \in \bar{F}'$, $x \gamma f = 0$. This can be tested for $x_1, \ldots, x_h, \bar{y}_{h+1}, \ldots, \bar{y}_r$:

$$x_i(x_1 + y_1)(x_2 + y_2)\ldots(x_h + y_h)y_1 y_2 \ldots y_r = x_i x_1 x_2 \ldots x_h y_1 \ldots y_r = 0$$

and since the $\bar{y}_i$, $i \geq h + 1$, can be expressed in terms of the y_j, $j \geq h + 1$, the same holds for them.   ∎

Lemma 10.3.6 *If W is an h-dimensional complex subspace of the m.t.i.s. F' and $\bar{W} = W$, a basis $y_1, y_2, \ldots, y_h$ of W can be found such that $y_1 = \bar{y}_1, \ldots, y_h = \bar{y}_h$.*

Proof. Consider $z_1 \in W$. If $\bar{z}_1 = \lambda z_1$, $\lambda \in \mathbf{C}$, $\lambda = \exp(i\theta)$, taking $y_1 = \exp(i\theta/2)z_1$, one obtains $y_1 = \bar{y}_1$.

If $(y_1, \bar{y}_1)$ is linearly independent, we form $y_1 = z_1 + \bar{z}_1$ and $y_2 = i(z_1 - \bar{z}_1)$, then $y_1 = \bar{y}_1$, $y_2 = \bar{y}_2$.

Suppose that we have obtained a linearly independent system $y_1, y_2, \ldots, y_s$, where $y_1 = \bar{y}_1, \ldots, y_s = \bar{y}_s$ and let z_{s+1} not belong to the space generated by these s vectors. If the system $y_1, \ldots, y_s, z_{s+1}, \bar{z}_{s+1}$ is linearly independent, we set $y_{s+1} = z_{s+1} + \bar{z}_{s+1}$ and $y_{s+2} = i(z_{s+1} - \bar{z}_{s+1})$. If the system is linearly dependent,

$$\bar{z}_{s+1} = \sum_{i=1}^{s} a^i y_i + a^{s+1} z_{s+1}, \quad a^{s+1} \neq 0,$$

and we take $Z_{s+1} = z_{s+1} - \sum_{i=1}^{s} a^i y_i$ different from zero. Choosing the a^i such that

$$a^i a^{s+1} - \bar{a}^i + a^i = 0,$$

$\bar{Z}_{s+1} = y_{s+1}/a^{s+1}$ and the previous situation is obtained. ∎

Proposition 10.3.2 *If F' is a m.t.i.s. of the space $E_{\mathbf{C}} = E'$ and if $\dim(F' \cap \bar{F}') = r - h$, a special Witt basis*

$$(x_1, x_2, \ldots, x_r, y_1, y_2, \ldots, y_r)$$

can be found, such that the $y_1, y_2, \ldots, y_r$ generate F' where :

1. $y_{h+1} = \bar{y}_{h+1}, y_{h+2} = \bar{y}_{h+2}, \ldots, \bar{y}_r = y_r$,
2. $y_1 = \delta \bar{x}_1, y_2 = \delta \bar{x}_2, \ldots, y_h = \delta \bar{x}_h$, δ *being either 1 or (-1),*
3. $x_{h+1} = \bar{x}_{h+1}, \ldots, x_r = \bar{x}_r$.

Proof.

1 follows from Proposition (10.3.1) and Lemma (10.3.6).

2 has already been proved.

3 can be proved using the construction method for a special Witt basis.

If the first $2h$ vectors of a Witt basis of the indicated type have been constructed, we choose $y_{h+1} = \bar{y}_{h+1}$ and x_{h+1}, then

$$B(\bar{x}_{h+1}, x_i) = B(\bar{x}_{h+1}, \bar{y}_i) = B(x_{h+1}, y_i) = 0, \quad i = 1, \ldots, h$$
$$B(\bar{x}_{h+1}, y_i) = B(\bar{x}_{h+1}, \bar{x}_i) = B(x_{h+1}, x_i) = 0, \quad i = 1, \ldots, h$$
$$B(\bar{x}_{h+1}, y_{h+1}) = B(x_{h+1}, \bar{y}_{h+1}) = B(x_{h+1}, y_{h+1}) = 1/2$$
$$B(\bar{x}_{h+1}, x_{h+1}) = 0.$$

x_{h+1} can be replaced by $(x_{h+1} + \bar{x}_{h+1})/2$ etc. ∎

Corollary 10.3.1 *In every m.t.i.s. F' of the complexification of a real space, r isotropic vectors $y_1, y_2, \ldots, y_r$ can be found and completed by r isotropic vectors $x_1, x_2, \ldots, x_r$ to a 'real' Witt basis in the sense of section 2.1.*

Proof. Obvious, invoking Proposition (10.3.2). ∎

Corollary 10.3.2 *If f is an isotropic r-vector determining the m.t.i.s. F' of the complexification of E, $\bar{f}f \neq 0$ if and only if Q is definite and $F' \cap \bar{F}' = 0$.*

Proof. This follows from Corollary (10.3.1) and Proposition (10.3.2). ∎

The computation of ϵ' where $\rho f = \epsilon' f$, $\rho = \bar{\gamma}\gamma$.

If $\delta = 1$ in Proposition (10.3.2), we form

$$e_j = (x_j + \bar{x}_j), \quad e'_j = i(x_j - \bar{x}_j), \quad j = 1, 2, \ldots, h$$
$$f_j = (x_j + y_j), \quad f'_j = (x_j - y_j), \quad j = h+1, \ldots, r$$
$$(e_j)^2 = (e'_j)^2 = (f_j)^2 = 1, \quad (f'_j)^2 = -1.$$

The Sylvester decomposition of Q then contains $2h + r - h = h + r$ positive terms and $r - h$ negative terms. If p is the number of positive terms, $p = h + r$.

If $\delta = -1$, an analogous method leads to $h + r$ negative terms and $p = r - h$ positive terms. This is exactly the basis given at the beginning of section 2.1, where $p \leq n - p$ was assumed.

Note that if $(p, n-p)$ is the signature of Q where p positive terms occur, $h = p - r$ if $2p \geq n$, or $r - p$ if $2p \leq n$.

Take $\gamma = (x_1 + y_1)(x_2 + y_2) \ldots (x_h + y_h)$, $N(\gamma) = 1$, $\gamma f = x_1 x_2 \ldots x_h f$,

$$\bar{\gamma} = (\bar{x}_1 + \delta x_1)(\bar{x}_2 + \delta x_2) \ldots (\bar{x}_h + \delta x_h)$$

$$\bar{\gamma}\gamma f = \bar{x}_1 \bar{x}_2 \ldots \bar{x}_h x_1 x_2 \ldots x_h f = (-1)^{h(h-1)/2} \bar{x}_1 \bar{x}_2 \ldots \bar{x}_h x_h x_{h-1} \ldots x_1 f.$$

If $\delta = 1$, $y_1 = \bar{x}_1, \ldots, y_h = \bar{x}_h$, $h = p - r$.

$$\bar{\gamma}\gamma f = (-1)^{h(h-1)/2} y_1 y_2 \ldots y_h x_h x_{h-1} \ldots x_1 f = (-1)^{h(h-1)/2} f,$$

$$\epsilon' = (-1)^{(p-r)(p-r-1)/2}$$

p being the number of positive terms in the Sylvester decomposition of Q.

If $\delta = -1$, $y_1 = -\bar{x}_1, \ldots, y_h = -\bar{x}_h$, $h = r - p$,

$$\epsilon' = (-1)^{h(h-1)/2+h} = (-1)^{h(h+1)/2},$$

and p being the number of positive terms,

$$\epsilon' = (-1)^{(r-p)(r-p+1)/2} = (-1)^{(p-r)(p-r-1)/2}.$$

We have proved the

Proposition 10.3.3 *If γf is conjugation pure spinor, $\gamma \in \mathrm{Pin}(Q')$, $\rho f = \bar{\gamma}\gamma f = \epsilon' f$ where*

$$\epsilon' = (-1)^{(p-r)(p-r-1)/2}, \tag{1}$$

p being the number of positive terms in the Sylvester decomposition of Q, $0 \le p \le 2r$. Furthermore $\gamma f = x_1 x_2 \ldots x_h f$ can be assumed, $h = |p - r|$.

For example, if Q is neutral, $p = r$ and $\epsilon' = 1$. If Q is positive definite, $p = 2r$ and $\epsilon' = (-1)^{r(r-1)/2} = \epsilon$. If (E, Q) is a Minkowski space of signature $(+ - --)$, $p = 1$, $r = 2$, $\epsilon' = -1$; if it has signature $(+ + +-)$, $p = 3$, $r = 2$ and $\epsilon' = 1$.

10.3.2   The fundamental hermitian sesquilinear form.

By the definition and the properties of pure spinors, we may write

$$\beta(\bar{u}\bar{f}) = a\mathcal{H}(uf, vf)\gamma f,$$

a and $\mathcal{H}(uf, vf)$ being scalar, a fixed.

$\mathcal{H}$ is a non-degenerate sesquilinear form, as can easily be verified, and we will choose a such that $\mathcal{H}$ is hermitian.

From $\beta(\bar{u}\bar{f})uf = a\mathcal{H}(vf, uf)\gamma f$, $\beta(\bar{u}\bar{f})vf\bar{\gamma} = \epsilon\bar{a}\overline{\mathcal{H}(vf, uf)}\bar{f}$ if $N(\gamma) = 1$, so $\beta(\bar{u}\bar{f})vf = \epsilon\epsilon'\bar{a}\overline{\mathcal{H}(vf, uf)}\gamma f$ and for $\mathcal{H}$ to be hermitian, $\epsilon\epsilon'\bar{a} = a$ must hold; so we can choose $a = \exp(i\theta)$ such that $a^2 = \epsilon\epsilon'$. The definition can be reformulated as follows :

Definition 10.3.1 *For the even dimensions, we define a hermitian sesquilinear form $\mathcal{H}$ on the complex spinor space, by*

$$\beta(\bar{u}\bar{f})vf = a\mathcal{H}(uf, vf)\gamma f, \quad a^2 = \epsilon\epsilon', \tag{2}$$

(where $\bar{f} = \gamma f \gamma^{-1}$, $\gamma \in G_0(Q')$, $\bar{\gamma}\gamma f = \epsilon' f$ and $\beta(f) = \epsilon f$).

Note that the definition of $\mathcal{H}$ only depends on γ by a factor of ± 1 (γ can be replaced by $\gamma\tau$, $\tau \in H'$). a can be chosen as 1 or i. Obviously, $\mathcal{H}$ is invariant under G_0.

Remarks. We could take $\gamma \in G'$ and absorb the coefficient a in γ; $\mathcal{H}$ would then be hermitian under a certain condition on γ.

Clearly $\mathcal{H}$ can be replaced by any non-zero real multiple.

We will find the conditions for the positive definiteness of $\mathcal{H}$ in section 3.5.

Using $\tilde{\beta}$ instead of β yields a form $\tilde{\mathcal{H}}$ which is invariant under G_0^+.

Clearly $\alpha_1\mathcal{H} + \beta_1\tilde{\mathcal{H}}$, $\forall\alpha_1, \beta_1 \in \mathbf{R}$ is a hermitian sesquilinear form, invariant under G_0^+.

10.3.3 'Charge conjugation' and the Dirac adjoint.

Consider the sequence of bijective mappings :

$$uf \to \bar{u}\bar{f} \to \bar{u}\bar{f}\gamma = \bar{u}\gamma f.$$

If we define

$$\mathcal{C}(uf) = \exp(i\theta)\bar{u}\gamma f, \tag{3}$$

θ being a fixed arbitrary real number, it can be verified that up to an unimportant sign change, the choice of $\gamma \in \mathrm{Pin}(Q')$ does not affect $\mathcal{C}(uf)$ (γ can be replaced by $\gamma\tau$, where $\tau \in H'$ (cf. 3.1), $\tau f = \mu f$, $\mu = \pm 1$ or $\pm i$). $\mathcal{C}$ commutes with the action of $g \in G$.

$\mathcal{C}$ is semi-linear and (whatever γ was chosen),

$$\mathcal{C}^2 = \epsilon' \,\mathrm{Id} \tag{4}$$

so that $\mathcal{C}^2 = \mathrm{Id}$ *if and only if* $p - r = 0, 1 \,(\mathrm{mod}\,4)$, *i.e. in the notations of Chapter 3, 2.1,* $p - q = 0, 2 \,(\mathrm{mod}\,8)$.

It is easy to verify that

$$\begin{aligned}
\mathcal{H}(\mathcal{C}(uf), vf) &= \epsilon\mathcal{H}(\mathcal{C}(vf), uf),\\
\mathcal{H}(\mathcal{C}(uf), \mathcal{C}(vf)) &= \epsilon\epsilon'\mathcal{H}(vf, uf).
\end{aligned} \tag{5}$$

As before, the Dirac adjoint $\mathcal{A}$ is defined by

$$\mathcal{A} \circ \mathcal{C} = \beta|_S,$$

$$\mathcal{A}(uf) = \epsilon'\beta(\mathcal{C}(uf)) \tag{6}$$

and

$$\mathcal{A} \circ \mathcal{C} = \mathcal{C} \circ \mathcal{A}. \tag{7}$$

where the meaning of the second member is obvious.

This is the sign table for $\epsilon\epsilon'$ depending on the values of r and p modulo 4 :

r	p			
	0	1	2	3
0	+	+	-	-
1	-	+	+	-
2	+	+	-	-
3	-	+	+	-

10.3.4 An application to quantum mechanics.

Definition 10.3.2 *Let* $w \in C(Q')$, *then by left multiplication, w defines a linear operator, which is called* hermitian *if*

$$\mathcal{H}(uf, wvf) = \mathcal{H}(wuf, vf), \quad uf, vf \in S.$$

Obviously this condition is equivalent to :

$$\beta(\bar{w}) = w.$$

Note that if A is this operator, $\mathcal{H}(A\varphi, \psi) = \overline{\mathcal{H}(A\psi, \varphi)}$, $\varphi = uf$, $\psi = vf$, so $\mathcal{H}(A \cdot, \cdot)$ is a new hermitian form.

In the Minkowski space of signature $(+++-)$, $\epsilon' = 1$, $f'f = 0$ and $\epsilon\epsilon' = -1$. Take an orthonormal basis :

$$(e_1)^2 = -1, \quad (e_2)^2 = (e_3)^2 = (e_4)^2 = 1,$$
$$x_1 = \frac{e_4 + e_1}{2}, \quad x_2 = \frac{ie_2 + e_3}{2}$$
$$y_1 = \frac{e_4 - e_1}{2}, \quad y_2 = \frac{-ie_2 + e_3}{2},$$

(x_1, x_2, y_1, y_2) form a special Witt basis, $y_2 = \bar{x}_2$ and we can choose $\gamma = -i(x_2 + y_2) = e_2$, $N(\gamma) = 1$, $\bar{\gamma}\gamma = 1$.

Writing (g_{kj}) for the components of the pseudo-metric tensor, we define

$$e^k = g^{kj}e_j.$$

Then e^k defines a hermitian operator A^k. $\beta(\bar{u}\bar{f})vf = i\mathcal{H}(uf, vf)e_2f$, setting $\psi = uf$,

$$\beta(\bar{\psi})e^k\psi = i\mathcal{H}(\psi, A^k\psi)e_2f.$$

The four real numbers $\mathcal{H}(\psi, A^k\psi)$ written $\langle \psi, A^k\psi \rangle = X^k$ define a vector X called the 'current vector' relative to $\mathcal{H}$ (cf. ultra, Chapter 16). (5) implies that

$$\mathcal{H}(\mathcal{C}(\varphi), \mathcal{C}(\psi)) = \epsilon\epsilon'\overline{\mathcal{H}(\varphi, \psi)}$$
$$\langle \psi, A^k\psi \rangle = -\langle \mathcal{C}\psi, A^k\mathcal{C}\psi \rangle.$$

The charge conjugation multiplies the current vector by -1.

Note that with the other signature, $\epsilon' = -1$ and the current vector is preserved.

10.3.5 The index of $\mathcal{H}$.

We compute

$$\beta(\bar{u}\bar{f})uf$$

for an element $(x_{i_1}x_{i_2} \ldots x_{i_s}f) = uf$, $1 \le i_1 < i_2 < \ldots < i_s \le r$ of a basis of $C(Q')f$, taking into account that $\gamma f = x_1x_2 \ldots x_h f$ and $y_i = \delta\bar{x}_i$, $i = 1, \ldots, h$.

If Q is definite, $\bar{x}_i = \delta y_i$, $i = 1, 2, \ldots, r$.

$$\beta(\bar{u}\bar{f})uf = \epsilon\bar{f}\bar{x}_{i_s} \ldots \bar{x}_{i_2}\bar{x}_{i_1}x_{i_1}x_{i_2} \ldots x_{i_s}f = \epsilon(\delta)^s\bar{f}f.$$

- If Q is positive definite, $\delta = 1$, $h = r$, $\gamma f = \bar{f}f$ and $\mathcal{H}$ is a definite hermitian form.

- If Q is negative definite : $\delta = -1$, $h = r$, $\gamma f = (-1)^r \bar{f}f$, $(\delta)^s = (-1)^s$.

  The linear isomorphism of the spinor space with an exterior algebra guarantees the existence of 2^r terms of each parity in the expression of a spinor in terms of the basis formed by the $x_{i_1} x_{i_2} \ldots x_{i_s} f$.

  $\mathcal{H}$ is a neutral hermitian form (i.e. of index 2^{r-1}).

- If Q is of signature $(p, n - p)$, $p \neq 0, n$, then $h = |p - r|$, $\gamma f = (\delta)^h \bar{y}_1 \bar{y}_2 \ldots \bar{y}_h f$.

  $$\beta(\bar{u}\bar{f})uf = \epsilon \bar{y}_1 \bar{y}_2 \ldots \bar{y}_h y_{h+1} \ldots y_r \bar{x}_{i_s} \ldots \bar{x}_{i_1} x_{i_1} \ldots x_{i_s} f,$$

  $x_i = \bar{x}_i$ or $\bar{x}_i = \delta y_i$; and only the latter case could produce a non-vanishing term if it occurred for $i = i_1, \ldots, i_s$; however, $h < r$ implies that a product $y_{h+1} \ldots y_r$ will then give 0. So the basis vectors of $C(Q')f$ are all isotropic for $\mathcal{H}$. $\mathcal{H}$ being non-degenerate, this means that this basis gives rise to a maximal Witt decomposition with a (2^r)-dimensional m.t.i.s. (relative to $\mathcal{H}$) : $\mathcal{H}$ is therefore neutral.

This result generalizes the special case of the Minkowski space, for which Dirac already introduced a neutral sesquilinear form. Summarizing,

Proposition 10.3.4 *The hermitian form $\mathcal{H}$ is positive definite if Q is positive definite, otherwise it is neutral.*

Remark. For $\tilde{\beta}$, $\tilde{\mathcal{H}}$ is definite if Q is negative definite, otherwise it is neutral.

10.3.6   The construction of a positive definite sesquilinear form.

1. Note that the results in section 3.1 could also be obtained for the composition of the complex conjugate with an automorphism of the Clifford algebra. In particular, if the automorphism α' is the natural extension of the linear mapping $e_j \to e_j$ if $Q(e_j) = 1$, $e_j \to -e_j$ if $Q(e_j) = -1$, $j = 1, 2, \ldots, n = 2r$, where the e_j form a fixed orthonormal basis. We will write $\mathring{u} = \alpha'(\bar{u})$, $u \in C(Q')$.

   If $\gamma' \in \text{Pin}(Q')$ determines the pure spinor $\gamma'f$ for the generalized complex conjugate, i.e. $\overset{c}{f} = \gamma'f\gamma'^{-1}$, we see, just as in 3.1, that there exists a scalar ϵ_0 equal to ± 1 such that $\mathring{\gamma}'\gamma'f = \epsilon_0 f$ and ϵ_0 is well-defined. If, for instance, $p \leq n - p$, we choose

   $$\gamma' = (x_1 + y_1)(x_2 + y_2) \ldots (x_r + y_r),$$
   $$\gamma'f = x_1 x_2 \ldots x_r f = \overset{c}{f}f,$$
   $$\overset{c'}{\gamma}{}' = \gamma',$$
   $$\overset{c'}{\gamma}{}'\gamma'f = (-1)^{r(r-1)/2} f$$

   and $\epsilon_0 = (-1)^{r(r-1)/2} = \epsilon$.

2. Extending the definition of section 3.2 and noticing that $\epsilon\epsilon_0 = 1$, we define

$$\epsilon\beta(uf)^c vf = \mathcal{H}^+(uf, vf)\gamma'f \tag{8}$$

or, if $\overset{c}{f}f = x_1 x_2 \ldots x_r f = \gamma' f$,

$$\epsilon\beta(uf)^c vf = \mathcal{H}^+(uf, vf)\overset{c}{f}f, \tag{9}$$

and it is easily verified that $\mathcal{H}^+$ is hermitian and that

$$\mathcal{H}^+(x_{i_1} x_{i_2} \ldots x_{i_s} f, x_{i_1} x_{i_2} \ldots x_{i_s} f) = 1,$$

so

Proposition 10.3.5 $\mathcal{H}^+$ *is a positive definite hermitian sesquilinear form.*

Remark. Formally, this is the same situation as in subsection 10.2.2, part 2, since $\overset{c}{f}f \neq 0$.

The definition of $u \to \overset{\circ}{u}$ depends on the choice of the orthonormal basis, but the conjugation itself is the same for all orthonormal bases determining the same direct decomposition

$$x = x^+ \oplus x^-, \quad x \in E, \quad Q(x^+) > 0, \quad Q(x^-) < 0.$$

$\mathcal{H}^+$ is invariant under the subgroup of elements in G_0 which preserve this direct decomposition.

10.3.7 Majorana spinors.

We have noted in section 1.2 that if a conjugation $\mathcal{C}$ exists such that $\mathcal{C}^2 = \mathrm{Id}$ in a space W, the realified space rW of W is the direct sum of two real spaces E_1 and E_2 such that

$$E_1 = (I + \mathcal{C})(rW), \quad E_2 = (I - \mathcal{C})(rW).$$

These two real space have the same dimension.

If we consider the space of spinors *such that the constant* $\epsilon' = 1$ (cf. section 3.1), this space gives rise to such a property; using standard notations from theoretical physics such a spinor, written ψ, is the sum of two spinors ψ_1 and ψ_2 where

$$\psi_1 = (\frac{1 + \mathcal{C}}{2})(\psi), \quad \psi_2 = (\frac{1 - \mathcal{C}}{2})(\psi). \tag{10}$$

The spinors ψ_1 and ψ_2 are called Majorana spinors; the product by i exchanges them :

$$\mathcal{C}\psi_1 = \psi_1, \quad \mathcal{C}\psi_2 = -\psi_2. \tag{11}$$

If the complex dimension of the spinor space under consideration is 2^r, the real dimension of the Majorana spinors is also 2^r.

$\mathcal{C}$ commuting with the action of G, the two Majorana spinor spaces are globally invariant by the left action of G (which is obvious) : the representation of G in the spinor space is therefore the sum of two real representations and we note that $p - q = 0, 2 \,(\mathrm{mod}\ 8)$.

Remark. If $\mathcal{C}^2 = -\,\mathrm{Id}$ we have seen (in section 1.2) that the spin representation is 'quaternionic'. It is still possible to write $\psi = \psi_1 + \psi_2$ uniquely, where

$$\psi_1 = (\frac{1 + i\mathcal{C}}{2})(\psi), \quad \psi_2 = (\frac{1 - i\mathcal{C}}{2})(\psi), \tag{12}$$

$$\mathcal{C}(\psi_1) = i\psi_2, \quad \mathcal{C}\psi_2 = -i\psi_1. \tag{13}$$

This does not lead to a decomposition of the spinor space as a direct sum of two complex subspaces, since $(\frac{1 + i\mathcal{C}}{2})(1 - i\mathcal{C}) = 1$ and every element ψ can be written as

$$\psi = (\frac{1 + i\mathcal{C}}{2})(1 - i\mathcal{C})\psi = \frac{1 + i\mathcal{C}}{2}\psi',$$

the 'ψ_1-form', or also

$$\psi = (\frac{1 - i\mathcal{C}}{2})(1 + i\mathcal{C})\psi = \frac{1 - i\mathcal{C}}{2}\psi'',$$

the 'ψ_2-form'.

This situation occurs when $p - q = 4, 8 \,(\mathrm{mod}\ 8)$ (cf. the table in Chapter 3, 2.1).

10.3.8 Weyl spinors.

Let us consider the product $e_N = e_1 e_2 \ldots e_n$ of the elements of an orthonormal basis. If $n = 2r$ we know that e_N anticommutes with all $x \in E$. By a change of orthonormal basis, e_N transforms in $\pm e_N$. We have proved that if $\varphi_g(x) = gxg^{-1}$, $x \in E$, $g \in G$, $\varphi(e_N) = -\,\mathrm{Id}_E$. Hence $\varphi(e_N)$ extends naturally to the main automorphism of the Clifford algebra.

It is obvious that $e_N f = (-1)^r f e_N$,

$$e_N f e_N^{-1} = (-1)^r f = \alpha(f)$$

and e_N is a pure spinor, determining $\alpha(f)$ (α replaces the complex conjugation).

As $\alpha(e_N) = (-1)^n e_N$, $\alpha(e_N)e_N f = \check{\epsilon} f$, $\check{\epsilon} = \pm 1$ and $(-1)^n (e_N)^2 = (e_N)^2 = \check{\epsilon}$ since $n = 2r$.

$$\check{\epsilon} = (-1)^{n(n-1)/2}(-1)^{n-p} = (-1)^{r-p}.$$

By analogy with section 3.3, the transformation $\mathcal{W}$, obtained as the composition of

$$uf \to \alpha(uf) \to \alpha(uf)e_N \to \alpha(u)e_N f = e_N uf,$$

is called the 'Weyl conjugation'. We should then set :

$$\mathcal{W}(uf) = (e_N)uf,$$

but to obtain $\mathcal{W}^2 = \text{Id}$ (in particular for the Minkowski space), $(e_N)^2 = (-1)^h$ we will introduce a scalar factor η such that $\eta^2 = (-1)^h$ and we define the 'Weyl conjugation'

$$\mathcal{W}(uf) = \eta e_N uf. \tag{14}$$

$\mathcal{W}$ commutes with the action of G^+. Putting

$$\varphi_1 = (\frac{1+\mathcal{W}}{2})\psi, \quad \varphi_2 = (\frac{1-\mathcal{W}}{2})\psi, \tag{15}$$

a spinor can be written as the direct sum of two 'Weyl spinors' such that

$$\mathcal{W}(\varphi_1) = \varphi_1, \quad \mathcal{W}(\varphi_2) = -\varphi_2. \tag{16}$$

Now it is obvious that a spinor common to the Majorana and the Weyl approach will exist *if and only if C commutes with the product by ηe_N*, i.e. if $r = p \,(\text{mod } 2)$; but as $p = r$ or $r + 1$ modulo 4, for a Majorana spinor, this condition reduces to $p = r$ modulo 4, and is not satisfied for Minkowski spaces (where $p = 1$ or 3).

10.3.9   A Weyl-Dirac charge conjugation or 'chiral' conjugation.

e_N defines an automorphism of the Clifford algebra whose composition with the usual complex conjugation yields a new complex conjugation (cf. 3.6). Using the notations of section 3.6 we will obtain a new factor ϵ_0, equal to

$$\epsilon'(-1)^{r-p},$$

and a Weyl-Dirac conjugation follows, using $\alpha(\bar{u})$ (it commutes with the action of G^+); the so-called Majorana condition will then be satisfied if

$$\epsilon'(-1)^{r-p} = 1$$

i.e. if $(-1)^{(p-r)(p-r+1)/2} = 1$, $p = r$ or $r - 1$ modulo 4 (or $p - q = 0, 6 \,(\text{mod } 8)$).

Using this definition, a Weyl spinor will be a Majorana spinor if $p = r$ modulo 4.

10.3.10   General conjugations.

To end this study of the even-dimensional case, let us try to find all conjugations for the representations of the Clifford group G or of its subgroup G^+, in the spinor space $C(Q')f$.

Given a conjugation C, all other conjugations J can be obtained by taking the composition of C with a linear isomorphism which commutes with G (or G^+). By the general study of Clifford algebras over even-dimensional spaces with neutral signature, every linear isomorphism of $C(Q')f$ is a left multiplication by some invertible element of $C(Q')$.

G is generated by the non-isotropic vectors in E. If a linear isomorphism of $C(Q')f$ commutes with all elements of G it must be a scalar, so that relying in $J^2 = \pm \mathrm{Id}$, the scalar will be of the form $\exp(i\theta)$, $\theta \in \mathbf{R}$: in this way the conjugation C of section 3.3 is reobtained.

If we replace G by G^+, the linear isomorphism under consideration is the left multiplication by an invertible element of the center of $C^+(Q')$, G^+ being generated by the products of an even number of non-isotropic vectors. Therefore a conjugation J can be obtained by applying C (defined in 3.3) and a multiplication by $\lambda + \mu e_N$, where

$$\begin{cases} \lambda^2 + \mu^2(-1)^h \neq 0 \\ |\lambda|^2 + |\mu|^2(-1)^h = \pm 1 \\ \lambda\bar{\mu} + \mu\bar{\lambda} = 0 \end{cases} \tag{17}$$

$\lambda, \mu \in \mathbf{C}$. We have proved :

Proposition 10.3.6 *If the dimension $n = 2r$,*

1. *the conjugations C which commute with the action of G are of the form $C(uf) = \exp(i\theta)\bar{u}\gamma f$, $\theta \in \mathbf{R}$, γf being the complex conjugation pure spinor defined in Proposition (10.3.3). If p is the number of positive terms and q is the number of negative terms in the Sylvester decomposition of Q, there exist Majorana spinors for C if and only if $p = r$ or $r + 1$ modulo 4 (or $p - q = 0, 2 \,(\mathrm{mod}\,8)$); Weyl-Majorana spinors exist if $p = r$ modulo 4 $(p - q = 0 \,(\mathrm{mod}\,8))$.*

2. *the conjugations J which commute with the action of G^+ are products of C defined in 1 and the left action of an element $\lambda + \mu e_N$, $e_N = e_1 e_2 \ldots e_n$, under the conditions expressed by (17). $J^2 = \epsilon'(|\lambda|^2 + |\mu|^2(-1)^h)\,\mathrm{Id}$, where ϵ' is defined in Proposition (10.3.3) and $h = |r - p|$.*

G can be replaced by Γ and G^+ by Γ^+ without affecting these results.

Remarks.

1. If h is even, $J^2 = \epsilon'\,\mathrm{Id}$, so if $h = |r - p|$ is even it is impossible to define Majorana spinors for a conjugation which commutes with G^+, if $|r - p| = 2\,(\mathrm{mod}\,4)$ or $p - q = 4\,(\mathrm{mod}\,8)$.

   The spin representations obtained for $p - q = 4\,(\mathrm{mod}\,8)$ are of 'quaternionic' type. In all other cases, they are of 'real' type.

2. To each of these conjugations C there corresponds a Dirac adjoint $\mathcal{A}$ by formula (7) of 3.3.

10.4 Invariant hermitian sesquilinear forms.

We are still considering the case where $n = 2r$. Let $\mathcal{H}_1$ be a sesquilinear form, $\mathcal{H}_1(uf, vf)$ is semi-linear in uf and linear in vf. C being the conjugation of section 3.3, where

$$Cuf = \bar{u}\gamma f, \quad \gamma \in \mathrm{Pin}(Q');$$

we define

$$\lambda\mathcal{B}_1(uf, vf) = \mathcal{H}_1(C(uf), vf), \quad \lambda \in \mathbf{C}^*, \tag{1}$$

λ arbitrary but fixed.

$\mathcal{B}_1$ clearly is a C-bilinear form and if $\mathcal{H}_1$ is non-degenerate, the same holds for $\mathcal{B}_1$. If $\mathcal{H}_1$ is invariant under the elements of the real Clifford group G, the action of G commuting with C, $\mathcal{B}_1$ will also be invariant. Conversely, given $\mathcal{B}_1$, (1) determines $\mathcal{H}_1$. We say that $\mathcal{B}_1$ and $\mathcal{H}_1$ are associated.

We will study the sesquilinear forms invariant under the action of G_0^+.

In particular, the forms $\mathcal{B}$ and $\mathcal{H}$ introduced before are associated, since

$$\beta(C(uf))vf = \beta(\bar{u}\gamma f)vf = \beta(\bar{u}\bar{f}\gamma)vf = \beta(\gamma)\beta(\bar{u}\bar{f})vf,$$

and γ can be chosen in $G_0(Q')$ so that $N(\gamma) = 1$, and $\gamma\beta(C(uf))vf = \beta(\bar{u}\bar{f})vf$, or

$$\frac{\epsilon'}{\lambda}\mathcal{H}(uf, vf)\gamma f = \beta(\bar{u}\bar{f})vf$$

which is of the form introduced in 3.2 to define $\mathcal{H}$.

By the previous results, every C-bilinear form $\mathcal{B}_1$, invariant under the action of G_0^+, can be written as

$$\mathcal{B}_1 = \alpha_1\mathcal{B} + \beta_1\tilde{\mathcal{B}}, \quad \alpha_1, \beta_1 \in \mathbf{C}.$$

We have obtained :

Proposition 10.4.1 *Every sesquilinear form on the spinors which is invariant under the action of G_0^+ is of the form*

$$\mathcal{H}_1 = \alpha_1\mathcal{H} + \beta_1\tilde{\mathcal{H}}, \quad \alpha_1, \beta_1 \in \mathbf{C},$$

where $\mathcal{H}$ and $\tilde{\mathcal{H}}$ are associated to $\mathcal{B}$ and $\tilde{\mathcal{B}}$.

Let us now assume that the sesquilinear forms are hermitian, the condition is expressed on $\mathcal{B}_1$ by

$$\lambda\mathcal{B}_1(C(uf), vf) = \overline{\lambda\mathcal{B}_1(C(vf), uf)},$$

it will be satisfied by $\mathcal{B}$ and $\tilde{\mathcal{B}}$ for a suitable choice of λ. Indeed, the preceding computations and the results of 3.2 show that we must take

$$\epsilon\epsilon'\bar{a} = a,$$

a being ϵ'/λ, or $\lambda\epsilon\epsilon' = \bar{\lambda}$, which can always be done if λ is 1 or i. The same holds for $\tilde{\mathcal{B}}$, where $\lambda\tilde{\epsilon}\epsilon' = \bar{\lambda}$ and $\tilde{\epsilon} = (-1)^{r(r+1)/2}$. Therefore :

Proposition 10.4.2 *Every hermitian sesquilinear form on the spinors which is invariant under the action of G_0^+ is of the form*

$$\mathcal{H}_1 = \alpha_1 \mathcal{H} + \beta_1 \tilde{\mathcal{H}}, \quad \alpha_1, \beta_1 \in \mathbf{R},$$

$\mathcal{H}$ *and* $\tilde{\mathcal{H}}$ *being the hermitian sesquilinear forms defined by*

$$\beta(\bar{u}\bar{f})vf = a\mathcal{H}(uf, vf)\gamma f, \quad \tilde{\beta}(\bar{u}\bar{f})vf = a\tilde{\mathcal{H}}(uf, vf)\gamma f,$$

$a = \epsilon \epsilon' \bar{a}$ *(so $a = 1$ or i) and γf being the conjugation pure spinor.*

This Proposition completes a remark in section 3.2.

Remark. The hermitian sesquilinear forms which are invariant under the action of all G_0 (instead of just G_0^+) are of the form $\alpha_1 \mathcal{H}$. These are also the forms for which the action of $x \in E$ is a self-adjoint operator. The forms $\beta_1 \tilde{\mathcal{H}}$ are these for which $x \in E$ is anti-self-adjoint.

If uf and vf are Majorana, a $k = \pm 1, \pm i$ exists such that $\mathcal{H}_1(uf, vf) = k\mathcal{B}_1(uf, vf)$. If $\mathcal{H}_1$ is hermitian, $\mathcal{H}_1(vf, uf) = \bar{k}\overline{\mathcal{B}_1(uf, vf)} = \pm k\mathcal{B}_1(uf, vf)$ if we assume also that $\mathcal{B}_1$ is either symmetric or antisymmetric. Hence $\mathcal{B}_1(uf, vf)$ is either real or pure imaginary, for all pairs of Majorana spinors.

10.5   Conjugations and hermitian sesquilinear forms if $n = 2r + 1$.

The real $(n = 2r + 1)$-dimensional space E is given a non-degenerate quadratic form Q of signature (p, q).

By Theorem (3.3.2), if $z_0 \in E$, z_0 being non-isotropic, $C^+(Q)$ is isomorphic to the Clifford algebra $C(Q_1)$ of $(z_0)^\perp = E_1$ given the quadratic form $Q_1 = -aQ$ where $Q(z_0) = a$. The study of $C^+(Q)$ can be reduced to that of the simple central algebra $C(Q_1)$.

Consider the complexified space (E', Q') of (E, Q) and its Witt decomposition

$$E' = F \oplus F' \oplus (z_0),$$

$F \oplus F'$ is also a Witt decomposition for the complexified space (E'_1, Q'_1) of (E_1, Q_1) and we can consider the space $S = \{x_{i_1} x_{i_2} \ldots x_{i_h} f, f = y_1 y_2 \ldots y_r\}$ (cf. Chapter 9), as the spinor space for the Clifford algebra $C(Q'_1)$, isomorphic to $C^+(Q')$.

We may assume that $a = \pm 1$.

Defining $\bar{f} = \gamma f \gamma^{-1}$, $\gamma \in \text{Pin}(Q'_1)$, we set $\mathcal{C}_1(uf) = \bar{u}\gamma f$ and it is obvious that $\mathcal{C}_1^2 = \epsilon' \text{Id}$, ϵ' being computed for (E_1, Q_1).

This conjugation allows us to define Majorana spinors if and only if $\epsilon' = 1$, i.e. if $a = -1$, $p = (p + q - 1)/2$ or $(p + q - 1)/2 + 1$ modulo 4, or $p - q = \pm 1 \, (\text{mod} \, 8)$.

If $a = 1$, p and q are exchanged and the same condition is obtained.

A conjugation C on the spinor space of $C^+(Q')$ will be defined by C_1 and the isomorphism $j : C(Q_1) \to C^+(Q)$ which associated $z_0 y \in C^+(Q)$ to $y \in E_1$ (cf. Chapter 9), i.e.

$$C(j(uf)) = j(C_1(uf)).$$

In the spinor space for $C(Q_1')$ the notion of Weyl spinor is well-defined, j can be used to define it in $C^+(Q')$.

There exists a second method to define a conjugation, if we consider, using the notations of Chapter 9, 1.1, the irreducible representation of $C^+(Q')$ by left multiplication in the space of

$$u^+ f = (u_1^+ + z_0 u_2^-)f.$$

By a result proved in Chapter 4, 1.6, the algebras $C^+(Q)$ and $C^+(-Q)$ are isomorphic. Therefore we can assume that $q > 1$.

We choose $Q(z_0) = -1$, $(z_0)^\perp$ is a space of signature $(p, q-1)$ and if $\bar{f} = \gamma f \gamma^{-1}$, there exists a $\gamma \in \mathrm{Pin}(Q')$ having the same parity as $|r-p|$ (by Proposition (10.3.3)).

1. If $h = |r - p|$ is even, we take

$$C(u^+ f) = \bar{u}^+ \gamma f.$$

2. If h is odd, $\gamma' = \gamma z_0$ is such that $\gamma' f \gamma'^{-1} = (-1)^r \bar{f}$.

   (a) If r is even, we define

   $$C(u^+ f) = \bar{u}^+ \gamma z_0 f.$$

   (b) If r is odd, we consider in $(z_0)^\perp$ the orthonormal basis $e_1, e_2, \ldots, e_{2r}$, $\gamma'' = \gamma z_0 e_1 e_2 \ldots e_{2r}$ then satisfies $\gamma'' f \gamma''^{-1} = \bar{f}$ and we define

   $$C(u^+ f) = \bar{u}^+ \gamma z_0 e_1 e_2 \ldots e_{2r} f,$$

   C then commutes with the action of G^+.

Majorana spinors will exist for this conjugation if and only if $p = r$ or $r + 1$, modulo 4 (cf. section 3.3), i.e. if $p - q = \pm 1 \,(\mathrm{mod}\,8)$.

Once the pure spinor γf is constructed, the latter method allows the construction of a hermitian sesquilinear form $\mathcal{H}$ using a similar method to that of the even-dimensional case, viz.

$$\beta(\overline{u^+ f})v^+ f = a\mathcal{H}(u^+ f, v^+ f)\gamma f, \quad a^2 = \epsilon\epsilon'$$

where the notations are analogous to those of 3.2. $\mathcal{H}$ is invariant under G_0^+.

Other results, analogous to those obtained in the even-dimensional case, can then be established by routine computations.

Remark. We have obtained Majorana spinors when $p - q = 0, 1, 2 \,(\mathrm{mod}\,8)$, in accordance with the results of Chapter 3, 2.1. Also note that for $p - q = 4, 6 \,(\mathrm{mod}\,8)$ we have obtained spinors which correspond to the quaternionic representation of the real Clifford group G. These results can be summarized as follows :

$p - q \,(\mathrm{mod}\,8)$	0	1	2	3	4	5	6	7
G-Majorana spinors	yes		yes		no		no	
G^{+}-Majorana spinors	yes	yes	yes	no	no	no	yes	yes

10.6   Selected references.

- A. Crumeyrolle, *Dérivations, formes et opérateurs usuels sur les champs spinoriels des variétés différentiables de dimension paire*, Ann. Inst. H. Poincaré, Vol. 16, no. 3, 1972.

Chapter 11

SPINORIALITY GROUPS.

We use the hypotheses of Chapter 10, 2, dim $E = n = 2r$, E being real, E' denotes its complexification. In the notations of Chapter 10, 2.1, $f = y_1 y_2 \ldots y_r$ defines the m.t.i.s. F'.

11.1

Definition 11.1.1 H is the subgroup of $\gamma \in \mathrm{Spin}(Q)$ such that

$$\gamma f = \pm f$$

and $p(H) = \mathcal{G}$ is the spinoriality group associated to f.

Note that if another isotropic r-vector f' was chosen in F', $f' = \lambda f$, $\lambda \in \mathbf{C}$, and the definition of H and $\mathcal{G}$ is therefore independent of the choice of f.

If another m.t.i.s. of E' is chosen, H will change into a conjugated subgroup of $\mathrm{Pin}(Q')$, but also :

Lemma 11.1.1 Two subgroups H and H_1 associated to f and f_1 by Definition (11.1.1) are conjugate in $\mathrm{Pin}(Q)$.

Proof. By Proposition (10.2.2), every m.t.i.s. contains r vectors $y_1, y_2, \ldots, y_r$ to which vectors $(x_1, x_2, \ldots, x_r)$ can be added in order to obtain a 'real' Witt basis $\{x_i, y_j\}$; therefore an element of $O(Q)$ mapping one such basis to another one can always be found. ∎

In what follows, we will assume that the isotropic r-vector $f = y_1 y_2 \ldots y_r$ is built from vectors belonging to such a 'real' Witt basis.

Proposition 11.1.1 In the elliptic case, $\mathcal{G}$ can be identified with the special unitary group $\mathrm{SU}(r, \mathbf{C})$. $\mathcal{G}$ is the set of elements with determinant 1 in the stabilizer of a m.t.i.s. The dimension of $\mathcal{G}$ is $r^2 - 1$, $\mathcal{G}$ is connected and simply connected.

Proof. Take $f = y_1 y_2 \ldots y_r$, $y_j = (ie_j - e_{n-j+1})/2$, $j = 1, 2, \ldots, r$. $\gamma f = \pm f$ if and only if $\gamma f \gamma^{-1} = N(\gamma) f$ (by the Fundamental Lemma (10.3.1)). $\gamma \in C^+(Q)$ implies $N(\gamma) = 1$ and $\gamma f \gamma^{-1} = f$. If $y_{j'} = \gamma y_j \gamma^{-1}$, $y_{i'} = A^j_{i'} y_j$, $x_j = (ie_j + e_{n-j+1})/2$ implies $x_{j'} = \gamma x_j \gamma^{-1}$ and $x_{i'} = \bar{A}^j_{i'} x_j$.

The Witt basis $\{x_i, y_j\}$ is mapped to the Witt basis $\{x_{i'}, y_{j'}\}$ and

$$\sum_j A^j_{i'} \bar{A}^j_{k'} = \delta_{i'k'}.$$

Considering f as a r-vector in $\wedge^r(F')$ leads to $\det(p(\gamma)) = 1$.

The other results are either obvious or have already been proved.   ∎

Proposition 11.1.2 *In the signature* $(p, n - p)$, $p \leq n - p$, $r \geq 2$, $\mathcal{G}$ *is isomorphic to the subgroup of elements with determinant one, of the group of* $n \times n$-*matrices of the form*

$$\begin{pmatrix} \alpha & -\bar{\mu} & \lambda & \mu \\ 0 & \beta & \nu & 0 \\ 0 & 0 & \rho & 0 \\ 0 & 0 & \bar{\nu} & \bar{\beta} \end{pmatrix}$$

where

$$\begin{cases} \alpha \in M_p(\mathbf{R}), \quad \det(\alpha) = \pm 1 \\ \beta \in M_{r-p}(\mathbf{C}), \quad \beta \bar{\beta}^T = \mathrm{Id}, \quad \det(\beta) = \pm \det(\alpha) \\ \alpha \rho^T = \mathrm{Id}, \quad \lambda \in M_k(\mathbf{R}), \quad \mu \in \mathbf{C}^{p(r-p)}, \quad \nu \in \mathbf{C}^{(r-p)p} \\ \nu = -\beta \mu^T \rho, \quad \rho^T \lambda + \lambda^T \rho = \nu^T \bar{\nu} + \bar{\nu}^T \nu. \end{cases}$$

$\mathcal{G}$ *has 4 connected components and can be identified with the set of elements with determinant 1 in the stabilizer of a m.t.i.s.*

$$\dim \mathcal{G} = r^2 - 2 + \frac{p(p-1)}{2}.$$

Proof. Write

$$y_j = \frac{e_j - e_{n-j+1}}{2}, \quad x_j = \frac{e_j + e_{n-j+1}}{2}, \quad j \leq p,$$

$$y_j = \frac{ie_j - e_{n-j+1}}{2}, \quad x_j = \frac{ie_j + e_{n-j+1}}{2}, \quad j > p.$$

Define $\sigma \in \mathcal{G}$ by its matrix

$$y_{i'} = \sum_{j \leq k} A^j_{i'} y_j + \sum_{j > k} A^j_{i'} y_j,$$
$$x_{i'} = \sum_{j \leq k} B^j_{i'} x_j + \sum_{j > k} B^j_{i'} x_j + \sum_{j \leq k} C^j_{i'} y_j + \sum_{j > k} C^j_{i'} y_j.$$

Expressing that σ is real, we obtain that α is real, that $\bar{\beta}$ is conjugated to β, that ρ is real, that λ is real, and two conjugation relations; then, expressing that $B(x_{i'}, y_{j'}) = \frac{1}{2} \delta_{i'j'}$ and $B(x_{i'}, x_{j'}) = 0$,

$$\alpha \rho^T = \mathrm{Id}, \quad \beta \bar{\beta}^T = \mathrm{Id}, \quad \nu = -\beta \mu^T \rho,$$

and

$$\rho^T \lambda + \lambda^T \rho = \nu^T \bar{\nu} + \bar{\nu}^T \nu.$$

If we consider a differentiable path starting at the identity in $\mathcal{G}$, this last condition implies

$$l + l^T = 0, \quad l = (\frac{d\lambda(t)}{dt})_{t=0}.$$

Finally, $\gamma f \gamma^{-1} = N(\gamma) f$ if $p(\gamma) = \sigma$.

These facts prove the proposition, $SL(p)$ and $SU(r-p)$ being connected. ∎

Proposition 11.1.3 *If Q is neutral, $p = r$, $\mathcal{G}$ is isomorphic to the subgroup of elements in $SL(n, \mathbf{R})$ of the form*

$$\begin{pmatrix} \alpha & \lambda \\ 0 & \rho \end{pmatrix}$$

where $\alpha \in M_r(\mathbf{R})$, $\det(\alpha) = \pm 1$, $\alpha \rho^T = \mathrm{Id}$ and $\rho^T \lambda + \lambda^T \rho = 0$.

$\mathcal{G}$ consists of two connected components and can be identified with the set of elements with determinant 1 in the stabilizer of an m.t.i.s.

$$\dim \mathcal{G} = \frac{(r-1)(3r+2)}{2}$$

Proof. In this case we do not use the complexification; the proof remains similar to that of Proposition (11.1.2). ∎

Definition 11.1.2 *H_e is the subgroup of $\gamma \in \mathrm{Spin}(Q)$ such that*

$$\gamma f = \chi \exp(i\theta) f,$$

$\chi \exp(i\theta)$ being an arbitrary element of $\mathbf{C}^*$. $p(H_e) = \mathcal{G}_e$ is called an enlarged spinoriality group.

$\mathcal{G}_e$ is the stabilizer of an m.t.i.s. under the action of $SO(Q)$.

In the elliptic case, the proof of Proposition (11.1.1) shows that $\chi = 1$ so that $p(H_e) = \mathcal{G}_e$ must be isomorphic to $U(r, \mathbf{C})$.

In the general case, χ may differ from 1. These groups are of dimension $r^2 + p(p-1)/2$ when $0 \le p \le r$.

If $p \ne 0$ the enlarged spinoriality group is not isomorphic to a generalized unitary group (their dimensions differ).

$\mathcal{G}_e$ is connected if $p = 0$ and has two connected components if $0 < p \le r$.

$\mathcal{G}$ is an invariant subgroup of $\mathcal{G}_e$ if both groups are associated to the same isotropic r-vector.

The complex spinoriality groups. Let us consider in $SO(Q')$ the subgroup $\mathcal{G}'$ of the stabilizer of an m.t.i.s. whose elements have determinant equal to ± 1. Since, by Chapter 10, 3.1, $\gamma f = \pm \mu f$, $\mu \in \mathbf{C}^*$ if and only if $\gamma f \gamma^{-1} = N(\gamma)\mu^2 f$, an element of $\mathcal{G}'$ must be of the form

$$\begin{pmatrix} A & C \\ 0 & B \end{pmatrix}$$

where

$$\begin{cases} A \in \mathrm{GL}(\mathbf{C}, r), & \det A = \pm 1, \\ A^T B = \mathrm{Id}, \\ B^T C + C^T B = 0, \end{cases}$$

so it is the product of a change of Witt bases for $F \oplus F'$:

$$\begin{pmatrix} \alpha & 0 \\ 0 & \beta \end{pmatrix}$$

where $\det(\alpha) = \pm 1$ and $\alpha^T \beta = \mathrm{Id}$, and a matrix of the form

$$m = \begin{pmatrix} 1 & \theta \\ 0 & 1 \end{pmatrix}$$

where θ is an antisymmetric complex matrix, $m \in p(\exp(\wedge^2(F')))$. Conversely, every element of $p(\exp(\wedge^2(F')))$ is of the form given for m (cf. Chapter 8, 1.2).

The definition of the group H' in $\mathrm{Spin}(Q')$ such that

$$p(H') = \mathcal{G}'$$

is obvious.

Similarly we can define groups H'_e and $\mathcal{G}'_e$, analogous to the H_e and $\mathcal{G}_e$ in Definition (11.1.2).

More results concerning the structure of spinoriality groups can be found in the thesis of J. Timbeau, (Toulouse, 1985).

11.2 Selected references.

- A. Crumeyrolle, *Groupes de spinorialité*, Ann. Inst. H. Poincaré, Vol. XIV, no. 4, 1971.

- A. Crumeyrolle, *Algèbres de Clifford et spineurs*, Cours de l'Université de Toulouse III, 1974.

Chapter 12

COVERINGS OF THE COMPLETE CONFORMAL GROUP—TWISTORS.

12.1 The complete conformal group.

E is a real n-dimensional vector space with a non-degenerate quadratic form of signature (p, q) (p positive and q negative terms). B is the associated symmetric bilinear form, it satisfies $B(x, x) = Q(x)$.

Let us consider the set of continuously differentiable homeomorphisms f from an open set $\mathcal{O}$ of E to some other open set in E, such that if $x \in \mathcal{O}$, and if $\tilde{f}_x$ is the tangent linear transformation in x, there exists a continuous function λ which has no zeroes and such that :

$$B(\tilde{f}_x(u), \tilde{f}_x(v)) = \lambda^2(x)B(u, v), \quad \forall u, v \in E. \tag{1}$$

This set forms a pseudo-group of transformations called (abusively) the *conformal group* of (E, Q) and denoted by $C_n(p, q)$.

By a theorem of Liouville (which we accept without proof), when $n \geq 3$, every element of $C_n(p, q)$ is a product (in arbitrary order) of a number of translations, isometries, homotheties and inversions. Only the inversions can cause some problems for their domain of definition. One can easily verify that all these transformations have the property expressed by (1).

If we include the inversions, we will speak of the 'complete' conformal group; if we do not include them, a subgroup called the 'restricted' conformal group is obtained; its elements are defined in all points of E.

The notations of the previous chapters are still used : E is also written $E(p, q)$ and has an orthonormal basis of the form

$$e_1, e_2, \ldots, e_n, \quad (e_i)^2 = 1, \quad i = 1, 2, \ldots, p, \quad (e_i)^2 = -1, \quad i = p+1, \ldots, p+q.$$

It will prove useful to consider $E(p, q)$ as a subspace of $E(p + 1, q + 1)$, an $(n + 2)$-dimensional vector space with a quadratic form of signature $(p + 1, q + 1)$ and a basis

$$e_0, e_1, \ldots, e_n, e_{n+1}, \quad (e_0)^2 = 1, \quad (e_{n+1})^2 = -1.$$

149

The Clifford algebras of $E(p,q)$ and $E(p+1, q+1)$ are denoted by $\mathrm{Cl}(p,q)$ and $\mathrm{Cl}(p+1, q+1)$ respectively. When we will need them, the Clifford and spin groups will be written $G(p,q)$, $G(p+1, q+1)$, $\mathrm{Spin}(p,q)$ etc.

12.2   Coverings of the complete conformal group.

12.2.1

Lemma 12.2.1 *There exists an injective mapping u from $E(p,q)$ to the isotropic cone C of $E(p+1, q+1)$, defined by :*

$$u(x) = \tfrac{1}{2}(e_{n+1} + e_0)x^2 + x + \tfrac{1}{2}(e_{n+1} - e_0), \quad x \in E(p,q). \tag{1}$$

It is easily verified that $u(x) = u(x')$ implies $x = x'$ and that $(u(x))^2 = 0$ in $\mathrm{Cl}(p+1, q+1)$.

The *isotropic transformation* (1) can also be written as :

$$u(x) = x_0 x^2 + x - y_0, \tag{2}$$

x_0, y_0 belonging to a special Witt basis of $E(p+1, q+1)$.

Note that the vector $u(x)$ can have any isotropic direction except that of x_0 (unless we would assign this direction as the limit for x^2 tending to infinity). Indeed, every isotropic vector of $E(p+1, q+1)$ is of the form

$$y = \lambda x + ae_0 + be_{n+1}, \quad \lambda^2 x^2 = b^2 - a^2$$

and has the same direction as $u(x)$ if we choose

$$\lambda = b - a, \quad x^2 = \frac{b+a}{b-a}, \quad b \neq a$$

since we need that $(x^2 - 1)/(2a) = (x^2 + 1)/(2b) = \lambda$.

Proposition 12.2.1 *If $g \in \mathrm{Pin}(p+1, q+1)$,*

$$\alpha(g)u(x)g^{-1} = \sigma_g(x)u(f(x)) \tag{3}$$

some special values of x possibly excepted; here $x \in E(p,q)$, f is a conformal transformation and $\sigma_g(x) \in \mathbf{R}$.

Proof. Every element of $\mathrm{Pin}(p+1, q+1)$ can be factored in terms of non-isotropic elements of $E(p+1, q+1)$. The proof consists of two steps :

- If (3) holds for $g_1, g_2 \in \mathrm{Pin}(p+1, q+1)$,

$$\alpha(g_i)u(x)g_i^{-1} = \sigma_{g_i}(x)u(f_i(x)), \qquad i = 1, 2,$$
$$\alpha(g_1 g_2)u(x)(g_1 g_2)^{-1} = \alpha(g_1)\alpha(g_2)u(x)g_2^{-1}g_1^{-1}$$
$$= \sigma_{g_1}(f_2(x))\sigma_{g_2}(x)u(f_1 \circ f_2(x))$$

and (3) holds for $g_1 g_2$ where $\sigma_{g_1 g_2}(x) = \sigma_{g_1}(f_2(x))\sigma_{g_2}(x)$ and $f_1 \circ f_2$ is associated with $g_1 g_2$.

Therefore it is sufficient to prove (3) for non-isotropic elements of $E(p+1, q+1)$.

- Take $g = v \in E(p+1, q+1)$ and set $N(v) = \epsilon = \pm 1$. By Chapter 4,

$$\alpha(v)u(x)v^{-1} = u(x) - 2B(v, u(x))\frac{v}{N(v)}.$$

Set $v = ae_0 + be_{n+1} + w^k e_k$, $e_k \in E(p, q)$, then $N(v) = w^2 + a^2 - b^2 = \epsilon$. Equating $\alpha(v)u(x)v^{-1}$ and $\sigma_v(x)u(f(x))$ leads to :

$$\sigma_v(x) = \epsilon((a - b)x + w)^2 \qquad (4)$$

and if $a \neq b$:

$$f(x) = \frac{\epsilon}{(a - b)^2(x + w/(a - b))} - \frac{w}{a - b} \qquad (5)$$

whereas if $a = b$:

$$f(x) = x - 2\epsilon B(w, x)w + 2a\epsilon w. \qquad (6)$$

(5) is a conformal transformation defined whenever

$$(x + w/(a - b))^2 \neq 0$$

and (6) is the composition of a translation and an isometry. ∎

Proposition 12.2.2 If $f = \psi(g)$ as in Proposition (12.2.1), ψ is a surjective homomorphism from $\mathrm{Pin}(p+1, q+1)$ to $C_n(p, q)$.

Proof. The factorization of $g \in \mathrm{Pin}(p+1, q+1)$ proves that

$$\psi(g_1 g_2) = \psi(g_1) \circ \psi(g_2)$$

for all $g_1, g_2 \in \mathrm{Pin}(p+1, q+1)$; the proof of Proposition (12.2.1) then establishes that every element of $C_n(p, q)$ can be reached by a product of a finite number of suitably chosen v. ∎

Proposition 12.2.3 *The kernel of* $\psi : \mathrm{Pin}(p+1, q+1) \to C_n(p,q)$ *is*

$$\{1, -1, e_N, -e_N\}$$

where $e_N = e_0 e_1 \ldots e_n e_{n+1}$.

Proof. If $g \in \ker \psi$, $\alpha(g)u(x)g^{-1} = \sigma_g(x)u(x)$ for all $x \in E(p,q)$. By the remark following Lemma (12.2.1), the action of g fixes all isotropic directions of $E(p+1, q+1)$, hence $p(g) = \pm \mathrm{Id}_E$ or

$$\alpha(g)yg^{-1} = \pm y, \quad \forall y \in E(p+1, q+1)$$

by the Lemma (12.2.2) following this proposition. The study of spin groups then leads to the four possible values for g.   ∎

We still have to prove the

Lemma 12.2.2 *If* (E, Q) *has isotropic elements and if* $\sigma \in O(Q)$ *fixes all isotropic lines,* $\sigma = \pm \mathrm{Id}_E$ *(*$\dim E = n \geq 3$*).*

Proof. E contains a hyperbolic pair (x, y) and $(x, y)^{\perp}$ is non-isotropic. Let $z \in (x, y)^{\perp}$. If $Q(z) = 0$, $\sigma(z) = cz$ by assumption; if $Q(z) \neq 0$, (x, y, z) is a Witt decomposition of the space E_3 generated by x, y, z and (z) is the only line in E_3 which is orthogonal to (x, y). It is sufficient to prove that σ maps E_3 to itself, then $\sigma(z) = cz$ follows since, by assumption, $\sigma(x) = ax$ and $\sigma(y) = by$. But

$$u = x - \tfrac{1}{2}Q(z)y + z$$

is isotropic,

$$\sigma(u) = du = dx - \tfrac{1}{2}Q(z)dy + dz = ax - \tfrac{1}{2}bQ(z)y + \sigma(z)$$

hence $\sigma(z) \in E_3$, $a = b = c = d$, every vector in (x, y) and in $(x, y)^{\perp}$ is multiplied by a, $\sigma(X) = aX$, $\forall X \in E$ and the lemma follows from Proposition (1.2.6).   ∎

12.2.2   Some remarkable results.

If $v = ae_0 + be_{n+1} + w$, $N(v) = \epsilon = \pm 1$, $w \in E(p,q)$, it follows from formula (6) that $\psi(v)$ is the symmetry relative to $(w)^{\perp}$ if $a = b = 0$ and (4) leads to $\sigma_v(x) = 1$, hence

1. The vectorial isometries, i.e. the elements of $O(p,q)$, are obtained from products g of non-isotropic vectors in $E(p,q)$, for which $N(g) = \pm 1$, $g \in \mathrm{Pin}(p,q)$ and the coefficient $\sigma_g(x)$ is 1.

2.
$$g = 1 + x_0 y = 1 + \frac{e_0 + e_{n+1}}{2} y, \quad y \in E(p,q)$$

   corresponds to the translation $x \to x + y$ in $E(p,q)$. $N(g) = 1$ can be verified by direct computation, and $\sigma_g(x) = 1$.

3.

$$g = 1 - y_0 y = 1 + \frac{e_{n+1} - e_0}{2} y, \quad y \in E(p, q), \quad N(g) = 1$$

corresponds to $x \to x(1 + yx)^{-1}$, which is called a special conformal transformation. In this case, $\sigma_g(x) = N(1 + yx)$ where $N(1 + yx)$ stands for the product $(1 + yx)\beta(1 + yx) = 1 + 2B(x, y) + x^2 y^2$. This transformation is defined unless $N(1 + yx) = 0$.

4.

$$g = \exp(\frac{t}{2} e_0 e_{n+1}), \quad t \in \mathbf{R}$$

corresponds to the homothety $x \to \exp(-t)x$ and $\sigma_g(x) = \exp t$.

5.

$$g = e_0$$

corresponds (by (4) and (5) for $\epsilon = 1$, $a = 1$, $b = 0$, $w = 0$) to the inversion $x \to 1/x$, defined whenever $x^2 \neq 0$; similarly $g = e_{n+1}$ corresponds to the inversion $x \to -1/x$, $x^2 \neq 0$. In the first case, $\sigma_g(x) = x^2$, in the second case it equals $-x^2$.

To establish 2, 3 and 4 one can rely on the concept of one-parameter subgroup (note that $\exp(tx_0 y) = 1 + tx_0 y$ and $\exp(-ty_0 y) = 1 - ty_0 y$) whose tangent is known for $t = 0$.

12.2.3 The connectedness of the conformal group.

Let $C_n^0(p, q)$ denote the connected component of the identity in the conformal group $C_n(p, q)$. If an element of $\mathrm{Pin}(p + 1, q + 1)$ is mapped by ψ into $C_n^0(p, q)$ the same holds for all other elements in its connected component. Conversely every element of $C_n^0(p, q)$ has a pre-image under ψ in one of the connected components of $\ker \psi$, ψ being a covering. Hence the set of connected components of $\ker \psi$ in $\mathrm{Pin}(p + 1, q + 1)$ is the pre-image of $C_n^0(p, q)$.

Let $G_0(p + 1, q + 1)$ (or G_0 for short) be the subgroup of elements in $\mathrm{Pin}(p + 1, q + 1)$ of spin norm 1 and let $G_0^+ = G_0 \cap \mathrm{Spin}(p + 1, q + 1)$. By Chapter 6, section 2, G_0^+ is a connected subgroup of index 2 in $\mathrm{Spin}(p + 1, q + 1)$. But $\mathrm{Pin}(p + 1, q + 1)$ has four connected components (as was proved in Chapter 6, 2), these being G_0^+, $\mathrm{Spin}(p + 1, q + 1) \setminus G_0^+$ and the components obtained from these two by the action of a non-isotropic vector of spin norm 1.

Note that by a classical result :

$$\frac{C_n(p, q)}{C_n^0(p, q)} \simeq \frac{\mathrm{Pin}(p + 1, q + 1)}{\psi^{-1}(C_n^0(p, q))}$$

and that $\pm 1 \in G_0^+$.

1. If $p + q$ is even

    (a) if pq is even (so both p and q are even) :

$$N(e_N) = -1, \quad \pm e_N \in \mathrm{Spin}(p+1, q+1) \setminus G_0^+.$$

$$\psi^{-1}(C_n^0(p, q)) = \mathrm{Spin}(p+1, q+1) \text{ and since}$$

$$\frac{\mathrm{Pin}(p+1, q+1)}{\mathrm{Spin}(p+1, q+1)} \simeq \{-1, 1\},$$

    $C_n(p, q)$ has two connected components.

    (b) If pq is odd (so both p and q are odd) :

$$N(e_N) = 1, \quad \pm e_N \in G_0^+, \quad \psi^{-1}(C_n^0(p, q)) = G_0^+,$$

$$\frac{\mathrm{Spin}(p+1, q+1)}{G_0^+} \simeq \{-1, 1\}$$

    and $C_n(p, q)$ has four connected components.

2. If $p + q$ is odd :

    (a) if p is even and q is odd :

$$N(e_N) = 1, \quad \pm e_N \in G_0 \setminus G_0^+$$

    and there are two connected components.

    (b) if p is odd and q is even :

$$N(e_N) = -1$$

    and $\pm e_N$ belongs to the complement of G_0 in $\mathrm{Pin}(p+1, q+1)$, so there are two connected components.

    Summarizing :

Proposition 12.2.4 *If pq is odd the group $C_n(p, q)$ has four connected components; otherwise it has two connected components.*

12.2.4 The Möbius group.

$C_n(p, q)$ is isomorphic to $\dfrac{\mathrm{Pin}(p+1, q+1)}{\ker \psi}$, so applying to both the projection p on the orthogonal group, we see that

$$C_n(p, q) \simeq \frac{O(p+1, q+1)}{\mathbf{Z}_2}.$$

$O(p+1, q+1)/\mathbf{Z}_2$ is the orthogonal projective group $\mathrm{PO}(p+1, q+1)$, which is also called the 'Möbius group'.

    The complete conformal group is therefore isomorphic to the Möbius group.

12.3 Twistors.

12.3.1

Definition 12.3.1 *Every irreducible representation space for a Clifford algebra being called a spinor space, we will call twistor space all direct sums of spinor spaces.*

If the spinor space is a minimal left ideal, the twistor space will be a left ideal of the Clifford algebra under consideration.

In the following, *we will only consider the direct sums of two spinor spaces* and the term 'twistors' will only be used in this sense.

Essentially, we will only use real even-dimensional vector spaces $E(p, q)$ where $p + q = n = 2r$, and the notations of Chapter 10, 2. The complexified Clifford algebra and, more generally, all complexified spaces will be denoted with primes.

Let f and f_1 be two isotropic r-vectors such that the sum

$$\mathrm{Cl}'(p, q)f \oplus \mathrm{Cl}'(p, q)f_1$$

is direct and defines a twistor space.

Proposition 12.3.1 *There exists a representation of $\mathrm{Cl}'(p + 1, q + 1)$ in*

$$\mathrm{Cl}'(p, q)f \oplus \mathrm{Cl}'(p, q)f_1$$

for which $\mathrm{Cl}'(p, q)f$ and $\mathrm{Cl}'(p, q)f_1$ are isomorphic to the spaces of even, resp. odd spinors of $\mathrm{Cl}'(p + 1, q + 1)$.

Proof. Recall that

$$\mathrm{Cl}(p + 1, q + 1) = \mathrm{Cl}(p, q) \otimes \mathrm{Cl}(1, 1).$$

Let $\{x_i, y_j\}$ be a special Witt basis of $E'(p + 1, q + 1)$ and $\{x, y\}$ a special Witt basis of $E'(1, 1)$, then

$$\{x_1, x_2, \ldots, x_r, x, y_1, y_2, \ldots, y_r, y\}$$

is a special Witt basis of $E'(p + 1, q + 1)$ and a $\gamma \in \mathrm{Pin}(Q')$ can be found such that $x_{i'} = \gamma x_i \gamma^{-1}$, $y_{i'} = \gamma y_i \gamma^{-1}$ if $\{x_{i'}, y_{j'}\}$ is another special Witt basis. Set $\Phi = f \otimes y = fy$, then $\mathrm{Cl}'(p + 1, q + 1)\Phi$ is a spinor space for $\mathrm{Cl}'(p + 1, q + 1)$. Let $f = y_1 y_2 \ldots y_r$, $f_1 = \gamma f \gamma^{-1} = y'_1 y'_2 \ldots y'_r$ with f and f_1 not collinear, and assume for instance that γ is odd.

A linear isomorphism i from $\mathrm{Cl}'(p, q)f \otimes \mathrm{Cl}'(p, q)f_1$ to $\mathrm{Cl}'(p + 1, q + 1)$ is defined by associating

$$x_{i_1} \ldots x_{i_h} f \quad \rightarrow \quad \begin{cases} x_{i_1} \ldots x_{i_h} \Phi & \text{if } h \text{ is even,} \\ x_{i_1} \ldots x_{i_h} x \Phi & \text{if } h \text{ is odd,} \end{cases}$$

$$x'_{i_1} \ldots x'_{i_h} f_1 \quad \rightarrow \quad \begin{cases} x'_{i_1} \ldots x'_{i_h} \gamma \Phi & \text{if } h \text{ is even,} \\ x'_{i_1} \ldots x'_{i_h} \gamma x \Phi & \text{if } h \text{ is odd.} \end{cases}$$

The representation ρ of $\mathrm{Cl}'(p+1,q+1)$ in the sum of the spaces is then $\rho_w = i^{-1} \circ w \circ i$ (w being the left product by w in $\mathrm{Cl}'(p+1,q+1)\Phi$). The parity conditions can be verified if w is even.   ∎

Hence a twistor space for $\mathrm{Cl}(p,q)$ (or $\mathrm{Cl}'(p,q)$) is isomorphic to a spinor space for $\mathrm{Cl}(p+1,q+1)$ (or $\mathrm{Cl}'(p+1,q+1)$).

The four-fold covering of the conformal group $C_n(p,q)$ by $\mathrm{Pin}(p+1,q+1)$ is called the 'twistor' covering because of this property of $\mathrm{Cl}(p+1,q+1)$.

Remark. When $p+q$ is even, the four-fold covering of $C_n^0(p,q)$ is either $\mathrm{Spin}(p+1,q+1)$ or $G_0^+(p+1,q+1)$ by section 2.3. $\mathrm{Spin}(p+1,q+1)$ acts on the space $\mathrm{Cl}'^+(p+1,q+1)\Phi$ of semi-spinors isomorphic to $\mathrm{Cl}'(p,q)f$, which becomes the 'twistor representation space' for $\mathrm{Cl}'(p,q)$. This leads some authors to (wrongly) consider $\mathrm{Cl}'(p,q)f$ as the twistor space for $\mathrm{Cl}'(p,q)$; the correct choice for this space is, by Proposition (12.3.1), $\mathrm{Cl}'(p,q)f \oplus \mathrm{Cl}'(p,q)f_1$.

12.3.2 Neutral spin groups and unitary groups.

In Chapter 10 a hermitian form $\mathcal{H}$ was defined on the spinor space $C(Q')f$, satisfying

$$\beta(\overline{uf})vf = a\mathcal{H}(uf,vf)\gamma f,$$

a being 1 or i and $\bar{f} = \gamma f \gamma^{-1}$ defining the conjugation pure spinor γf.

If the quadratic form Q is not positive definite, $\mathcal{H}$ is neutral; this will therefore always be the case for the spinor space of $\mathrm{Cl}'(p+1,q+1)$.

Proposition 12.3.2 *If $p+q = 2r$, $r \geq 2$, the connected component of $\mathrm{Spin}(p+1,q+1)$ is a subgroup of $\mathrm{SU}(r,r)$.*

Proof. If g belongs to the connected component in $\mathrm{Spin}(p+1,q+1)$, $\beta(g)g = 1$ so that the natural action of this component on $\mathrm{Cl}'(p+1,q+1)\Phi$ by left multiplication is an isometry for $\mathcal{H}$.

The connected component of $\mathrm{Spin}(p+1,q+1)$ is generated by that of $\mathrm{Spin}(p,q)$ and by

$$\exp(\frac{e_0 + e_{n+1}}{2}z) = 1 + \frac{e_0 + e_{n+1}}{2}z, \qquad z \in E(p,q)$$
$$\exp(\frac{e_0 - e_{n+1}}{2}z) = 1 + \frac{e_0 - e_{n+1}}{2}z, \qquad z \in E(p,q)$$
$$\exp(\frac{e_0 e_{n+1}}{2}\eta), \qquad\qquad\qquad \eta \in \mathbf{R}^*.$$

$e_0 e_{n+1}$ defines an isomorphism of $\mathrm{Cl}'(p+1,q+1)$ sending a positive vector for $\mathcal{H}$ to a negative one, since

$$\beta(e_0 e_{n+1})e_0 e_{n+1} = -1.$$

$e_0 e_{n+1}$ commutes with the generators of the connected component of $\mathrm{Spin}(p+1,q+1)$ listed above.

In a reference frame where the matrix of g is diagonal, to every eigenvector which is positive for $\mathcal{H}$ there uniquely corresponds a negative one, by the action of $e_0 e_{n+1}$:

$$gu\Phi = \lambda u\Phi, \quad \lambda \in \mathbf{C}^*$$

$$g(e_0 e_{n+1} u\Phi) = e_0 e_{n+1} gu\Phi = \lambda(e_0 e_{n+1} u\Phi).$$

Note that if $\theta_g : u\Phi \to gu\Phi$, $g \in G$,

$$\det \theta_g = (N(g))^{2^{n+1}},$$

by a lemma of independent interest :

Lemma 12.3.1 *If x is a non-isotropic element of the n-dimensional vector space $E(p,q)$ and if $\theta_x : y \to xy$, then $\det(\theta_x) = (x)^{2^n}$.*

Proof. Let x be the element (e_i) of an orthogonal basis : $(e_i)^2 = \alpha_i$ and order the elements of a basis $e_{i_1} e_{i_2} \ldots e_{i_k}$ of $C(Q)$ lexicographically :

$$e_1, e_1 e_2, e_1 e_2 e_3, \ldots, e_1 e_2 \ldots e_n, e_1 e_3, e_1 e_3 e_4, \ldots, e_1 e_n$$

followed by the $e_{i_1 i_2 \ldots i_k}$ with $i_1 > 1$.

θ_{e_1} is bijective and sends the 2^{n-1} elements starting with e_1 to the 2^{n-1} other elements of the basis (up to the coefficient α_1). ϵ_{e_1} being the sign of the permutation

$$(e_{i_1} e_{i_2} \ldots e_{i_k}) \to e_1(e_{i_1} e_{i_2} \ldots e_{i_k}), \quad (\bmod \, \alpha_1)$$

it is readily seen that

$$\det \theta_{e_1} = (\alpha_1)^{2^{n-1}} \epsilon_{e_1} = (\alpha_1)^{2^{n-1}},$$

there being 2^{2n-2} inversions in the permutation. Hence

$$\det \theta_x = (x)^{2^n}.$$

∎

The result concerning the determinant of θ_g follows immediately from the factorization of $g \in \mathrm{Spin}(p+1, q+1)$ as a product of non-isotropic vectors.

In the special case under consideration, $N(g) = 1$ and $\det \theta_g = 1$ so that g defines a special unitary transformation on a space with neutral metric.

The dimension of $\mathrm{Spin}(p+1, q+1)$ is $(r+1)(2r+1)$, hence if $r \geq 2$,

$$(r+1)(2r+1) \leq (2r)^2 - 1$$

and the connected component of $\mathrm{Spin}(p+1, q+1)$ can be considered as a subgroup of $\mathrm{SU}(r, r)$. ∎

Corollary 12.3.1 *The connected component of the identity in $\mathrm{Spin}(2,4)$ is isomorphic to $\mathrm{SU}(2,2)$.*

Proof. $r = 2$, $p = 1$ and $(r+1)(2r+1) = (2r)^2 - 1 = 15$, and $\mathrm{SU}(2,2)$ is connected ∎

This corollary explains the importance of the group $\mathrm{SU}(2,2)$ in 'twistor' minkowskian geometry.

12.3.3 Linear and affine description of the properties of the spinors in a Minkowski space E.

Here $r = 2$, $p = 1$ and $q = 3$.

Using the notations of Chapter 10, $h = r - p = 1$ and $\epsilon' = -1$ so that the conjugation C defined by $C(uf) = \bar{u}\gamma f$ satisfies $C^2 = -\,\mathrm{Id}$ and the quaternionic case applies. We may choose $\gamma f = x_2 f = ie_2 f$.

The space S^- of odd spinors is $(x_1 f, x_2 f) = (\rho, \sigma)$ and, applying C, a basis $(f, x_1 x_2 f)$ for the space S^+ of even spinors is obtained.

We will write $(uf)^* = C(uf)$.

Using the notations of Chapter 8, 3.1, we set

$$\psi(uf \otimes vf) = uf\beta(v),$$

ψ being the classical isomorphism from $S \times S$ to $C(Q')$. If we set[1] $\chi = (1 + \beta) \circ \psi$, it is easy to verify :

$$\begin{cases} \chi(\rho \otimes \rho^*) = x_1, \\ \chi(\sigma \otimes \rho^*) = x_2, \\ \chi(\sigma \otimes \sigma^*) = y_1, \\ \chi(\rho \otimes \sigma^*) = -y_2, \end{cases} \tag{1}$$

and χ is a linear isomorphism from $S^- \otimes S^+$ to E', equivariant under the action of G_0^+.

Consider $uf \otimes vf$, or explicitly, $(a\rho + b\sigma) \otimes (-a'\sigma^* + b'\rho^*)$, then

$$\chi(uf \otimes vf) = ab'x_1 + bb'x_2 - a'by_1 + aa'y_2, \tag{2}$$

and $\chi(uf \otimes vf)$ is readily seen to be isotropic when $uf \in S^-$ and $vf \in S^+$.

This isotropic vector is real if and only if

$$aa' = -\overline{bb'}, \quad ab' = \overline{ab'}, \quad a'b = \overline{a'b},$$

which holds in particular if $b = -\bar{a}'$ and $a = \bar{b}'$ so that :

$$\chi(uf \otimes (uf)^*), \quad uf \in S^- \quad \text{is a real isotropic vector.} \tag{3}$$

Similar facts apply to

$$\chi(uf \otimes (vf)^*) + \chi(vf \otimes (uf)^*), \quad uf, vf \in S^-$$

being real; it is the image under χ of

$$(u + v)f \otimes ((u + v)f)^* - uf \otimes (uf)^* - vf \otimes (vf)^*.$$

---

[1]cf. the works of K. Bugajska.

Remark. Anticipating the results of Chapter 13, χ can be related to the generalized triality principle.

$f \circ x_1 f$ is obtained from $\mathcal{B}(xf, x_1 f) = \Lambda(f \circ x_1 f, x) = B(f \circ x_1 f, x)$ for all $x \in E' = E_{\mathbf{C}}$; taking x equal to x_1, y_1, y_2 successively, $f \circ x_1 f$ is seen to be collinear with y_2.

For $x = x_2$ we get

$$-f x_2 x_1 f = B(f \circ x_1 f, x_2) f = -f.$$

Finally[2],

$$\begin{cases} f \circ x_1 f = -2y_2 & \text{or} \quad \sigma^* \circ \rho = 2y_2, \\ f \circ x_2 f = 2y_1 & \text{or} \quad \sigma^* \circ \sigma = -2y_1, \\ x_1 x_2 f \circ x_1 f = -2x_1 & \text{or} \quad \rho^* \circ \rho = -2x_1, \\ x_1 x_2 f \circ x_2 f = -2x_2 & \text{or} \quad \rho^* \circ \sigma = -2x_2. \end{cases} \tag{4}$$

Hence the isomorphism χ is, up to a constant factor, the generalized triality principle in Minkowski space.

In what follows, many properties could be proved for the general case of a complexified four-dimensional real space with arbitrary metric. We will, however, restrict ourselves to the spaces of initial signature $(1, 3)$, i.e., the quaternionic case.

Proposition 12.3.3 *Every minkowskian semi-spinor is a pure spinor, and conversely.*

The converse was proved in Chapter 8.

Proof. If F is the plane (x_1, x_2), F' the plane (y_1, y_2), $E' = F \oplus F'$, by Proposition (8.1.4), uf is a pure spinor if and only if u is of the form

$$\lambda \exp(v) z_1 z_2 \ldots z_k, \quad \lambda \in \mathbf{C}^*, \quad v \in \wedge^2(F),$$

$z_1, z_2, \ldots, z_k$ being linearly independent vectors of F. This gives the following possibilities for u :

$$\lambda(1 + a x_1 x_2) f, \lambda(1 + a x_1 x_2)(b x_1 + c x_2)(b' x_1 + c' x_2) f, \lambda(1 + a x_1 x_2)(b x_1 + c x_2) f.$$

Clearly $(\alpha x_1 + \beta x_2) f$ and $(\alpha' + \beta' x_1 x_2) f$ belong to these. ∎

Proposition 12.3.4 *To each minkowskian semi-spinor of $\mathrm{Cl}'(1, 3) = C'$ a totally isotropic plane can be associated.*

Proof. The spinor space $C'f$ chosen, there exists a bijection between the set of m.t.i.s. and the set of pure spinor fields, up to a non-zero complex coefficient (cf. Chapter 8, 1.1); the m.t.i.s. are two-dimensional in this case. ∎

---

[2]Here 'o' should not be mistaken for the composition of functions.

Consequence. To every spinor in $C'f$ written as the direct sum

$$uf = u^+f + u^-f$$

of two semi-spinors, a pair of totally isotropic planes of different parity can be associated. These planes intersect along an isotropic line D (cf. Proposition (8.1.2)). Hence : to every non-zero spinor without parity, an isotropic line of the vector space E can be associated. (This is called 'construction (A)'.)

Proposition 12.3.5 *The isotropic line associated to* $(u^+f, (u^+f)^*)$ *is real.*

Proof. By Lemma (8.1.3), the points x of the plane associated to u^+f are characterized by $x(u^+f) = 0$. u^+f and $(u^+f)^*$ correspond to planes intersecting along the line Δ, and a point $x \in \Delta$ is characterized by

$$xu^+f = 0, \quad x(u^+f)^* = 0$$

or $x\bar{u}^+\gamma f = 0$. A point $y \in \bar{\Delta}$ is characterized by

$$y\overline{u^+f} = 0, \quad y\overline{u^+\gamma f} = 0$$

or $y\bar{u}^+\gamma f\gamma^{-1} = 0$ and $yu^+\bar{\gamma}\gamma f\gamma^{-1} = 0$, but $\bar{\gamma}\gamma f = \epsilon'f$ leads to

$$y\bar{u}^+\gamma f = 0, \quad yu^+f = 0$$

hence $\Delta = \bar{\Delta}$. ∎

Proposition 12.3.6 *The line D associated to* $uf = u^+f + u^-f$ *by construction (A) can be identified with the line determined by* $\chi(u^+f \otimes u^-f)$.

Proof. If $u^-f = (ax_1 + bx_2)f$, $u^+f = (a' + b'x_1x_2)f$,

$$x = \chi(u^+f \otimes u^-f) = ab'x_1 + bb'x_2 - a'by_1 + aa'y_2;$$

it is immediately verified that

$$x(u^+f) = 0, \quad x(u^-f) = 0.$$

So, given $uf = u^+f + u^-f$ up to a scalar factor (such an object will be called a projective spinor), two real isotropic lines can be associated to it, determined by

$$\chi((u^+f) \otimes (u^+f)^*) \quad \text{and} \quad \chi((u^-f) \otimes (u^-f)^*).$$

Using the above formula for $\chi(uf \otimes vf)$, the necessary and sufficient condition for these lines to coincide is

$$\mathcal{H}(u^-f, u^+f) = -\bar{a}a' - \bar{b}b' = 0$$

(which implies that $\mathcal{H}(uf, uf) = 0$).

Now it is possible to associate to every spinor a pair of affine lines, one of them containing $\chi(u^-f \otimes (u^-f)^*)$ and parallel to $\chi(u^+f \otimes (u^+f)^*)$, the other one defined by the same conditions with both vectors exchanged. We have already noted the possibility to associate an isotropic line through the origin in the affine space E' to every pair of semi-spinors, i.e. a homogeneous isotropic line to every spinor.

Proposition 12.3.7 *Construction (A) establishes a bijection between the set of pairs of projective pure spinors of different parity and the set of homogeneous isotropic lines in E'.*

Proof. This follows at once from Corollary (8.1.4) (each $(r-1)$-dimensional totally isotropic subspace of E' is contained in exactly one even m.t.i.s. and exactly one odd m.t.i.s.)

A direct proof could be given here, using Proposition (12.3.6) and formula (2); it is sufficient to compute a, b, a', b' associated with a vector

$$\alpha x_1 + \beta x_2 - \gamma y_1 + \delta y_2$$

where $\alpha\gamma = \beta\delta$. ∎

Every vector in E' being the sum of two isotropic vectors, every vector in E' can be obtained as the image under χ of a sum $uf \otimes vf + u'f \otimes v'f$ involving only two terms.

Construction (B). To generalize construction (A), we determine the set of $x \in E'$ such that :

$$x(u^+f) = \lambda(u^-f), \quad x(u^-f) = \lambda'(u^+f) \tag{5}$$

where λ, λ' are non-zero fixed scalars. Note that

$$x^2(u^+f) = \lambda x(u^-f) = \lambda\lambda'(u^+f)$$

so that $x^2 = \lambda\lambda' = k$, a constant, and (5) is equivalent to

$$\begin{cases} x(u^+f) &= \lambda(u^-f) \\ x^2 &= k \end{cases}$$

If we set $x = x_0 + x'$, x_0 being a particular solution of $x(u^+f) = \lambda(u^-f)$ (if such x_0 exists), we see that $x'(u^+f) = 0$. x' is a general element of a totally isotropic plane. The first condition in (5) means that x lies in a totally isotropic affine plane and the second condition has a similar meaning. Hence, once an x_0 has been found, satisfying (5), the solution set of (5) is an affine isotropic line Δ. But from the equivalent formulation, it follows that Δ is the intersection of a m.t.i.s. and a quadric $x^2 = k$; such an intersection must contain the line at infinity of the m.t.i.s. and another line

Δ (which coincides with the intersection of the two affine planes of different parity obtained in (5)).

An explicit solution to (5) is easily obtained if we set

$$x = \alpha x_1 + \beta x_2 + \gamma y_1 + \delta y_2, \quad u^+ f = (a' + b' x_1 x_2) f, \quad u^- f = (a x_1 + b x_2) f,$$

the first equation in (5) leads to :

$$x = (\frac{b' x_1}{a'} + y_2)\delta + (y_1 - \frac{b' x_2}{a'})\gamma + \frac{\lambda(a x_1 + b x_2)}{a'}, \quad a' \neq 0$$

and the second one to :

$$x = (\frac{b x_2}{a} + x_1)\alpha + (y_2 - \frac{b}{a} y_1)\delta + \frac{\lambda'(a' y_1 - b' x_2)}{a}, \quad a \neq 0,$$

from which the points X of Δ are determined :

$$X = (ab' x_1 + bb' x_2 - a' by_1 + aa' y_2)\frac{\delta}{aa'} + \frac{\lambda a^2 x_1 + \lambda abx_2 + \lambda' a'^2 y_1 - \lambda' a' b' y_2}{aa'}. \quad (6)$$

If two pairs $(u^+ f, u^- f)$ and $(u_1^+ f, u_1^- f)$ are sent by construction (B) to the same line $\Delta = \Delta_1$, from the identification of their directions we see that :

$$\frac{a}{a_1} = \frac{b}{b_1}, \quad \frac{a'}{a_1'} = \frac{b'}{b_1'},$$

and expressing that the point of Δ obtained when $\delta = 0$ lies on Δ_1 :

$$\lambda(\frac{a'}{a} - \frac{a_1'}{a_1}) = 0, \quad \lambda(\frac{b}{a'} - \frac{b_1}{a_1'}) = \lambda'(\frac{b'}{a} - \frac{b_1'}{a_1}), \quad \lambda'(\frac{a}{a'} - \frac{a_1}{a_1'}) = 0.$$

If both λ and λ' differ from 0, $\Delta = \Delta_1$ is equivalent to

$$(u_1^+ f, u_1^- f) = (tu^+ f, tu^- f), \quad t \in \mathbf{C}^*.$$

If $\lambda = \lambda' = 0$, $\Delta = \Delta_1$ is equivalent to

$$(u_1^+ f, u_1^- f) = (tu^+ f, t'u^- f) \quad t, t' \in \mathbf{C}^*.$$

Conversely, let Δ be an isotropic line belonging to the quadric $X^2 = k$, can a spinor $uf = u^+ f + u^- f$ be found such that construction (B) associates Δ to it ? The direction of Δ determines a homogeneous line Δ_0 and therefore the pair $(u^+ f, u^- f)$, up to two factors t and t'; the line Δ intersects $(y_2)^\perp$ (the exceptional case will be treated separately) in a point $Ax_1 + Bx_2 + Cy_1$ such that $AC = k$, from which λ and λ' are deduced (fixing a, b and b'/a', $A = \lambda a/a'$, $B = \lambda b/a' - \lambda' b'/a$, $C = \lambda' a'/a$ yielding λ, λ' and b'/a, with a single arbitrary coefficient; if $\lambda' \neq 0$, $\lambda \neq 0$, b'/a is fixed). This proves :

Proposition 12.3.8 *There exists a bijection between the set of affine isotropic lines intersecting the quadric $X^2 = k$, $k \neq 0$, and the set of projective spinors.*

Conditions for Δ to be real ($\lambda\lambda' \neq 0$). Expressing that there exists a complex scalar u such that the product of u and the coefficient of δ in X equals its complex conjugate (since Δ contains a real vector), $u = a\bar{b}$ will do and we get $a\bar{a}' + b\bar{b}' = 0$, meaning that $\mathcal{H}(u^- f, u^+ f) = 0$.

Next one expresses that the intersection of Δ and y_1^+ is a real point, yielding :

$$\lambda' f' \bar{a}' = \bar{\lambda} ab$$

which is equivalent to λ' being of the form $(-\bar{\lambda}\rho\bar{\rho})$, since $\mathcal{H}(u^- f, u^+ f) = 0$, and, in the general case, $\lambda\lambda'$ real and negative,

$$\rho\bar{\rho} = -\frac{\lambda\lambda'}{|\lambda|^2}.$$

Note that $\mathcal{H}(uf, uf) = 0$.

Conditions for real Δ and Δ_1 to intersect. There exists at least one point $x \in E'$ such that

$$x(u^+ f) = \lambda(u^- f), \qquad x(u^- f) = \lambda'(u^+ f),$$
$$x(u_1^+ f) = \lambda(u_1^- f), \qquad x(u_1^- f) = \lambda'(u_1^+ f).$$

These four conditions are equivalent to

$$x(u^+ f) = \lambda(u^- f), \quad x(\bar{u}_1 f) = \lambda'(u_1^+ f), \quad x^2 = \lambda\lambda' \tag{7}$$

or to

$$x(u^- f) = \lambda'(u^+ f), \quad x(u_1^+ f) = \lambda(u_1^- f), \quad x^2 = \lambda\lambda'. \tag{8}$$

Considering the first condition, we see that, since the isotropic planes associated to $(u^+ f)$ and $u_1^- f$ must intersect, there exists a line determined by

$$x(u^+ f) = \lambda(u^- f) \quad \text{and} \quad x(u_1^- f) = \lambda'(u_1^+ f);$$

to express the existence of $x \in \Delta \cap \Delta_1$ we can write $x^2 = \lambda\lambda'$, Δ and Δ_1 are real and if they have the same direction,

$$\mathcal{H}(u^- f, u^+ f) = \mathcal{H}(u_1^- f, u_1^+ f) = \mathcal{H}(u^- f, u_1^+ f) = \mathcal{H}(u^+ f, u_1^- f) = 0 \tag{9}$$

Conversely, the explicit form of these conditions implies

$$a\bar{a}' + b\bar{b} = a\bar{a}_1' + b\bar{b}_1' = a_1\bar{a}' + b_1\bar{b}' = a_1\bar{a}_1' + b_1\bar{b}_1' = 0,$$

$a/a_1 = b/b_1$ and $a'/a_1' = b'/b_1'$, meaning that the homogeneous lines of directions Δ and Δ_1 coincide. Hence *(9) expresses that Δ and Δ_1 have the same direction.*

We shall henceforth assume that Δ and Δ_1 have different directions. If there exists an $x \in \Delta \cap \Delta_1$, computing $\mathcal{H}(uf, u_1 f)$ and applying (7), (8), immediately leads to

$$\mathcal{H}(u^+ f, u_1^- f) = \bar{\lambda} \lambda^{-1} \mathcal{H}(u^- f, u_1^+ f), \tag{10}$$

$$\mathcal{H}(u^- f, u_1^+ f) = \bar{\lambda}' \lambda'^{-1} \mathcal{H}(u^+ f, u_1^- f), \tag{11}$$

$\lambda \lambda'$ being non-zero and real.

Conversely, both lines having different directions, the left-hand sides of (10) and (11) cannot both vanish; if $\mathcal{H}(u^+ f, u_1^- f) \neq 0$, if $x(u^+ f) = \lambda(u^- f)$ and if $x(u_1^- f) = \lambda'(u_1^+ f)$, ($x$ being real, take the intersection of Δ and the plane $x(u_1^- f) = \lambda'(u_1^+ f)$, this point generally exists), then (10) implies that $x^2 = \lambda \lambda'$, so $x \in \Delta \cap \Delta_1$. Hence *(10) and (11) express that the real lines Δ and Δ_1 intersect either in a finite point or 'at infinity'.*

For special λ, λ', a suggestive condition is obtained : if λ is pure imaginary (and therefore λ' too),

$$\mathcal{H}(uf, u_1 f) = \mathcal{H}(u^- f, u_1^+ f) + \mathcal{H}(u^+ f, u_1^- f) = 0,$$

and Δ *and* Δ_1 *intersect in a finite point or at infinity if and only if their representing spinors are orthogonal under* $\mathcal{H}$.

Finally, note that if $\lambda \lambda' = 0$ and only one of the factors vanishes, the same conclusions hold, although the proofs will not.

In fact, one can use the continuity for $\lambda' = \epsilon$, $\epsilon \to 0$, condition (10) remains.

12.4   Selected references.

- A. Crumeyrolle, *Twisteurs sans twisteurs*, Geometro-dynamic Proc. 1983, Tecnoprint, Bologna 1984.

- P. Lounesto, E. Latvamaa, *Conformal transformations and Clifford algebras*, Report HTKK-MAT, A-123 Helsinki, 1978.

- R. Penrose, R. Ward, *Twistors for flat and curved space-time*, in General Relativity and Gravitation, Inst. Phys. th. Bern, t. 2, 1979.

Chapter 13

THE TRIALITY PRINCIPLE, THE INTERACTION PRINCIPLE AND ORTHOSYMPLECTIC GRADED LIE ALGEBRAS.

Notations and basic notions. (E, Q) is a vector space of even dimension $n = 2r$ with a non-degenerate quadratic form Q and B is the associated bilinear form.

$C(Q)$ is the Clifford algebra over (E, Q).

G, G_0, G_0^+, $\text{Pin}(Q)$ and $\text{Spin}(Q)$ are the classical Clifford and spin groups.

If (x_i, y_i), $i = 1, 2, \ldots, r$ is a Witt basis of the complexified space (E', Q') of (E, Q) and if we set $y_1 y_2 \ldots y_r = f$, the spinor space is $S = C(Q')f$, where $C(Q')$ is the complexified Clifford algebra of $C(Q)$, i.e. the Clifford algebra of (E', Q'); a spinor has the form uf, $u \in C(Q')$ and $S = S^+ \oplus S^-$ where $S^+ = C^+(Q')f$ and $S^- = C^-(Q')f$.

β is the main antiautomorphism of $C(Q)$ and α its main automorphism, $\tilde{\beta} = \beta \circ \alpha$.

13.1   E. Cartan's triality principle.

13.1.1

E is an eight-dimensional vector space over a field $\mathbf{K}$ of characteristic zero (usually, $\mathbf{K} = \mathbf{R}$ or $\mathbf{C}$) with a non-degenerate quadratic form of maximal index 4. The Witt bases and the spinor spaces are those defined in the previous chapters.

A G_0-invariant bilinear form $\mathcal{B}$ is defined by

$$\beta(uf)vf = \mathcal{B}(uf, vf)f, \tag{1}$$

$\mathcal{B}$ is non-degenerate and, since $2r = 8$, $r = 4$, $\mathcal{B}$ is symmetric, zero on $S^+ \times S^-$ (cf. Chapter 8, 2) and we recall that :

$$\mathcal{B}(guf, gvf) = N(g)\mathcal{B}(uf, vf), \quad \forall g \in G, \tag{2}$$

$$\mathcal{B}(xuf, xvf) = Q(x)\mathcal{B}(uf, vf), \quad \forall x \in E, \tag{3}$$

165

$$B(xuf, vf) = B(uf, xvf), \quad \forall x \in E. \tag{4}$$

A quadratic form γ exists on S such that :

$$\gamma(uf) = B(uf, uf).$$

$A = E \times S$ is given a symmetric bilinear form Λ such that :

$$\Lambda(x + uf, x' + u'f) = B(x, x') + B(uf, u'f), \quad x, x' \in E, \quad uf, u'f \in S.$$

Λ is non-degenerate and the subspaces E, S^+ and S^- are non-isotropic for Λ; the orthogonal complement of any of these spaces is clearly the sum of the two others.

A cubic form F_0 on A is defined by :

$$F_0(x + uf + u'f) = B(xuf, u'f) = B(uf, xu'f), \quad x \in E, \quad uf \in S^+, \quad u'f \in S^-.$$

Through 'polarization' a symmetric trilinear form Φ_0 on $A \times A \times A$ is deduced :

$$\Phi_0(\xi, \eta, \zeta) = F_0(\xi + \eta + \zeta) + F_0(\xi) + F_0(\eta) + F_0(\zeta) - F_0(\xi + \eta) - F_0(\xi + \zeta) - F_0(\eta + \zeta),$$

$$\xi, \eta, \zeta \in A.$$

Proposition 13.1.1 *There exists a commutative, non-associative algebra structure on $A = E \times S$ such that :*

$$\xi \circ \eta = \eta \circ \xi = \omega \in A, \quad \forall \xi, \eta \in A, \tag{5}$$

where $\Lambda(\omega, \zeta) = \Phi_0(\xi, \eta, \zeta), \forall \zeta \in A$.

Proof. The commutativity follows from the symmetry of Φ_0 and the existence from the non-degeneracy of Λ. ∎

Proposition 13.1.2 $\xi \circ \eta = 0$ *if ξ and η each belong to exactly one of the subspaces E, S^+ and S^-.*

Proof. Note that $F_0(\omega) = 0$ if ω belongs to any of the subspaces $E + S^+$, $E + S^-$ and $S^+ + S^-$, hence

$$\Phi_0(\xi, \eta, \zeta) = F_0(\xi + \eta + \zeta)$$

if $\xi \in E, \eta \in S^+$ and $\zeta \in S^-$. Under the hypotheses of the proposition, $\Phi_0(\xi, \eta, \zeta) = 0$ for all ζ (decompose ζ as a sum of three terms with zero contribution). ∎

Proposition 13.1.3 *The inclusions*

$$E \circ S^+ \subseteq S^-, \quad S^+ \circ S^- \subseteq E, \quad S^- \circ E \subseteq S^+ \tag{6}$$

hold.

Proof. If $\xi \in E$ and $\eta \in S^+$, $\Phi_0(\xi, \eta, \zeta)$ is zero when $\zeta \in E + S^+$, so $\xi \circ \eta$ belongs to the orthogonal space of $E + S^+$, relative to Λ, i.e. S^-. The other inclusions have similar proofs. ∎

Proposition 13.1.4 *For all* $x \in E$ *and* $uf \in S$:

$$x \circ uf = xuf, \quad \gamma(x \circ uf) = Q(x)\gamma(uf), \quad x \circ (x \circ uf) = Q(x)uf. \qquad (7)$$

Proof. The first formula need only be proved for $uf \in S^+$ (or $uf \in S^-$). It is an immediate consequence of the definition and of the properties of $\mathcal{B}$. For example, if $x \circ uf \in S^-$, $u'f \in S^-$,

$$\mathcal{B}(x \circ uf, u'f) = \Lambda(x \circ uf, u'f) = \Phi_0(x, uf, u'f) = F_0(x + uf + u'f) = \mathcal{B}(xuf, u'f)$$

and $x \circ uf = xuf$ since $\mathcal{B}$ is non-degenerate on $S^- \times S^-$. The other formulas are obvious. ∎

There exists a natural representation μ of the Clifford group G in A, if $\omega = x + uf + u'f$ where $x \in E$, $uf \in S^+$ and $u'f \in S^-$, μ is defined by :

$$\mu(g)(\omega) = gxg^{-1} + g(uf + u'f),$$

and we have :

Proposition 13.1.5

$$\Phi_0(\mu(g)\xi, \mu(g)\eta, \mu(g)\zeta) = N(g)\Phi_0(\xi, \eta, \zeta), \quad \xi, \eta, \zeta \in A, \qquad (8)$$

$$guf \circ gvf = N(g)g(uf \circ vf)g^{-1}, \quad g \in G, \quad uf, vf \in S, \qquad (9)$$

$\mu(g)$ is an automorphism of A if $g \in G_0$.

Proof. These results all follow from routine computations. ∎

Remark. In general, every automorphism of the vector space A for which Λ and F_0 are invariant gives rise to an algebra automorphism of A. This is the case for the action of elements in the group G_0 and (9) expresses this result.

Note that the spaces E, S^+ and S^- are all of the same dimension 8 and that this follows from the choice of $r = 4$ (we need $2^{r-1} = 2r$). An important result holds in this case :

The triality principle. *There exists an automorphism* $\mathcal{J}$ *of the vector space* $A = E \times S = E \oplus S^+ \oplus S^-$ *such that* $\mathcal{J}^3 = \mathrm{Id}$. *Both the bilinear form* Λ *and the cubic form* F_0 *are invariant under* $\mathcal{J}$. *Furthermore,* $\mathcal{J}$ *maps* E *to* S^+, S^+ *to* S^- *and* S^- *to* E.

A complete proof can be found in Chevalley's book; we will only indicate its main points.

Choose $x_1 \in E$ such that $Q(x_1) = 1$, $x_1 \in G_0$.

Choose $u_1 f \in S^+$ such that $\gamma(u_1 f) = 1$.

If $x \in E$ we set $\tau(x) = u_1 f \circ x = x u_1 f \in S^-$. τ is an isomorphism from E to S^-. If $u' f \in S^-$ we set $\tau(u' f) = x \in E$. τ is thereby defined on $E \oplus S^-$.

If $uf \in S^+$, we set

$$\tau(uf) = \mathcal{B}(uf, u_1 f) u_1 f - uf \in S^+,$$

τ is an automorphism of S^+. Finally we define $\mathcal{J} = \mu(x_1)\tau$.

Then the Cayley numbers or octonions can be constructed if we choose x_1 and $u_1 f$ as before, $u'_1 = x_1 \circ u_1 f = x_1 u_1 f$. If $x, y \in E$, $x u'_1 f \in S^+$, $y u_1 f \in S^-$ and $x u'_1 f \circ y u_1 f \in E$, we define

$$x * y = (x u'_1 f) \circ (y u_1 f)$$

and prove that if $\mathbf{K} = \mathbf{R}$ the octonion algebra is obtained. This non-associative algebra can be constructed more directly as follows : consider the set $E = \mathbf{H} \times \mathbf{H}$ where $\mathbf{H}$ is the usual skewfield of Hamilton's quaternions. E is an eight-dimensional vector space over $\mathbf{R}$.

If $x = (a, b)$ and $x' = (a', b')$, we define

$$x + x' = (a + a', b + b'), \quad \alpha x = (\alpha a, \alpha b), \quad \alpha \in \mathbf{R}$$

and if $\bar{a}$ denotes the conjugate of $a \in \mathbf{H}$,

$$xx' = (aa' - \bar{b}'b, b'a + b\bar{a}')$$

defines a multiplication which satisfies the bilinearity conditions for algebras; the pair $(1, 0)$ is its neutral element.

Setting $\bar{x} = (\bar{a}, -b)$ and $x\bar{x} = |x|^2$,

$$x\bar{x} = (a\bar{a} + b\bar{b}, 0)$$

so that $|x|^2 = (\alpha^2, 0)$, $\alpha \in \mathbf{R}$ and $|x| = 0$ is equivalent to $x = (0, 0)$. If $x \neq 0$, $\bar{x}/|x|^2$ is the inverse of x.

A direct computation shows that $|xy| = |x||y|$ and that $xy = 0$ implies that $x = 0$ or $y = 0$.

13.2   The generalized triality principle.

13.2.1

Now E is a vector space of even[1] dimension $n = 2r$ over the field $\mathbf{K}$ ($\mathbf{K} = \mathbf{R}$ or $\mathbf{C}$), with a non-degenerate quadratic form Q of maximal index r. Using the previous notations,

$$S^+ = C^+(Q)f, \quad S^- = C^-(Q)f.$$

---

[1]These results can be adapted to hold in the odd-dimensional case as well.

On $S \times S$ a bilinear form $\mathcal{B}$ can be defined as in Chapter 8, 2 (or as in the previous section when $r = 4$) using the anti-automorphism β; its composition $\tilde{\beta} = \beta \circ \alpha$ with the main automorphism gives rise to another bilinear form, $\tilde{\mathcal{B}}$. $\mathcal{B}$ is invariant under G_0, $\tilde{\mathcal{B}}$ is invariant under G_0^+ (cf. Chapter 8, 2).

$\mathcal{B}$ and $\tilde{\mathcal{B}}$ are non-degenerate and either symmetric or antisymmetric.

$$\mathcal{B}(uf, vf) = (-1)^{r(r-1)/2} \mathcal{B}(vf, uf), \quad \tilde{\mathcal{B}}(uf, vf) = (-1)^{r(r+1)/2} \tilde{\mathcal{B}}(vf, uf).$$

If $r = 2, 3 \, (\text{mod } 4)$, $\mathcal{B}$ is antisymmetric.
If $r = 0, 1 \, (\text{mod } 4)$, $\mathcal{B}$ is symmetric.
If $r = 1, 2 \, (\text{mod } 4)$, $\tilde{\mathcal{B}}$ is antisymmetric.
If $r = 0, 3 \, (\text{mod } 4)$, $\tilde{\mathcal{B}}$ is symmetric.

$\mathcal{B}$ vanishes on $S^+ \times S^-$ and $S^- \times S^+$ if r is even, it vanishes on $S^+ \times S^+$ and $S^- \times S^-$ if r is odd.

$\tilde{\mathcal{B}}$ coincides with $\mathcal{B}$ on $S^+ \times S$ and with $-\mathcal{B}$ on $S^- \times S$.

These facts lead us to consider different cases. In the following, F will be a vector subspace of $C(Q)$ on which $O(Q)$ can act naturally, F is given a quadratic form extending Q : if, for instance, F is the Lie algebra of a spin group, the Killing form can be chosen. The intersections $F \cap C^\pm(Q)$ will be written $F^\pm$.

13.2.2   The generalized triality principle or interaction principle. ($n = 2r$, $r = 1, 2, 3 \, (\text{mod } 4)$.)

If $r = 2, 3 \, (\text{mod } 4)$, $\mathcal{B}$ is antisymmetric.

$A = F \times S = F \oplus S$ is given a bilinear form Λ :

$$\Lambda(x + uf, x' + u'f) = B(x, x') + \mathcal{B}(uf, u'f), \quad x, x' \in F, \quad uf, u'f \in S, \quad (1)$$

where B stands for the extension to F of the symmetric bilinear form associated to Q. Λ is non-degenerate.

Using obvious notations, we define :

$$F_0(x + u^+ f + u^- f) = \mathcal{B}(xu^+ f, u^- f), \quad (2)$$

$x \in F^+$ if r is odd, $x \in F^-$ if r is even. Next we define for ξ, η, ζ a trilinear form Φ_0 as in section 1.

If $r = 1, 2 \, (\text{mod } 4)$, $\tilde{\mathcal{B}}$ is antisymmetric.

Similar definitions are used for $\tilde{\Lambda}$, $\tilde{F}_0$ and $\tilde{\Phi}_0$.

Now we are in an analogous situation to that leading to Propositions (13.1.1), (13.1.2) and (13.1.3). The results are summarized in the :

Proposition 13.2.1

1. If r is even, defining $A = F^- \oplus S$, there exists a non-associative commutative

algebra structure over A such that

$$\xi \circ \eta = \eta \circ \xi = \omega \in A, \quad \forall \xi, \eta \in A,$$

where $\Lambda(\omega, \zeta) = \Phi_0(\xi, \eta, \zeta), \quad \forall \zeta \in A. \ \xi \circ \eta = 0$ if ξ and η each belong to one of the subspaces F^-, S^+ and S^-.

$$F^- \circ S^+ \subseteq S^-, \quad S^+ \circ S^- \subseteq F^-, \quad S^- \circ F^- \subseteq S^+. \tag{3}$$

2. *If r is odd, defining $A = F^+ \oplus S$, analogous results hold, but*

$$F^+ \circ S^+ \subseteq S^+, \quad S^+ \circ S^- \subseteq F^+, \quad S^- \circ F^+ \subseteq S^-. \tag{4}$$

A similar proposition holds for $\tilde{\Lambda}$ and $\tilde{\Phi}_0$.

Since we can no longer consider the automorphism $\mathcal{J}$ or call this 'E. Cartan's triality principle', these results are called a generalized triality principle or, relying on the inclusions in (3) and (4), an *interaction principle*.

It can be verified at once that $x \circ uf = xuf$ and $guf \circ gvf = N(g)g(uf \circ vf)g^{-1}$, $g \in G^+$.

The remark following Proposition (13.1.5) still holds.

The case where B and $\tilde{B}$ are symmetric could also be considered; but the anti-symmetric case will prove useful and essential in the sequel.

13.2.3   Orthosymplectic graded Lie algebras.

V is a $(\mathbf{Z}_2)$-graded vector space if it is the direct sum of two subspaces V_0 and V_1, where V_0 is called the even component and V_1 the odd component.

$$V = \oplus_{i=0}^1 V_i, \quad i \in \mathbf{Z}_2.$$

An element of V_i is said to be of degree i, and the degree of v will be written $|v|$. Of course, the degree is only defined for homogeneous elements, i.e. those belonging to V_0 or V_1.

A Lie algebra V is said to be graded if $V = V_0 \oplus V_1$ and

$$[a, b] = -(-1)^{|a||b|}[b, a], \quad \sum_{\text{(cycl.)}} (-1)^{|a||c|}[a, [b, c]] = 0$$

where the second requirement is called the 'graded Jacobi identity'. The set $\text{End}(V)$ of endomorphisms of V is a graded Lie algebra with

$$[a, b] = ab - (-1)^{|a||b|}ba$$

$\text{End}_0(V)$ sends V_i to V_i, $i = 0, 1$, whereas $\text{End}_1(V)$ sends V_0 to V_1 and V_1 to V_0.

Definition 13.2.1 *Let V be a graded vector space and h a non-degenerate bilinear form on $V \times V$ such that $h|_{V_0 \times V_0}$ is antisymmetric, $h|_{V_1 \times V_1}$ is symmetric and $h|_{V_0 \times V_1} = h|_{V_1 \times V_0} = 0$;*

$$\mathcal{G}_0 = \{a \in \mathrm{End}_0(V) | h(ax, y) + h(x, ay) = 0, \forall x, y \in V_i\} \tag{5}$$

$$\mathcal{G}_1 = \{a \in \mathrm{End}_1(V) | h(ax, y) = h(x, ay), \forall x \in V_1, \forall y \in V_0\} \tag{6}$$

$\mathcal{G}_0 \oplus \mathcal{G}_1$ is called an orthosymplectic graded Lie algebra.

We will use the notation $\mathcal{G}_0 \oplus \mathcal{G}_1 = \mathrm{Osp}(V_0, V_1)$.
The verification of this definition may be carried out by the reader.

13.2.4 The interaction algebra and the construction of some orthosymplectic graded Lie algebras.

Considering the situation of section 1.2, if $r \neq 0 \, (\bmod 4)$, a non-degenerate antisymmetric bilinear form invariant under the action of G_0^+ on the spinor space is given by $\mathcal{B}$ or $\tilde{\mathcal{B}}$, so that S is a symplectic space.
Using the notations of the previous subsection, set

$$V_0 = S, \quad V_1 = \begin{cases} F^+ & \text{if } r \text{ is odd}, \\ F^- & \text{if } r \text{ is even}, \end{cases}$$

and take Λ or $\tilde{\Lambda}$ for h. G_0^+ acts on $F^\pm$ by an isometric extension of its action on E and by left multiplication on S as a symplectic transformation.
We define

$$uf(vf) = \begin{cases} uf \circ vf, & \text{if } uf \text{ and } vf \text{ have different parity}, \\ 0, & \text{if } uf \text{ and } vf \text{ have the same parity}, \end{cases} \tag{7}$$
$$uf(x) = xuf, \quad x \in F^\pm.$$

Now we can verify that

$$\Lambda(uf(x), u'f) = \Lambda(x, uf(u'f)),$$

which is condition (6), since for uf and $u'f$ of different parity, it is equivalent to (assuming $uf \in S^+$, $u'f \in S^-$):

$$\mathcal{B}(xu^+f, u^-f) = \mathcal{B}(x, u^+f \circ u^-f) = F_0(x + u^+f + u^-f).$$

The graded Lie algebra generated by the (odd) elements of S and by the (even) endomorphisms obtained from the representation of the Lie algebra of G_0^+ (or $\mathrm{Spin}(Q)$) in $F^\pm$ and S is an orthosymplectic algebra which we will denote by $\mathcal{G}_0 \oplus \mathcal{G}_1$.

The graded brackets corresponding to odd elements will be written using $\{\cdot,\cdot\}$, the other ones using $[\cdot,\cdot]$.

$$\{uf, vf\}(wf) = uf \circ (vf \circ wf) + vf \circ (uf \circ wf) \in S$$

$$\{uf, vf\}(x) = xuf \circ vf + xvf \circ uf \in F^{\pm}, \quad x \in F^{\pm}.$$

$\{uf, vf\}$ acts in S as an element of the Lie algebra of the symplectic group of S and in $F^{\pm}$ as an element of the Lie algebra of the orthogonal group.

For $g \in G_0^+$ we obtained :

$$guf \circ gvf = g(uf \circ vf)g^{-1}, \tag{8}$$

from which, taking a derivative along a path starting at the identity in G_0^+ and tangent to $a \in \mathcal{L}(\mathrm{Spin}(Q))$, we get

$$auf \circ vf + uf \circ avf = a(uf \circ vf) - (uf \circ vf)a. \tag{9}$$

But (8) still holds for g in the group of transformations preserving Λ and F_0, so that (9) still holds for any even element a of the orthosymplectic graded Lie algebra $\mathcal{G}_0 \oplus \mathcal{G}_1$ just defined.

Let a be an even element of $\mathcal{G}_0 \oplus \mathcal{G}_1$.

$$a(x) = ax - xa, \quad x \in F^{\pm},$$
$$a(vf) = avf, \quad vf \in S,$$
$$[uf, a](x) = (ax - xa)uf - axuf = -xauf,$$

$$[uf, a](x) = -(auf)(x) \tag{10}$$

$$\begin{aligned}[uf, a](vf) &= uf \circ avf - a(uf \circ vf)\\ &= uf \circ avf - a(uf \circ vf) + (uf \circ vf)a,\end{aligned}$$

applying (9) gives :

$$[uf, a](vf) = -(auf)(vf) \tag{11}$$

and from (10) and (11) we get :

$$[uf, a] = -auf. \tag{12}$$

Remark. If $a \in F^{\pm}$, $\{uf, vf\}(a) = auf \circ vf + avf \circ uf$ and if a is an even element of $\mathcal{G}_0 \oplus \mathcal{G}_1$,

$$\{uf, vf\}(a) = a(uf \circ vf) - (uf \circ vf)a. \tag{13}$$

Similarly, if $\{uf, vf\}$ is also an element of $F^{\pm}$, (12) yields

$$[uf, \{uf, vf\}] = -\{uf, vf\}(uf),$$

and the graded Jacobi identity

$$\sum_{\text{(cycl.)}} [uf, \{vf, wf\}] = 0$$

becomes

$$\sum_{\text{(cycl.)}} vf \circ (wf \circ uf) + wf \circ (vf \circ uf) = 0$$

or

$$\sum_{\text{(cycl.)}} (uf \circ wf) \circ vf = 0 \tag{14}$$

if $\{uf, vf\}$ is an element of $F^{\pm}$ for any choice of uf, vf. This also means that

$$\{uf, vf\}(wf) = -(uf \circ vf)(wf). \tag{15}$$

Relying on (13) and (15), $\{uf, vf\}$ can be identified with the natural action of $-(uf \circ vf)$.

Note that $\{uf, vf\}$ being an infinitesimal symplectic mapping in the space S, given the bilinear form $\mathcal{B}$ (for instance), it can be identified with an element $z \in C(Q)$ by the classical isomorphism between $C(Q)$ and the endomorphism space of S. This element z satisfies $\beta(z) + z = 0$ in order to satisfy (5); if z is decomposed in terms of homogeneous components relative to the products of orthonormal basis vectors e_i, z only contains terms of degree $2, 3 \,(\text{mod } 4)$.

If, for example, $r = 3$ and if we choose $F^+ = \wedge^2(E) + \wedge^6(E)$, $\{uf, vf\}$ can always be identified with an even z and $\{uf, vf\}$ can be identified with $-(uf \circ vf)$ as indicated above.

13.2.5 An example of an orthosymplectic graded Lie algebra.

The algebra which we are about to construct will be called *complex conformosymplectic minkowskian.*

The Minkowski space E is given an orthonormal basis e_1, e_2, e_3, e_4 with $(e_1)^2 = 1$, $(e_2)^2 = (e_3)^2 = (e_4)^2 = -1$. A two-dimensional space E_0 with orthonormal basis $\{e_0, e_5\}$ satisfying $(e_0)^2 = 1$, $(e_5)^2 = -1$ will be added to it. Its complexification yields a six-dimensional space (E_1', Q_1') with Clifford algebra $C(Q_1')$. The following special Witt basis is chosen on E_1' :

$$x_0 = \frac{e_0 + e_5}{2} \qquad x_1 = \frac{e_1 + e_4}{2} \qquad x_2 = \frac{ie_2 + e_3}{2}$$

$$y_0 = \frac{e_0 - e_5}{2} \qquad y_1 = \frac{e_1 - e_4}{2} \qquad y_2 = \frac{ie_2 - e_3}{2}$$

$$B(x_\alpha, y_\alpha) = \tfrac{1}{2}, \quad x_\alpha y_\alpha + y_\alpha x_\alpha = 1, \quad 0 \le \alpha \le 2.$$

Note that :

$$J_0 = x_0 y_0 - y_0 x_0 = -e_0 e_5,$$
$$J_1 = x_1 y_1 - y_1 x_1 = -e_1 e_4,$$
$$J_2 = x_2 y_2 - y_2 x_2 = -i e_2 e_3.$$

A basis for the Lie algebra $\mathcal{L}(O(2,4))$, isomorphic to that of the conformal group $C_4(1,3)$ is :

$e_i e_j, \quad 1 \leq i < j \leq 4,$ (corresponding to the rotations),
$e_0 e_5,$ (corresponding to the homotheties),
$x_0 e_k, \quad 1 \leq k \leq 4,$ (corresponding to the translations),
$y_0 e_k, \quad 1 \leq k \leq 4,$ (corresponding to the special conformal transformations).

We will set $S = C(Q_1')f$, $f = y_0 y_1 y_2$.

Since $r = 3$, we choose $F^+ = \wedge^2(E_1') \oplus C e_N$ where $e_N = e_0 e_1 e_2 e_3 e_4 e_5$, and the form B on S.

Let $\hat{B}_1$ be the natural extension of the quadratic form of E_1' to F^+. We choose

$$\hat{B}_1(e_r e_s, e_i e_j) = 2(g_{rj} g_{is} - g_{sj} g_{ir}),$$

the (g_{ij}) being the coefficients of the 'metric' on E_1'.

To compute $f \circ x_0 f$, we consider

$$B(zf, x_0 f) = \hat{B}_1(f \circ x_0 f, z), \quad \forall z \in F^+$$

yielding $f \circ x_0 f = 2y_1 y_2$. The other computations are similar.

A suitable choice of $\hat{B}_1(e_N, e_N)$ gives the following table for the 'anti-brackets' $\{\cdot, \cdot\}$ of the algebra :

$\circ$	$x_0 f$	$x_1 f$	$x_2 f$	$x_0 x_1 x_2 f$
f	$2y_1 y_2$	$-2y_0 y_2$	$2y_0 y_1$	$-\frac{J_0 + J_1 + J_2 + 3i e_N}{2}$
$x_0 x_1 f$	$2x_0 y_2$	$2x_1 y_2$	$\frac{-J_0 - J_1 + J_2 + 3i e_N}{2}$	$2x_0 x_1$
$x_0 x_2 f$	$-2x_0 y_1$	$\frac{J_0 - J_1 + J_2 - 3i e_N}{2}$	$-2x_2 y_1$	$2x_0 x_2$
$x_1 x_2 f$	$\frac{J_0 - J_1 - J_2 + 3i e_N}{2}$	$2x_1 y_0$	$2x_2 y_0$	$2x_1 x_2$

$\qquad\qquad\qquad\qquad\qquad\qquad\qquad\qquad\qquad\qquad\qquad\qquad\qquad\qquad\qquad\qquad$ (16)

u being a linear combination of 1, x_1 and x_2 in the following formulas, we distinguish between the spinors of the form uf and these of the form $x_0 uf$ to obtain the following table :

$$
\begin{aligned}
&[x_0 uf, a] = -a x_0 uf, \quad a \in \mathcal{L}(O'(1,3)) = \wedge^2(E_1'),\\
&[uf, a] = -auf,\\
&[x_0 uf, x_0 e_k] = [uf, y_0 e_k] = 0,\\
&[x_0 uf, y_0 e_k] = e_k uf, \quad [uf, x_0 e_k] = e_k x_0 uf,\\
&[x_0 uf, y_0 x_0 - x_0 y_0] = x_0 uf,\\
&[uf, y_0 x_0 - x_0 y_0] = -uf.
\end{aligned}
$$

$\qquad\qquad\qquad\qquad\qquad\qquad\qquad\qquad\qquad\qquad\qquad\qquad\qquad\qquad\qquad\qquad$ (17)

The reader may also consider the action of e_N and the brackets of even elements.

The real case. Let $\mathcal{H}$ be a hermitian sesquilinear form on the spinor space $C(Q_1')f$ ($f = y_0 y_1 y_2$) such that :

$$\tilde{\beta}(uf)f = \epsilon\epsilon'\mathcal{H}(uf, vf)\gamma f.$$

In this signature and dimension, $\epsilon = \epsilon' = -1$ and $\gamma f = -ix_2 f = e_2 f$ is a possible choice.

The charge conjugation C satisfies $C^2 = -\,\mathrm{Id}$.

$$C(f) = -ix_2 f, \quad C(x_1 x_2 f) = ix_1 f,$$
$$C(x_0 x_1 f) = -ix_0 x_1 x_2 f, \quad C(x_0 x_2 f) = ix_0 f.$$

Note that C changes the parity.

Definition 13.2.2 *A spinor* $uf = u^+ f + u^- f$ *is said to satisfy a Majorana type condition if :*

$$C(u^+ f) = u^- f$$

or, equivalently, $C(u^- f) = -(u^+ f)$. *A spin basis* $\{x_{i_1} x_{i_2} \ldots x_{i_k} f\}$ *is said to be of Majorana type[2] if the conjugates of the even elements in the basis coincide with its odd elements.*

In such a basis, a 'Majorana spinor' has odd components given by the complex conjugates of its even components.

If uf satisfies the conditions of the definition, so does auf if a is even and real, and $gufg^{-1}$ if $g \in \mathrm{Pin}(Q)$.

The Majorana type properties are preserved by these natural real transformations.

From a 'Majorana type' basis, another basis can be deduced, in which the components of these spinors are real. For instance, the following basis :

$$\frac{f - ix_2 f}{\sqrt{2}} = E_1, \qquad \frac{x_0 x_1 f - ix_0 x_1 x_2 f}{\sqrt{2}} = E_2,$$
$$\frac{x_1 x_2 f + ix_1 f}{\sqrt{2}} = E_3, \qquad \frac{x_0 x_2 f + ix_0 f}{\sqrt{2}} = E_4,$$
$$\frac{-x_2 f + if}{\sqrt{2}} = E_5, \qquad \frac{-x_0 x_1 x_2 f + ix_0 x_1 f}{\sqrt{2}} = E_6, \qquad (18)$$
$$\frac{x_1 f + ix_1 x_2 f}{\sqrt{2}} = E_7, \qquad \frac{x_0 f + ix_0 x_2 f}{\sqrt{2}} = E_8.$$

The basis (18) is a symplectic basis for a form σ obtained from minus the imaginary part of $\mathcal{H}$.

$$\sigma(f, x_0 x_1 f) = 1,$$
$$\sigma(x_1 f, x_0 f) = 1,$$
$$\sigma(x_2 f, x_0 x_1 x_2 f) = 1, \qquad (19)$$
$$\sigma(x_1 x_2 f, x_0 x_2 f) = 1.$$

---

[2]This notion differs slightly from that introduced in Chapter 10, 3.7

σ vanishes on $C^+(Q_1')f \times C^-(Q_1')f$, but

$$\sigma(auf, vf) = \sigma(avf, uf)$$

when uf and vf are of the same parity and a belongs to $\mathcal{L}(O(2,4))$, to $\mathbf{R}i$ or equals e_N ($e_N f = -if$).

In the real case a generalized triality principle is constructed with σ instead of $\mathcal{B}$:

$$\sigma(zuf, vf) = \hat{B}_1(z, uf \circ vf)$$
$$z \circ uf = zuf,$$

$uf, vf \in S$ of the same parity, $z \in F^+$ and the other compositions are $S^+ \circ S^- = 0$, $F^+ \circ F^+ = 0$.

We choose $F^+ = \mathcal{L}(O(2,4)) \oplus \mathbf{R}e_N$ in order to identify $\{uf, vf\}$ and $-(uf \circ vf)$. The following table is obtained; we have not written the upper triangular part :

$\circ$	E_1	E_2	E_3	E_4	E_5	E_6	E_7	E_8
E_1	$2y_0y_1$							
E_2	$\frac{-(J_0+J_1)}{2}$	$2x_0x_1$						
E_3	e_3y_0	$-e_3x_1$	$-2x_1y_0$					
E_4	$-e_3y_1$	$-e_3x_0$	$\frac{-J_0+J_1}{2}$	$2x_0y_1$				
E_5	0	$\frac{-iJ_2+3e_N}{2}$	e_2y_0	$-e_2y_1$	$2y_0y_1$			
E_6	$\frac{iJ_2-3e_N}{2}$	0	$-e_2x_1$	$-e_2x_0$	$-\frac{J_0+J_1}{2}$	$2x_0x_1$		
E_7	$-e_2y_0$	e_2x_1	0	$\frac{iJ_2+3e_N}{2}$	e_3y_0	$-e_3x_1$	$-2x_1y_0$	
E_8	e_2y_1	e_2x_0	$-\frac{iJ_2+3e_N}{2}$	0	$-e_3y_1$	$-e_3x_0$	$\frac{-J_0+J_1}{2}$	$2x_0y_1$

$$(20)$$

13.2.6 Minkowskian graded Lie algebras.

The signature will be $(+ - - -)$ as in the previous section. We will use the fundamental hermitian form $\mathcal{H}$ and the symplectic form σ for which

$$\sigma(f, x_1f) = 1, \quad \sigma(x_1x_2f, x_2f) = 1,$$

($f = y_1y_2$ here), and the conjugation $\mathcal{C}$

$$\begin{cases} \mathcal{C}(f) = -ix_2f, & \mathcal{C}(x_1x_2f) = ix_1f, \\ \mathcal{C}(x_2f) = if, & \mathcal{C}(x_1f) = -ix_1x_2f, \end{cases}$$

used in the construction of the symplectic Majorana type basis. The brackets $\{uf, vf\}$ which, relying on the remark in section 4, can be trivially identified with

$-(uf \circ vf)$, are listed in the following table :

$\circ$	$f - ix_2f$	$x_1f + ix_1x_2f$	$-x_2f + if$	$x_1x_2f + ix_1f$
$f - ix_2f$	$-2e_2y_1$	$-J_1$	$2e_3y_1$	iJ_2
$x_1f + ix_1x_2f$	$-J_1$	$2e_2x_1$	$-iJ_2$	$-2e_3x_1$
$-x_2f + if$	$2e_3y_1$	$-iJ_2$	$2e_2y_1$	$-J_1$
$x_1x_2f + ix_1f$	iJ_2	$-2e_3x_1$	$-J_1$	$-2e_2x_1$

$$(21)$$

The ordinary brackets and the brackets similar to (17) should be added to this table; together, they define a real minkowskian orthosymplectic graded Lie algebra.

13.2.7   The relationship with the Wess-Zumino heuristic formalism and the work of other authors.

In a large number of publications from 1970 on, many authors have introduced graded Lie algebra tables, generally without any logical justification. Using their notation, this is the table of Wess-Zumino :

$$[P_m, P_n] = 0, \quad [P_m, Q_\alpha] = [P_m, \bar{Q}_{\dot\alpha}] = 0,$$
$$\{Q_\alpha, \bar{Q}_{\dot\beta}\} = 2\sigma^m_{\alpha\dot\beta} P_m,$$
$$\{Q_\alpha, Q_\beta\} = \{\bar{Q}_{\dot\alpha}, \bar{Q}_{\dot\beta}\} = 0. \tag{22}$$

P_m is a translation operator in Minkowski space, the Q_α define the spinors and the $\bar{Q}_{\dot\alpha}$ the conjugate spinors. In terms of coordinates, the space is described by x^m, θ^α and $\bar{\theta}^{\dot\alpha}$, where the coordinates with latin indices commute whereas those with greek indices anticommute ('graded analysis'). Then one defines :

$$P_m = i\frac{\partial}{\partial x^m},$$

$$Q_\alpha = \frac{\partial}{\partial \theta^\alpha} - i\sigma_{\alpha\dot\beta}\bar{\theta}^{\dot\beta}\left(\frac{\partial}{\partial x^m}\right), \tag{23}$$

$$\bar{Q}_{\dot\alpha} = -\frac{\partial}{\partial \bar{\theta}^{\dot\alpha}} + i\theta^\beta\sigma^m_{\beta\dot\alpha}\left(\frac{\partial}{\partial x^m}\right).$$

From table (16) in section 5, we can extract :

$$\begin{aligned}
f \circ x_1f &= 2y_2y_0, \\
f \circ x_2f &= -2y_1y_0, \\
x_1x_2f \circ x_1f &= 2x_1y_0, \\
x_1x_2f \circ x_2f &= 2x_2y_0
\end{aligned} \tag{24}$$

(and a second isomorphic table in which $(-x_0)$ replaces y_0 in the right-hand sides).

We can think of (24) as isomorphic to the table :

$$
\begin{aligned}
f \circ x_1 f &= -2y_2, \\
f \circ x_2 f &= 2y_1, \\
x_1 x_2 f \circ x_1 f &= -2x_1, \\
x_1 x_2 f \circ x_2 f &= -2x_2
\end{aligned}
\tag{25}
$$

(f in the left-hand sides now being $y_1 y_2$).

More generally, from (16) and (17) we can obtain the table ($a = xe_k$ or ye_k) :

$$
\begin{aligned}
[uf, a] &= 0, & \{u^+ f, v^- f\} &= -(u^+ f \circ v^- f), \\
[a, b] &= 0, & \{u^+ f, v^+ f\} &= \{u^- f, v^- f\} = 0.
\end{aligned}
\tag{26}
$$

The analogy between (26) and (22) is obvious and can be made even more explicit. Using notations which are familiar to physicists, we choose :

$$
\sigma^1_{1\dot{1}} = \sigma^1_{2\dot{2}} = -1, \qquad \sigma^2_{1\dot{2}} = -i, \qquad \sigma^2_{2\dot{1}} = i,
$$

$$
\sigma^3_{1\dot{1}} = 1, \qquad\qquad \sigma^3_{2\dot{2}} = -1, \qquad \sigma^4_{1\dot{2}} = \sigma^4_{2\dot{1}} = 1,
$$

and easily verify that

$$
\begin{aligned}
\{Q_2 - Q_1, \bar{Q}_{\dot{2}} + \bar{Q}_{\dot{1}}\} &= 8y_2, \\
\{Q_2 + Q_1, \bar{Q}_{\dot{2}} + \bar{Q}_{\dot{1}}\} &= -8y_1, \\
\{Q_2 - Q_1, \bar{Q}_{\dot{2}} - \bar{Q}_{\dot{1}}\} &= -8x_1, \\
\{Q_2 + Q_1, \bar{Q}_{\dot{2}} - \bar{Q}_{\dot{1}}\} &= -8x_2.
\end{aligned}
$$

Defining :

$$
\begin{aligned}
Q_1 &= x_2 f - x_1 f, & Q_2 &= x_2 f + x_1 f, \\
\bar{Q}_{\dot{1}} &= -f - x_1 x_2 f, & \bar{Q}_{\dot{2}} &= -f + x_1 x_2 f
\end{aligned}
$$

we can identify (25) and the second line of (22). Note that the $\bar{Q}_{\dot{1}}$ and $\bar{Q}_{\dot{2}}$ defined this way are the charge conjugates of Q_1 and Q_2 respectively, if the charge conjugation $\mathcal{C}$ is chosen such as to satisfy $\gamma f = ie_2 f$. $(Q_1, Q_2, \bar{Q}_{\dot{1}}, \bar{Q}_{\dot{2}})$ can be taken as a 'Majorana type' basis for this conjugation.

Many tables for graded Lie algebras relevant to mathematical physics and used in the literature, could be found along similar lines.

13.3   Selected references.

- A. Crumeyrolle, *Algèbres de Clifford et spineurs*, Cours et séminaires du Département de Mathématique, Toulouse III, 1974.

- A. Crumeyrolle, *Constructions d'algèbres de Lie graduées orthosymplectiques et conformosymplectiques minkowskiennes*, Lecture Notes in Mathematics n. 1165, Springer-Verlag.

  This paper contains more explanations and results about (13.2.7) above.

- C. Chevalley, *The algebraic theory of spinors*, Columbia University Press, New-York, 1954.

- J. Wess, B. Zumino, Nucl. Phys. 70, (B. 39) 1974.

- L. Corwin, Y. Neeman, S. Sternberg, *Graded Lie Algebras in Mathematics and Physics*, Rev. of Mod. Physics Vol. 47 n. 3, July 1975, p. 573.

Chapter 14

THE CLIFFORD ALGEBRA AND THE CLIFFORD BUNDLE OF A PSEUDO-RIEMANNIAN MANIFOLD. EXISTENCE CONDITIONS FOR SPINOR STRUCTURES.

14.1 The Clifford algebra of a manifold.

14.1.1

V is a real n-dimensional C^∞ manifold with a pseudo-riemannian structure of signature $(p, n - p)$ (we will assume that $p \leq n - p$).

Q is, just as before in the case of $E = \mathbf{R}^n$, a non-degenerate quadratic form of signature $(p, n - p)$. V is 'modeled' on $(\mathbf{R}^n, Q)$. X stands for a smooth vector field of V, and the set of all such fields is denoted by $D^1(V)$. The field of quadratic forms is also denoted by Q.

Definition 14.1.1 *The Clifford algebra $\mathrm{Clif}_V(Q)$ of V is the quotient of the tensor algebra $\otimes D^1(V)$ of differentiable vector fields on V by the two-sided ideal J generated by the elements*

$$X \otimes X - Q(X), \quad X \in D^1(V).$$

Formally, this definition coincides with the definition given in Chapter 3 for vector spaces. $D^1(V)$ is a module over the ring $C^\infty(V)$ of C^∞ differentiable functions.

If U is an open set in V, $\mathrm{Clif}_V(Q)$ induces an algebra $\mathrm{Clif}_U(Q)$ on U which can be identified with the quotient of $\otimes D^1(U)$ by the two-sided ideal J_U (where $X \in D^1(U)$).

If the coordinates (x^α) are defined on U, $\partial/\partial x^1, \partial/\partial x^2, \ldots, \partial/\partial x^n$ form a basis of $D^1(U)$, which is a free module of dimension n, and $\mathrm{Clif}_U(Q)$ is a (2^n)-dimensional module over $C^\infty(U)$: the proofs are the same as for vector spaces.

If $x \in V$, $\mathrm{Clif}_V(Q)$ induces a Clifford algebra $\mathrm{Clif}_x(Q)$ at the point x, this algebra is a (2^n)-dimensional real vector space that can also be constructed directly on the tangent space by the well-known procedure.

The definition of Clifford fields is similar to that of vector fields.

14.1.2 Derivations in the Clifford algebra of a manifold.

Let $x \to Y_x$ be a vector field Y on V which induces a local one-parameter transformation group $t \to \varphi_t$. If $\tilde{\varphi}_t$ stands for the natural extension of the tangent map $d\varphi_t$ to tensor fields, the Lie derivative L_Y of the tensor field K is defined by :

$$(L_Y K)_x = \lim_{t \to 0} \frac{1}{t}(K_x - (\tilde{\varphi}_t(K))_x).$$

L_Y is a derivation in the tensor algebra of V. From this it follows that :

$$L_Y(X \otimes X - Q(X)) = L_Y(X) \otimes X + X \otimes L_Y(X) - L_Y(Q(X)),$$

and if g stands for the symmetric bilinear form associated to the field Q on V, by an easy contraction :

$$L_X(Q(X)) = 2g(L_Y X, X) + (L_Y Q)(X).$$

$X \otimes X' + X' \otimes X - 2g(X, X')$ belonging to J, L_Y is seen to globally preserve J if and only if $L_Y(Q) = 0$, in which case Y induces a local isometry group (Y is a Killing field). We have proved :

Proposition 14.1.1 *The Lie derivative L_Y extends to the Clifford algebra of V if and only if $L_Y(Q) = 0$, Y inducing a local isometry group.*

Now let ∇ be a linear connection. The same formal computations yield :

$$\nabla_Y(X \otimes X - Q(X)) = \nabla_Y(X) \otimes X + X \otimes (\nabla_Y X) - \nabla_Y(Q(X))$$
$$\nabla_Y(Q(X)) = 2g(\nabla_Y X, X) + (\nabla_Y g)(X, X)$$

and the result :

Proposition 14.1.2 *The covariant derivative ∇_Y extends to the Clifford algebra of V if and only if $\nabla_Y g = 0$ i.e. the connection ∇ is euclidean.*

14.1.3 The Clifford bundle of V.

The group $O(Q)$ has a natural extension to $C(Q)$ of its action in $E = \mathbf{R}^n$, as can be shown if we verify that $O(Q)$ preserves the ideal generated by the $x \otimes x - Q(x)$, or also by the results of Chapter 8, 3.2, using the representation θ such that :

$$\theta_{\varphi(g)}(w) = gwg^{-1}, \quad g \in \text{Pin}(Q)$$

(or $\theta_{p(g)(w)} = \alpha(g)wg^{-1}$).

To the principal bundle of orthonormal bases for V a vector bundle of fiber $C(Q)$ can be associated by a classical procedure; its structural group is the previously defined extension of $O(Q)$. This bundle is the *Clifford bundle* of V, denoted by $\text{Clif}(V, Q)$ (or $\text{Clif}(V)$ for short). It is a bundle of rank 2^n over $\mathbf{R}$ and its fibers are real Clifford algebras.

A section of $\text{Clif}(V)$ is a Clifford field by the definition in section 1.

Proposition 14.1.3 *The bundles* $\mathrm{Clif}(V)$ *and* $\wedge(T(V))$ *are linearly isomorphic when considered as vector bundles.*

Proof. $\wedge(T(V))$ is the exterior algebra bundle associated to the tangent bundle $T(V)$ of V. This follows at once from the linear identification between $C(Q)$ and $\wedge(E)$. ∎

Note that the space of sections of $\mathrm{Clif}(V)$ has a Clifford algebra structure.

Algebraic spin subbundles.

Definition 14.1.2 *Every vector subbundle* $\mathrm{Spin}(V)$ *of* $\mathrm{Clif}(V)$ *such that* $\forall x \in V$, $\mathrm{Spin}_x(V)$ *is a minimal left ideal of* $\mathrm{Clif}_x(V)$ *is called an amorphic algebraic spin subbundle. Every local or global section of* $\mathrm{Spin}(V)$ *is called an amorphic spinor field. Every section of* $\mathrm{Clif}(V)$ *such that* e_x *is an idempotent element of* $\mathrm{Clif}_x(V)$ *for all* $x \in V$ *will be called an idempotent field. Every section* $x \to f(x)$ *of* $\mathrm{Clif}(V)$ *such that* $f(x)$ *is an isotropic r-vector is called a field of isotropic r-vectors.*

It is easy to give *sufficient* conditions for the existence of an amorphic spin subbundle :

1. there exists a global primitive idempotent field

2. there exists a global isotropic r-vector field (if Q is neutral, $n = 2r$).

By the results of Chapter 5, 1.7 and 2.4, the first condition is stronger than the second if Q is neutral. But both conditions clearly can be weakened since there may exist different idempotent elements which determine the same left ideal and since the isotropic r-vector may be replaced by λf, $\lambda(x) \neq 0$, $\forall x \in V$.

We will prove an important result, assuming that the second condition holds :

Proposition 14.1.4 *The covariant derivative defined by an euclidean connection* ∇ *does not extend naturally to the amorphic spinors.*

Proof. Consider the case where Q is neutral, $n = 2r$, and where there exists an isotropic r-vector field $x \to f(x)$ defining the amorphic spin subbundle.

Take a Witt basis over a suitable open set $\mathcal{U}$, $x \to \{\xi_\alpha(x), \xi_{\beta^*}(x)\}$, $\alpha, \beta = 1, 2, \ldots, r$, such that $f(x) = (\xi_1 \cdot \xi_2 \cdot \ldots \xi_{r^*})(x)$ and that the spinor field is given by

$$x \to \psi(x) = (\xi_{\alpha_1} \xi_{\alpha_2} \ldots \xi_{\alpha_h} f)(x).$$

If ∇ could be naturally extended to the spinors,

$$\nabla_{\xi_\beta} \psi = \nabla_\beta \psi = \sum_i \xi_{\alpha_1} \ldots (\nabla_\beta \xi_{\alpha_i}) \ldots \xi_{\alpha_h} f + \xi_{\alpha_1} \xi_{\alpha_2} \ldots \xi_{\alpha_h} \nabla_\beta f$$

would hold.

But if we define the connection coefficients by

$$\nabla_\beta \xi_{\alpha^*} = \Gamma^\lambda_{\alpha^* \beta} \xi_\lambda + \Gamma^{\lambda^*}_{\alpha^* \beta} \xi_{\lambda^*},$$

we get

$$\nabla_\beta f = \sum_{\alpha^*} \Gamma^\lambda_{\alpha^* \beta} \xi_{1^*} \dots \xi_\lambda \dots \xi_{r^*} + \Gamma^{\lambda^*}_{\lambda^* \beta} f.$$

For $\nabla_x f$ to be an element of $\mathrm{Clif}(V)f$,

$$\Gamma^\lambda_{\alpha^* \beta} = 0, \quad \Gamma^\lambda_{\alpha^* \beta^*} = 0$$

should hold, meaning that, if the ω^i_j are the connection forms for the chosen Witt bases,

$$\omega^\alpha_{\beta^*} = 0.$$

However, if we express that ∇ is euclidean in terms of the Witt basis, we obtain

$$\omega^\beta_\alpha + \omega^{\alpha^*}_{\beta^*} = 0, \quad \omega^{\alpha^*}_\beta + \omega^{\beta^*}_\alpha = 0, \quad \omega^\alpha_{\beta^*} + \omega^\beta_{\alpha^*} = 0, \tag{1}$$

and these conditions do not imply that $\omega^\alpha_{\beta^*} = 0$.

Remarks. A similar result holds for the Lie derivative.

In Chapter 11, we introduced the group H of elements γ such that $\gamma f = \pm f$, $p(H) = \mathcal{G}$ and $\mathcal{L}(\mathcal{G}) = \mathcal{L}(H)$ consists of elements u such that $uf = 0$.

If (x_α, x_{α^*}) is a Witt basis for E, u can be defined by :

$$u = a^{\alpha \beta^*} x_\alpha x_{\beta^*} + \tfrac{1}{2} b^{\alpha^* \beta^*} x_{\alpha^* \beta^*}, \quad b^{\alpha^* \beta^*} = -b^{\beta^* \alpha^*},$$

where both summations are over all α and β from 1 to r.
$\beta(u) + u = 0$ implies that $\sum_1^r a^{\alpha \alpha^*} = 0$.

If, using a classical procedure, we compute

$$ux_\gamma - x_\gamma u = a^{\alpha \gamma^*} x_\alpha + b^{\alpha^* \gamma^*} x_{\alpha^*}$$
$$ux_{\gamma^*} - x_{\gamma^*} u = -a^{\gamma \alpha^*} x_{\alpha^*},$$

the conditions (1) are reobtained, along with

$$\sum_1^r \omega^\alpha_\alpha = 0 \tag{2}$$

and

$$\omega^\alpha_{\beta^*} = 0. \tag{3}$$

If an element of an enlarged spinoriality group is considered, only condition (3) arises. This proves :

Proposition 14.1.5 *If the signature is neutral, an euclidean connection will naturally induce a derivation on the local amorphic spinor fields (defined by a local isotropic r-vector field) if and only if its forms have values in the Lie algebra of an enlarged spinoriality group.*

Analogous results can be obtained for the case where the first sufficient condition is assumed to hold.

∇ *extends naturally to the amorphic spinors if*

$$e\nabla e = 0. \tag{4}$$

Indeed, $e^2 = e$ and $\nabla_X ue = \nabla_X(ue^2) = \nabla_X(ue) + ue\nabla_X e$, so that ∇ will not, in general, satisfy (4).

14.2   Existence conditions for spinor structures.

We assume that $n = 2r$ and use the notations and the assumptions of Chapter 10, 2.1 concerning the space (E, Q) and its complexification (E', Q').

14.2.1   Trivial $\text{Pin}(Q)$ spinor structures.

Let $\{x_\alpha, y_\beta\}$ denote the 'real' Witt basis W obtained from an orthonormal real basis using the method of Chapter 10, 2.1.

$$f = y_1 y_2 \ldots y_r,$$

$C(Q')f$ is a spinor space for the representation ρ of $C(Q')$ for which $\rho(u)vf = uvf$. The space $C(Q')f$ will be considered as the space of a faithful representation of $\text{Pin}(Q)$.

Consider another 'real' Witt basis $\Omega = \{\xi_\alpha, \eta_\beta\}$, associated to the orthonormal basis $\mathcal{R}$, such that

$$\Omega = \varphi_g(W), \quad g \in \text{Pin}(Q), \quad \varphi_{(g)} = \tau \in O(Q).$$

Then

$$S = \{\xi_{\alpha_1}\xi_{\alpha_2}\ldots\xi_{\alpha_h}gf\} = \{gx_{i_1}x_{i_2}\ldots x_{i_h}f\} \tag{1}$$

is a basis for $C(Q')f$.

If Ω is replaced by another real Witt basis Ω' associated to the orthonormal basis $\mathcal{R}'$ of E, where :

$$\Omega' = \{\xi_{\alpha'}, \eta_{\beta'}\}, \quad \Omega' = \varphi_\gamma(\Omega),$$

$g' = \gamma g$, S is replaced by

$$S' = \{\xi_{\alpha'_1}\xi_{\alpha'_2}\ldots\xi_{\alpha'_h}g'f\} = \gamma S. \tag{2}$$

If ψ is an element of $\mathbf{C}^{2^r}$ defining a spinor in the basis S and ψ' defines the same spinor in the basis S', we set

$$\psi' = \gamma^{-1}\psi. \tag{3}$$

The choice of S corresponds to that of the pair $(\mathcal{R}, g)$ (W being fixed). Hence $(\mathcal{R}, g)$ and (Ω, g) can be called 'spin bases' over $\mathcal{R}$ and Ω respectively. Since there exist Witt bases for $C(Q')f$ which are not associated to real orthonormal bases, we consider pairs (Ω, g) and (Ω', g') where Ω and Ω' are arbitrary Witt bases and $g, g' \in \mathrm{Pin}(Q')$, but $\gamma = g'g^{-1}$ belonging to $\mathrm{Pin}(Q)$, since we are only interested in the action of $\mathrm{Pin}(q)$ on $C(Q')f$ and on the bases.

Definition 14.2.1 Ω, Ω' *being Witt bases,* $g, g' \in \mathrm{Pin}(Q')$, *the pairs* (Ω, g) *and* (Ω', g') *are said to define a trivial* $\mathrm{Pin}(Q)$ *spinor structure if*

$$\Omega' = \sigma(\Omega), \quad \varphi_{(\gamma)} = \sigma, \quad \gamma = g'g^{-1} \in \mathrm{Pin}(Q).$$

Two triples (Ω, g, ψ) *and* (Ω', g', ψ'), $\psi, \psi' \in \mathbf{C}^{2^r}$ *then determine the same spinor if and only if*

$$\psi' = \gamma^{-1}\psi$$

in the sense of (3).

Note that if the basis $W = \{x_i, y_j\}$ and the ideal $C(Q')y_1 y_2 \ldots y_r$ are replaced by another Witt basis $\{x'_i, y'_j\}$ and the ideal $C(Q')y'_1 y'_2 \ldots y'_r$, a $\gamma_1 \in \mathrm{Pin}(Q')$ can be found such that it maps W' to W and (Ω, g) is replaced by $(\gamma_1 \Omega \gamma_1^{-1}, g\gamma_1) = (\Omega_1, g_1)$. (Ω, g, ψ) and (Ω_1, g_1, ψ') still determine the same spinor since

$$g'g^{-1} = (g'\gamma_1)(g\gamma_1)^{-1} = \gamma.$$

Hence the choice of the spinor ideal does not influence the definition.

Note that if we would use a primitive idempotent instead of an isotropic r-vector, we would not be certain that a $\gamma_1 \in \mathrm{Pin}(Q')$ can be found such that it maps one primitive idempotent to another (cf. Chapter 5, 2.4).

This situation can be naturally generalized if we replace $\mathrm{Pin}(Q)$ by a principal bundle, the set of Ω by a bundle of bases under the group $O(Q)$, the spinor space by a 'vector spin' bundle on which the group $\mathrm{Pin}(Q)$ acts, and the previous 'equivariant' actions by morphisms : all this leads us to the following

14.2.2

Definition 14.2.2 ($\mathrm{Pin}(Q)$ *spinor structure on a manifold,* $n = 2r$.) *Let* V *be a pseudo-riemannian (or a riemannian)* C^∞ *paracompact* $(n = 2r)$-*dimensional manifold, let* ξ *be the principal bundle of orthonormal bases, with group* $O(Q)$. *A* $\mathrm{Pin}(Q)$ *spinor structure is said to exist on* V *if a principal bundle* η *with group* $\mathrm{Pin}(Q)$ *and a principal morphism* h *from* η *to* ξ *can be constructed.*

This means that the following diagram must commute :

$$
\begin{array}{ccc}
\eta & \xrightarrow{\;R_\gamma\;} & \eta \\
{\scriptstyle h}\downarrow & & \downarrow{\scriptstyle h} \quad \searrow{\scriptstyle V} \\
\xi & \xrightarrow[{R_{\varphi(\gamma)}}]{} & \xi \quad \nearrow{\scriptstyle \Pi}
\end{array}
$$

where R_γ and $R_{\varphi(\gamma)}$, $\gamma \in \mathrm{Pin}(Q)$, stand for the 'right translations' and q, Π for the projections on the base.

More explicitly, since we can always define the bundles η and ξ using open sets $(U_\alpha)_{\alpha \in A}$ corresponding to a common trivialization, local sections z_α, $\mathcal{R}_\alpha$ with transition functions $\gamma_{\alpha\beta}$ and $\varphi(\gamma_{\alpha\beta})$ must exist, such that $\forall x \in U_\alpha \cap U_\beta$,

$$
z_\beta(x) = z_\alpha(x)\gamma_{\alpha\beta}(x), \quad \gamma_{\alpha\beta}(x) \in \mathrm{Pin}(Q),
$$
$$
h(z_\beta(x)) = \mathcal{R}_\beta(x) = h(z_\alpha(x)\varphi(\gamma_{\alpha\beta}(x))) = \mathcal{R}_\alpha(x)\varphi(\gamma_{\alpha\beta}(x)).
$$

Clearly, the orthonormal bases could be replaced by 'real' Witt bases.

By a standard procedure, a 'vector spin' bundle ζ ($\hat{q} : \zeta \to V$) with group $\mathrm{Pin}(Q)$ and fiber $\mathbf{C}^{2^r}$ can be associated to the principal bundle η, $\mathrm{Pin}(Q)$ acting on $\mathbf{C}^{2^r}$ by (3).

$\mathbf{C}^{2^r}$ will be identified with the standard space for which

$$
\{x_{i_1} x_{i_2} \dots x_{i_h} f\}
$$

is a basis.

If θ stands for the isomorphism from $C(Q')$ to $\mathbf{C}^{2^r}$ which is determined by the basis $\{x_i, y_j\}$, its restriction to $C(Q')$ allows us to set $\psi = \theta(u)$, $u \in C(Q')f$, $\psi \in \mathbf{C}^{2^r}$.

By classical results a field of $\mathrm{Pin}(Q)$ spinors ψ over an open subset U of V is a differentiable mapping

$$
\psi : z \to \psi(z),
$$

from η to $\mathbf{C}^{2^r}$, such that if $\psi(z) = \theta(u)$ and $\gamma \in \mathrm{Pin}(Q)$,

$$
\psi(z\gamma^{-1}) = \gamma\psi(z), \tag{4}
$$

or $\theta(\gamma u)$, in accordance with (3). Such a mapping defines a section ψ of ζ :

$$
x \to \psi_x \in \hat{q}^{-1}(x), \quad \psi(z(x)) = \psi_x,
$$

so that

$$
(\gamma\psi_x)^{i_1 \dots i_h} x_{i_1 \dots x_h} f = \psi_x^{i_1 \dots i_h}(\gamma x_{i_1} \dots x_{i_h} f). \tag{5}
$$

14.2.3 Necessary conditions for the existence of a $\text{Pin}(Q)$ spinor structure.

If $x \in U_\alpha$ and if $x \to \mathcal{R}_\alpha(x)$ is a differentiable field of orthonormal bases, this field can be identified with a field of 'real' Witt bases $x \to W_\alpha(x)$, so that

$$h(z_\alpha(x)) = W_\alpha(x) = \theta_\alpha^x(\{x_i, y_j\})$$

where the θ_α^x are local isomorphisms : if $x \in U_\alpha \cap U_\beta$ and if (θ_α^x), (θ_β^x) correspond to U_α and U_β respectively, the transition functions from θ_α^x to θ_β^x are of the form $\varphi(\gamma_{\alpha\beta})$, $\gamma_{\alpha\beta}(x) \in \text{Pin}(Q)$, where $\gamma_{\alpha\beta}$ is a differentiable function of x. For each $x \in U_\alpha$ the (θ_α^x) define an isomorphism from $C(Q)$ to $\text{Clif}_x(V)$ which can be extended to the complexifications $C(Q') \to \text{Clif}'_x(V)$.

The $\text{Pin}(Q)$ spinor structure on V then induces a trivial $\text{Pin}(Q)$ spinor structure (cf. section 1) on the tangent space at x : an open set U containing x can be found along with an equivalence class of pairs (Ω_x, g_x) where $x \to \Omega_x$ is a local section of the bundle ξ_C and $x \to g_x$ is a differentiable mapping sending $x \in U$ into $\text{Pin}(Q')$, (Ω'_x, g'_x) being equivalent to (Ω_x, g_x) if $g'_x g_x^{-1} \in \text{Pin}(Q)$.

Note that arbitrary Witt bases must be used in the complexified bundle ξ_C.

If $x \in U_\alpha \cap U_\beta$, two equivalent pairs (in the sense of Definition (14.2.1)) must be obtained :

$$(\Omega_\alpha^x = \theta_\alpha^x\{\lambda_\alpha(x)(x_i, y_j)\lambda_\alpha^{-1}(x)\}, g_\alpha(x)),$$
$$(\Omega_\beta^x = \theta_\beta^x\{\lambda_\beta(x)(x_i, y_j)\lambda_\beta^{-1}(x)\}, g_\beta(x)),$$

where $\lambda_\alpha(x), \lambda_\beta(x), g_\alpha(x), g_\beta(x) \in \text{Pin}(Q')$, λ_α and λ_β are defined on U_α and U_β respectively, and g_α, g_β on some open neighborhood of x in $U_\alpha \cap U_\beta$, $g_\beta g_\alpha^{-1}(x) \in \text{Pin}(Q)$ and $g_\beta(x) = g_{\alpha\beta}(x)g_\alpha(x)$, the $\varphi(g_{\alpha\beta})$ being the $O(Q)$-valued transition functions from Ω_α to Ω_β. Also note that $\lambda_\alpha g_{\alpha\beta} \lambda_\beta^{-1} = \gamma_{\alpha\beta}$.

We set

$$\theta_\alpha^x\{\lambda_\alpha(x)(x_i, y_j)\lambda_\alpha^{-1}(x)\} = \varphi_\alpha^x\{x_i, y_j\},$$

the φ_α^x determine an isomorphism from $C(Q')$ to $\text{Clif}'_x(V)$.

By the results of section 1, the space of spinors over x can be identified with an ideal of $\text{Clif}'_x(V)$ and the spin bases with the bases of this ideal, hence a spinor in x can be considered as a well-defined element of some minimal left ideal of $\text{Clif}'_x(V)$.

The identification of a basis for ζ_x with $\varphi_\alpha^x\{x_{i_1} \ldots x_{i_h} f\}$ commutes with the action of $\gamma \in \text{Pin}(Q)$, since a spinor in x has components $\alpha^{i_1 \cdots i_h}(x)$ and

$$\alpha^{i_1 \cdots i_h}(x)\varphi_\alpha^x(x_{i_1} \ldots x_{i_h} f) = \varphi_\alpha^x(\alpha^{i_1 \cdots i_h}(x)x_{i_1} \ldots x_{i_h} f)$$

so relying on (5),

$$\varphi_\alpha^x(\gamma^{-1}\alpha^{i_1 \cdots i_h}\gamma x_{i_1} \ldots x_{i_h} f) = \gamma^{-1}\alpha^{i_1 \cdots i_h}(x)\varphi_\alpha^x(\gamma x_{i_1} \ldots x_{i_h} f).$$

φ_α^x corresponding to the action of $\epsilon g_{\alpha\beta}(x)$, $\epsilon = \pm 1$, the spinor which is identified with $\varphi_\alpha^x(f)$ in the first basis is described by $\varphi_\beta^x(\epsilon g_{\alpha\beta}^{-1}(x)f)$ in the second one, so

$$f = \epsilon f g_{\alpha\beta}^{-1}(x),$$

and applying the main antiautomorphism,

$$g_{\alpha\beta}(x)f = \pm f, \tag{6}$$

which means that $g_{\alpha\beta}$ belongs to the subgroup H of $\text{Pin}(Q)$ introduced in Chapter 11 and that $\varphi(g_{\alpha\beta})$ belongs to the spinoriality group $\mathcal{G} \subset O(Q)$.

Next we obtain

$$\varphi^x_\beta(f) = \varphi^x_\alpha(g_{\alpha\beta}(x))\varphi^x_\alpha(f)\varphi^x_\alpha(g^{-1}_{\alpha\beta}(x))$$

which we will write as

$$f_\beta(x) = \hat{g}_{\alpha\beta}(x)f_\alpha \hat{g}^{-1}_{\alpha\beta}(x). \tag{7}$$

$\varphi^x_\alpha(g_{\alpha\beta}(x)) = \varphi^x_\beta(g_{\alpha\beta}(x)) = \hat{g}_{\alpha\beta}(x)$ and, relying on (6) :

$$f_\beta(x) = N(\hat{g}_{\alpha\beta}(x))f_\alpha(x), \tag{8}$$

where $N(\hat{g}_{\alpha\beta}(x)) = N(g_{\alpha\beta}(x))$.

Note that the $\varphi(g_{\alpha\beta})$ are transition functions for sections of the complexification of ξ, the cocycle $\varphi(\gamma_{\alpha\beta})$ which defines ξ and the cocycle $\varphi(g_{\alpha\beta})$ are *cohomologous* in $O(Q')$.

We have proved :

Proposition 14.2.1 *If a* $\text{Pin}(Q)$ *spinor structure exists on the manifold* V :

1. *Modulo a factor* (± 1) *there exists, on* V, *a field of isotropic* r-*vectors (a 'pseudo-field') i.e. a vector subbundle of* $\text{Clif}'(V, Q')$ *defined, for* $x \in U_\alpha$, *by* $f_\alpha(x)$.

2. *The complexified pseudo-riemannian bundle* $\xi_{\mathbf{C}}$ *of frames admits local sections*

$$x \in U_\alpha \to \Omega_\alpha(x),$$

such that if $\Omega_\beta(x) = \Omega_\alpha(x)\varphi(g_{\alpha\beta}(x))$, $x \in \nu_\alpha \cap \nu_\beta$, $\varphi(g_{\alpha\beta}(x)) \in O(Q)$,

$$f_\beta(x) = \hat{g}_{\alpha\beta}(x)f_\alpha(x)\hat{g}^{-1}_{\alpha\beta}(x),$$
$$f_\beta(x) = N(\hat{g}_{\alpha\beta}(x))f_\alpha(x).$$

3. *The structural group of the pseudo-riemannian bundle* ξ *of frames can be reduced, in* $O(Q')$, *to a spinoriality group.*

14.2.4   Sufficient conditions for the existence of a $\text{Pin}(Q)$ spinor structure.

Proposition 14.2.2 *Let* $(U_\alpha, \varphi_\alpha)_{\alpha \in A}$ *be an atlas of local trivializations of the complexified pseudo-riemannian bundle* $\xi_{\mathbf{C}}$, *its transition functions* $\varphi(g_{\alpha\beta})$ *being* $O(Q)$-*valued.*

If there exists a pseudo-field of isotropic r-*vectors on* V, *locally defined by*

$$x \in U_\alpha \to f_\alpha(x),$$

and such that for $x \in U_\alpha \cap U_\beta \neq \emptyset$,

$$f_\beta(x) = \hat{g}_{\alpha\beta}(x)f_\alpha(x)\hat{g}_{\alpha\beta}^{-1}(x),$$
$$f_\beta(x) = N(\hat{g}_{\alpha\beta})(x)f_\alpha(x),$$

$(g_{\alpha\beta}(x) \in \text{Pin}(Q))$, a $\text{Pin}(Q)$ spinor structure exists on V.

In this proposition, the $g_{\alpha\beta}$ are determined up to a sign change.
Proof.

1. Relying on (8), $f_\beta(x)\hat{g}_{\alpha\beta}(x) = \hat{g}_{\alpha\beta}(x)f_\alpha(x)$ leads to

$$N(\hat{g}_{\alpha\beta}(x))f_\alpha(x)\hat{g}_{\alpha\beta}(x) = \hat{g}_{\alpha\beta}(x)f_\alpha(x),$$

and by the results on pure spinors :

$$\hat{g}_{\alpha\beta}(x)f_\alpha(x) = \mu(x)f_\alpha(x), \quad \mu(x) \in \mathbf{C}^*.$$

But since

$$\hat{g}_{\alpha\beta}(x)f_\alpha(x)\hat{g}_{\alpha\beta}^{-1}(x) = N(\hat{g}_{\alpha\beta}(x))f_\alpha(x)$$

the Fundamental Lemma (10.3.1), ensures that

$$N(\hat{g}_{\alpha\beta}(x)) = \mu^2 N(\hat{g}_{\alpha\beta}(x)), \quad \mu^2 = 1$$

so that $\hat{g}_{\alpha\beta}(x)f_\alpha(x) = \pm f_\alpha(x)$ and $f_\beta(x)\hat{g}_{\alpha\beta}^{-1}(x) = \pm f_\alpha(x)$ i.e.

$$g_{\alpha\beta}(x)f = \pm f, \quad g_{\alpha\beta}(x) \in H.$$

2. Choose a standard f such that $\varphi_\alpha^x(f) = f_\alpha(x)$, $\varphi_\beta^x(f) = f_\beta(x)$ and introduce the local Clifford section $x \to \varphi_\alpha^x(x_{i_1} \ldots x_{i_h} f)$ over (U_α); we will write $(x_{i_1} \ldots x_{i_h} f)_\alpha^x$ for short.
Since $f_\beta(x)\hat{g}_{\alpha\beta}^{-1}(x) = \pm f_\alpha(x)$,

$$(x_{i_1} \ldots x_{i_h} f)_\beta^x = \pm \hat{g}_{\alpha\beta}(x)(x_{i_1} \ldots x_{i_h} f)_\alpha^x$$

and both Clifford sections being well-defined, for $x \in U_\alpha \cap U_\beta$ a $\hat{\gamma}_{\alpha\beta}(x)$ can be found such that $\varphi(\hat{\gamma}_{\alpha\beta}) = \varphi(\hat{g}_{\alpha\beta}(x))$, $\gamma_{\alpha\beta}(x) \in \text{Pin}(Q)$.

$$(x_{i_1} \ldots x_{i_h} f)_\beta^x = \hat{\gamma}_{\alpha\beta}(x)(x_{i_1} \ldots x_{i_h} f)_\alpha^x$$

and a $\text{Pin}(Q)$ spinor structure for V is obtained, since at every point $x \in U_\alpha \cap U_\beta \cap U_\gamma$, the coefficients of $\gamma_{\alpha\beta}(x)$, $\gamma_{\alpha\gamma}(x)$ and $\gamma_{\beta\gamma}(x)$ will satisfy the coherence condition for a bundle, $\varphi(\gamma_{\alpha\beta}(x)) = \varphi(g_{\alpha\beta}(x))$.

∎

Proposition 14.2.3 *A pseudo-riemannian manifold has a* $\mathrm{Pin}(Q)$ *spinor structure if and only if the structural group of the principal bundle of frames can be reduced to a spinoriality group in* $\mathrm{O}(Q')$.

Proof. Denoting the transition functions by $\varphi(g_{\alpha\beta}(x))$,

$$g_{\alpha\beta}(x)f = \pm f \quad \text{and} \quad \varphi_\alpha^x(g_{\alpha\beta}(x)f) = \pm f_\alpha(x),$$

the φ_α^x being the local isomorphisms :

$$\hat{g}_{\alpha\beta}(x)f_\alpha(x) = \pm f_\alpha(x)$$

leads to $f_\alpha(x)\hat{g}_{\alpha\beta}^{-1}(x) = \pm f_\alpha(x)$ and from $f_\beta(x) = \hat{g}_{\alpha\beta}(x)f_\alpha(x)\hat{g}_{\alpha\beta}^{-1}(x)$ we find

$$f_\beta(x) = \pm\hat{g}_{\alpha\beta}(x)f_\alpha(x) = \pm f_\alpha(x),$$

$$f_\beta(x)\hat{g}_{\alpha\beta}^{-1}(x) = \pm f_\alpha(x)$$

and the proof continues as in Proposition (14.2.2), part 2.   ∎

Corollary 14.2.1 *The set of* $\mathrm{Pin}(Q)$ *spinor structures on a manifold V on which at least one exists, is of the same cardinality as* $H^1(V, \mathbf{Z}_2)$.

Proof. This follows from part 2 in the proof of Proposition (14.2.2); the good choice of $\hat{\gamma}_{\alpha\beta}$ depends on a cocycle with values in $\mathbf{Z}_2 \simeq \{-1, 1\}$.   ∎

Remark.

1. If a $\mathrm{Pin}(Q)$ spinor structure exists, it is important to note that the reduction to a spinoriality group leads to a $\mathrm{SO}(Q)$-valued cocycle; from this, one cannot conclude that the bundle ξ is orientable, since this reduction is carried out in $\mathrm{O}(Q')$.

2. Under the same assumptions as 1, an algebraic spin subbundle is defined by the pseudo-field whose restriction to U_α is $x \to f_\alpha(x)$; this is a complex subbundle of the bundle $\mathrm{Clif}'(V, Q')$.

3. If V is orientable, the notion of $\mathrm{Spin}(Q)$ spinor structure should be clear to the reader, who may easily adapt the results of this subsection to that case.

14.3   Some particular results.

14.3.1   Spinor structures on spacetime $V_{1,3}$.

Let us compute a spinoriality group, using the special Witt basis

$$x_1 = \frac{e_1 + e_4}{2}, \quad x_2 = \frac{ie_2 + e_3}{2}, \quad y_1 = \frac{e_1 - e_4}{2}, \quad y_2 = \frac{ie_2 - e_3}{2}$$

and express that $uf = 0$, $f = y_1 y_2$, if $u \in \mathcal{L}(H)$, the Lie algebra of H. These elements are of the form $\alpha x_2 y_1 + \bar{\alpha} y_1 y_2$, or

$$B(e_1 - e_4)e_3 + C(e_1 - e_4)e_2, \tag{1}$$

B and C being real numbers. $\mathcal{L}(H)$ is easily seen to be a commutative Lie algebra. Consider the connected component of the identity in the Lorentz group $O(1,3)$, i.e. the restricted Lorentz group, sometimes written as L_4^0. The connected component $\mathcal{G}_0$ of the spinoriality group $\mathcal{G}$ must be a commutative group of real dimension 2, its Lie algebra being commutative. Formula (1) shows that every element of this group is generated by :

a hyperbolic rotation ψ in (e_1, e_3),
a hyperbolic rotation θ in (e_1, e_2),
an ordinary rotation θ in (e_3, e_4),
an ordinary rotation ψ in (e_2, e_4).

This group therefore is not compact, and being connected and abelian, a classical result implies that it is isomorphic to $\mathbf{R}^2$. Under these conditions a spinoriality group (connected or not) is homeomorphic to a finite product of copies of $\mathbf{R}^2$.

If a $\mathrm{Pin}(Q)$ spinor structure exists over $V_{1,3}$, a reduction to a spinoriality group must occur for the structural group $O(1,3)$. By a known result, the bundle obtained in this way is trivial, its structural group having the topological property established in the previous paragraph. Hence :

Proposition 14.3.1 *If a $\mathrm{Pin}(Q)$ spinor structure exists over spacetime, the bundle of frames is trivial.*[1]

14.3.2 Almost hermitian structure and spinor structure.

Let V be an $(n = 2r)$-dimensional almost complex manifold and let Q be a positive definite metric on Q endowing V with an almost hermitian structure. In Chapter 11, we saw that in this case a spinoriality group $\mathcal{G}$ is similar to $\mathrm{SU}(r, \mathbf{C})$ in $O(Q')$.

We recall some classical results.

A real $(n = 2r)$-dimensional manifold is called a *almost complex* manifold if the structural group of its frame bundle can be reduced to $\mathrm{GL}(n, \mathbf{C})$. This is equivalent to the existence of a section $J : x \to J_x$ of $T(V) \otimes T^*(V)$ on V for which $J^2 = -\mathrm{Id}$, or to the decomposition of the complexified tangent bundle $T^{\mathbf{C}}(V)$ as a direct sum of two complex conjugate bundles, i.e. $T^{\mathbf{C}}(V) = T_1 \oplus \bar{T}_1$, the trivialization cocycles of the vector bundles T_1 and $\bar{T}_1$ having complex conjugate transition functions. Such a V is orientable.

Then a hermitian metric can be defined on V, such that for the associated symmetric bilinear form, $g(JX, JY) = g(X, Y)$ holds for all vector fields X, Y on V.

---

[1]Note that the reduction is in $O(Q')$.

Next a regular antisymmetric 2-form F is defined on V by

$$F(X, Y) = g(X, JY),$$

giving V an 'almost symplectic' structure.

In each $x \in V$, the spaces $(T_1)_x$ and $(\bar{T}_1)_x$ are eigenspaces of J with eigenvalues i and $-i$; they are totally isotropic for the metric g. The subbundles T_1 and $\bar{T}_1$ therefore naturally define two amorphic spin subbundles of $\text{Clif}'(V)$—in the purely algebraic sense of Definition (14.1.2).

Definition 14.3.1 *A* $\text{Spin}(Q)$ *spinor structure is said to be naturally associated to the almost hermitian structure if the amorphic spin subbundle defined by it is the complex subbundle* $\theta(V)$ *determined by* $T(V)$ *in* $T^{\mathbf{C}}(V)$.

Note that the introduction of an almost hermitian structure reduces the structural group of the tangent bundle $T(V)$ to $\text{U}(r, \mathbf{C})$ and that $\theta(V)$ can be identified with T_1.

Under these conditions, we have :

Proposition 14.3.2 *The reduction to* $\text{SU}(r, \mathbf{C})$ *in* $\text{SO}(Q')$ *of the structural group of* $T(V)$, *V being a manifold with almost hermitian structure, is equivalent to the existence of a* $\text{Spin}(Q)$ *spinor structure which is naturally associated to the almost hermitian structure.*

Proof. Let $(U_\alpha, \theta_\alpha)$ be an atlas of local trivializations of $T(V)$, θ_α^x extends to the Clifford algebras. We set

$$f'_\alpha(x) = \theta_\alpha^x(f), \quad f = y_1 y_2 \ldots y_r \quad \text{(as before)},$$
$$f'_\beta(x) = \hat{g}'_{\alpha\beta}(x) f'_\alpha(x) \hat{g}'^{-1}_{\alpha\beta}(x), \quad \varphi(g'_{\alpha\beta}(x)) \in \text{U}(r, \mathbf{C}),$$
$$f'_\beta(x) = \mu^{(x)}_{\alpha\beta} f'_\alpha(x),$$

where $|\mu_{\alpha\beta}(x)| = 1$ since this coefficient corresponds to the determinant of a unitary matrix. If the group $\text{U}(r, \mathbf{C})$ reduces to $\text{SU}(r, \mathbf{C})$, the cocycle defined by the $(\mu_{\alpha\beta})$ is cohomologous to the identity and local isomorphisms θ_α exist for $T(V)$, with transition functions in $\text{SU}(r, \mathbf{C})$, for which

$$\hat{\theta}_\alpha(f) = \theta_\alpha^x(\lambda_\alpha(x) f \lambda_\alpha^{-1}(x)), \quad \lambda_\alpha(x) \in \text{Pin}(Q') \quad \text{(even } \text{Spin}(Q'))$$
$$= f_\alpha(x).$$

$f_\alpha(x)$ and $f'_\alpha(x)$ define the same m.t.i.s. at x, but

$$f_\beta(x) = \hat{g}_{\alpha\beta}(x) f_\alpha(x) \hat{g}_{\alpha\beta}^{-1}(x), \quad f_\beta(x) = f_\alpha(x)$$

and $N(g_{\alpha\beta}(x)) = 1$ completes the proof. The converse follows from the previous results.   ∎

In fact V is orientable and we could speak of $\text{Spin}(Q)$ spinor structures.

The following Lemma will be used in the proof of Proposition (14.3.3) :

Lemma 14.3.1 *If f is the standard isotropic r-vector and if $\lambda \in S^1$, there exists a $\delta \in \mathrm{Spin}(Q)$ such that $\lambda f = \delta f$.*

Proof. Let $f = y_1 y_2 \ldots y_r$ and $f' = y_1' y_2' \ldots y_r' = \lambda^2 f$, $\lambda = \exp i\theta$, θ a real number. f' can be determined by the 'real' Witt basis

$$\{x_{i'}, y_{j'}\} = (x_1, \ldots, x_{r-1}, \exp(-2i\theta)x_r, y_1, y_2, \ldots, y_{r-1}, \exp(2i\theta)y_r),$$

constructed from the 'real' Witt basis $\{x_i, y_j\}$.

Let $\delta \in \mathrm{Pin}(Q)$ sending $\{x_i, y_j\}$ to $\{x_{i'}, y_{j'}\}$, $\delta f \delta^{-1} = \lambda^2 f$ implies that $\delta f = \mu f$ and $\lambda^2 = \mu^2 N(\delta)$ (cf. Chapter 10, 3.1), but $\delta \in \mathrm{Spin}(Q)$ and $N(\delta) = 1$. so $\delta f = \pm f$ and δ may be chosen such that $\delta f = \lambda f$. ∎

Proposition 14.3.3 *A $\mathrm{Spin}(Q)$ spinor structure for the manifold V with almost hermitian structure exists if and only if the complex line bundle $\wedge^r(\theta(V))$ is isomorphic to the tensor square of another complex line bundle.*

Proof. Using the notations of the proof of Proposition (14.3.2), if $\mu_{\alpha\beta}(x) = (\nu_{\alpha\beta})^2$ for some cocycle $(\nu_{\alpha\beta})$, local sections $x \in U_\alpha \to \hat{\sigma}_\alpha(x)$ exist in the bundle of cocycle $(\nu_{\alpha\beta})$, such that $\hat{\sigma}_\beta^2(x) = \hat{\sigma}_\alpha^2 \mu_{\alpha\beta}(x)$.

$\hat{\delta}_\alpha(x)$ is constructed over U_α so that

$$\hat{\delta}_\alpha(x) f_\alpha'(x) = \frac{1}{\hat{\sigma}_\alpha(x)} f_\alpha'(x),$$

then

$$\hat{\delta}_\alpha(x) f_\alpha'(x) \hat{\delta}_\alpha^{-1}(x) = \frac{1}{\hat{\sigma}_\alpha^2(x)} f_\alpha'(x).$$

Equating this to f_α and relying on $f_\beta'(x) = \mu_{\alpha\beta}(x) f_\alpha'(x)$, $f_\beta(x) = f_\alpha(x)$,

$$f_\beta(x) = \gamma_{\alpha\beta}(x) f_\alpha(x) \hat{\gamma}_{\alpha\beta}^{-1}(x), \quad \gamma_{\alpha\beta} = \delta_\beta g_{\alpha\beta}' \delta_\alpha^{-1},$$

and there exists a $\mathrm{Spin}(Q)$ spinor structure.

Conversely, if V has a $\mathrm{Spin}(Q)$ spinor structure, we may assume that it is naturally associated to $\theta(V)$, $x \to f_\alpha'(x)$ defining $\theta(V)$ locally and $x \to f_\alpha(x)$ being a $\mathrm{Spin}(Q)$ spinor structure (so $f_\beta(x) = f_\alpha(x)$, $f_\alpha'(x)$ and $f_\alpha(x)$ are collinear).

$$f_\beta'(x) = \mu_{\alpha\beta}(x) f_\alpha'(x), \quad |\mu_{\alpha\beta}(x)| = 1,$$

choosing $\hat{\delta}_\beta$ such that $f_\beta(x) = \hat{\delta}_\beta(x) f_\beta'(x) \hat{\delta}_\beta^{-1}(x)$,

$$\hat{\delta}_\beta(x) f_\beta'(x) = \frac{1}{\hat{\sigma}_\beta(x)} f_\beta'(x),$$

then

$$f_\beta(x) = \frac{1}{\hat{\sigma}_\beta^2(x)} f'_\beta(x).$$

From $f_\beta(x) = f_\alpha(x)$ we then see that

$$\hat{\sigma}_\beta^2(x) = \hat{\sigma}_\alpha^2(x)\mu_{\alpha\beta}(x).$$

∎

Remark. $H^1(V, \mathbf{C}^*)$ is known to be isomorphic to $H^2(V, \mathbf{Z})$. The cocycle $(\mu_{\alpha\beta})$ defines the first Chern class $c_1(\theta(V))$; Proposition (14.3.3) means that a $\mathrm{Spin}(Q)$ spinor structure exists if and only if $c_1(\theta(V)) = 2\tilde{c}_1(\theta(V))$ for some $\tilde{c}_1(\theta(V)) \in H^2(V, \mathbf{Z})$. This condition implies the vanishing of the second Stiefel-Whitney class w_2, since it is the $(\bmod\, 2)$ reduction of $c_1(\theta(V))$.

14.3.3   The generalized twistor bundle naturally associated to a $\mathrm{Pin}(Q)$ spinor bundle.

Having defined a twistor of order q as any element of a vector space isomorphic to the direct sum of q spinor spaces for the complexified Clifford algebra over (E, Q), a twistor is an element of a left ideal of $C(Q')$ for the chosen representation.

If a $\mathrm{Pin}(Q)$ spinor structure exists on the manifold V, we know by the previous results that a pseudo-field of isotropic r-vectors exists in its complexified Clifford bundle.

The choice of the standard spinor space corresponds to that of an isotropic r-vector f. From $f = y_1 y_2 \ldots y_r$, where $\{x_i, y_j\}$ is a real Witt basis, 2^r isotropic r-vectors, of which no two are collinear, can be obtained if we exchange some x_k and y_k (cf. the proof of Proposition (8.1.5)).

Clearly, the choice of f does not affect the definition of the spinor vector bundle associated to the spin principal bundle (H is changed to gHg^{-1}, $g \in \mathrm{Pin}(Q)$). Therefore we have the :

Proposition 14.3.4 *If there exists a $\mathrm{Pin}(Q)$ spinor bundle over V, there also exists a twistor bundle of order 2^r over V, i.e. the Clifford bundle is the Whitney sum of 2^r spinor bundles.*

14.4   Enlarged spinor structures and others.

14.4.1

Definition 14.4.1 *The pseudo-riemannian (possibly riemannian) manifold V is said to have a enlarged $\mathrm{Pin}(Q)$ spinor structure if the structural group $O(Q)$ of the principal bundle of frames can be reduced in $O(Q')$ to an enlarged spinoriality group.*

Proposition 14.4.1 *If an enlarged* $\mathrm{Pin}(Q)$ *spinor structure exists on* V :

1. *There exists, up to a non-vanishing scalar factor, a field of isotropic r-vectors on V and, consequently, a vector subbundle of* $\mathrm{Clif}'(V, Q')$*, defined for* $x \in U_\alpha$ *by* $f_\alpha(x)$.

2. *The complexified pseudo-riemannian bundle of frames has local sections :*

$$x \in U_\alpha \to \Omega_\alpha(x), \quad \Omega_\beta(x) = \Omega_\alpha(x)\varphi(g_{\alpha\beta}(x)), \quad \varphi(g_{\alpha\beta}(x)) \in O(Q),$$

$$x \in U_\alpha \cap U_\beta, \text{ such that } f_\beta(x) = \hat{\mu}_{\alpha\beta} f_\alpha(x), \quad \hat{\mu}_{\alpha\beta}(x) \in \mathbf{C}^*,$$

$$f_\beta(x) = \hat{g}_{\alpha\beta}(x) f_\alpha(x) \hat{g}_{\alpha\beta}^{-1}(x),$$

and conversely.

Proof. If 1 and 2 hold, the proof given for Proposition (14.2.2) leads to

$$g_{\alpha\beta}(x)f = \lambda_{\alpha\beta}(x)f, \quad \lambda_{\alpha\beta}(x) \in \mathbf{C}^*.$$

Conversely, if $g_{\alpha\beta}(x)f = \lambda_{\alpha\beta}(x)f$, the proof of Proposition (14.2.3) can be used. ∎

Corollary 14.4.1 *If V has almost complex structure, an enlarged* $\mathrm{Spin}(Q)$ *spinor structure exists on it.*

Proof. Obvious, since $|\mu_{\alpha\beta}| = 1$ in this case.   ∎

Remark. If we define the group Ge $\subseteq C(Q')$ of all $\rho \exp(i\theta)g$, $g \in \mathrm{Pin}(Q)$, $\rho > 0$ and θ real, replacing $\mathrm{Pin}(Q)$ by Ge in Definition (14.2.2) leads to the necessary condition expressed by parts 1 and 2 of Proposition (14.4.1).

We could therefore speak of a Ge spinor structure when the condition of Definition (14.4.1) is met.

Topological considerations which the reader may find in the literature, allows the reduction of Ge to its subgroup obtained by fixing $\rho = 1$; when Q is positive definite, this is already so for algebraic reasons.

In this case the group is called a reduced enlarged spinoriality group (or a torogonal spinoriality group) and we speak of a torogonal spinor structure.

14.4.2   Enlarged spinor structures on the spacetime $V_{1,3}$.

The computation of the Lie algebra for a torogonal spinoriality group leads to an element $v = B(e_1 - e_4)e_3 + C(e_1 - e_4)e_2 + Ae_2e_3$, using the notations of subsection 3.1, A, B and C being real.

This Lie algebra clearly is not commutative. The simple reasoning proving Proposition (14.3.1) no longer holds.

The Lie algebra of the 'little group', i.e. the stabilizer of $(e_1 - e_4)$ can be identified with the Lie algebra of a torogonal spinoriality group. Hence :

The 'little group' associated to an isotropic vector and a torogonal spinoriality group are (at least) locally isomorphic.

14.4.3 Pin(Q') spinor structures.

We shall only consider complexified spaces. Pin(Q') can be projected onto O(Q'), with a kernel consisting of four elements, $N(g) = \pm 1$ leading to $\{1, -1, i, -i\}$.

The Definition (14.2.2) and the proof of Proposition (14.2.1) can be adapted at once. $\epsilon = \pm 1$ or $\epsilon = \pm i$, so $f_\beta(x) = \pm f_\alpha(x)$ replaces formula (8). An analogous proposition to (14.2.2) can be stated and proved.

Consider a subbundle of m.t.i.s. in the complexified bundle $T_\mathbf{C}(V)$, i.e. a complex subbundle with, in every $x \in V$, an m.t.i.s. of $T_\mathbf{C}^x(V)$ as fiber. By Proposition (2.2.5), such a subbundle τ corresponds to a field of isotropic r-vectors defined over V up to a non-vanishing scalar factor. If $(U_\alpha)_{\alpha \in A}$ is a set of trivializing open sets for this subbundle, we can define, for $x \in U_\alpha$, $x \to f_\alpha(x) = \varphi_\alpha^x(y_1 y_2 \ldots y_r)$, where $f_\beta(x) = \mu_{\alpha\beta}(x) f_\alpha(x)$, $x \in U_\alpha \cap U_\beta$, $\mu_{\alpha\beta} \in \mathbf{C}^*$. The $\mu_{\alpha\beta}$ define a cocycle for the complex line bundle $\wedge^r(\tau)$. Such a description was given before when V was assumed to be almost complex (cf. Proposition (14.3.3), with $\tau = \theta(V)$).

Lemma 14.4.1 *If f_α and f_β are local isotropic r-vector fields connected with the m.t.i.s. subbundle over the trivializing open sets U_α and U_β, new isotropic r-vectors f'_α and f'_β can be chosen such as to define the same bundle and that if $x \in U_\alpha \cap U_\beta$, $f'_\beta(x) = \pm f'_\alpha$.*

Proof. As in Lemma (14.3.1) (but $\lambda = \rho \exp i\theta$ here, $\rho > 0$) we introduce the Witt basis $f' = y'_1 y'_2 \ldots y'_r = \lambda^2 f$,

$$\{x'_i, y'_j\} = (\frac{x_1}{\rho^2}, x_2, \ldots, \exp(-2i\theta)x_r, \rho^2 y_1, y_2, \ldots, \exp(2i\theta)y_r).$$

Then there exist local sections σ_α^2 for the bundle with cocycle $(\mu_{\alpha\beta})$: $\sigma_\rho^2(x) = \sigma_\alpha^2(x)\mu_{\alpha\beta}(x)$. A $\delta_\alpha(x) \in$ Pin(Q') can be found such that

$$\frac{f_\alpha(x)}{\sigma_\alpha(x)} = \delta_\alpha(x) f_\alpha(x),$$

so that

$$f'_\alpha(x) = \delta_\alpha(x) f_\alpha(x) \delta_\alpha^{-1}(x) = \pm \frac{f_\alpha(x)}{\sigma_\alpha^2(x)},$$

$$f'_\beta(x) = \pm f'_\alpha(x).$$

∎

Proposition 14.4.2 *The existence of a Pin(Q') spinor structure on V is equivalent to that of a m.t.i.s. subbundle of $T_\mathbf{C}(V)$.*

Proof. One part of this proposition has already been proved. Conversely, if a m.t.i.s. subbundle of $T_C(V)$ exists, (f'_α) can be found such that

$$f'_\beta(x) = \pm f'_\alpha(x),$$
$$f'_\beta(x) = \hat{g}_{\alpha\beta}(x)f'_\alpha(x)\hat{g}_{\alpha\beta}^{-1}(x), \quad \hat{g}_{\alpha\beta}(x) = \varphi(g_{\alpha\beta}(x)),$$
$$g_{\alpha\beta}(x) \in \mathrm{Pin}(Q'),$$

and the analogous result to Proposition (14.2.2) can be applied. ∎

Remark. By Corollary (10.3.1), a 'real' Witt basis can be considered when determining a m.t.i.s. If, therefore, we have a $\mathrm{Pin}(Q')$ spinor structure determined by the conditions of Proposition (14.4.1), except that $\varphi(g_{\alpha\beta}(x)) \in O(Q')$, we can return to the case where

$$f_\beta(x) = \hat{\mu}_{\alpha\beta}(x)f_\alpha(x)$$
$$f_\beta(x) = \hat{\gamma}_{\alpha\beta}f_\alpha(x)\hat{\gamma}_{\alpha\beta}^{-1}(x), \quad \varphi(\gamma_{\alpha\beta}(x)) \in O(Q),$$

relying on the previous remark.

The notions of enlarged spinor structure and of $\mathrm{Pin}(Q')$ spinor structure are therefore equivalent. Since the existence condition for a $\mathrm{Pin}(Q')$ structure is independent of the signature of Q, the same holds for the existence condition for an enlarged spinor structure (except for the existence of pseudo-riemannian bundles).

14.5 Spinor structures in the odd-dimensional case.

Assume that V is $(2r+1)$-dimensional and *orientable*. The groups under consideration are $SO(Q)$, $\mathrm{Spin}(Q)$ etc. Our definitions from section 2 go through unchanged. The standard spinors being a representation space of $C^+(Q)$ (cf. Chapter 9), we consider a Witt basis $\{x_i, y_j, z_0\}$, where z_0 is non-isotropic, the x_i, y_j, z_0 determine a Witt decomposition and the space of $\{x_{i_1} \ldots x_{i_h}f, f = y_1y_2 \ldots y_r\}$ is the standard spinor space. The results of section 2 can now easily be adapted to this case.

14.6 Spinor structures on spheres and on projective spaces.

14.6.1 *Manifolds are considered here with C^∞ differentiable structure.*

We will establish the existence of a spinor structure on all spheres S^n, where $n = 2r$ or $2r+1$, $r \geq 1$, by showing that r global isotropic sections of the complexified tangent bundle can be defined such that they are linearly independent and generate an m.t.i.s. field at every point.

An atlas with two maps is defined for S^n by inversions, with centers given by the poles P and P' whose Cartesian coordinates are $(0, 0, \ldots, 1)$ and $(0, 0, \ldots, -1)$. If $(x_1, x_2, \ldots, x_{n+1})$ stands for the coordinates of a point in $\mathbf{R}^{n+1}$ and $(u_1, u_2, \ldots u_n)$

for those of a point in the equatorial hyperplane $H(x^{n+1} = 0)$, we get for the point $M \in S^n$:

$$x_i = \frac{2u_i}{\Delta^2 + 1}, \quad i = 1, 2, \ldots, n, \quad x_{n+1} = \epsilon \frac{\Delta^2 - 1}{\Delta^2 + 1}, \quad \Delta^2 = \sum_1^n (u_i)^2, \qquad (1)$$

where $\epsilon = \pm 1$ depending on the center P ($\epsilon = 1$) or P' ($\epsilon = -1$), under the condition that, for $\epsilon = 1$, $-1 \le x_{n+1} < 1 - \alpha$, $0 < \alpha < 1$.

To each differentiable path in H, with tangent $(\dot{u}_1, \dot{u}_2, \ldots, \dot{u}_n)$, there corresponds a differentiable path in S^n, whose tangent components are given by :

$$\dot{x}_i = \frac{2(\Delta^2 + 1)\dot{u}_i - 4u_i \sum u_k \dot{u}_k}{(\Delta^2 + 1)^2}, \quad i = 1, 2, \ldots, n, \quad \dot{x}_{n+1} = \epsilon \frac{4 \sum u_k \dot{u}_k}{(\Delta^2 + 1)^2}, \qquad (2)$$

from which we see that

$$\sum_1^{n+1} (\dot{x}_i)^2 = \frac{4}{(\Delta^2 + 1)^2} \sum (\dot{u}_i)^2,$$

displaying the conformal property of the stereographic projection, with coefficient $2/(\Delta^2 + 1)$ (cf. Chapter 12, 2).

We will also use the inversion with center $Q = (-1, 0, \ldots, 0)$ and the equatorial plane $K(x_1 = 0)$, with general point w denoted by $(w_2, w_3, \ldots w_{n+1})$. It can easily be verified that :

$$w_2 = \frac{2u_2}{D}, \ldots, w_n = \frac{2u_n}{D}, w_{n+1} = 1 - \frac{2(u_1 + 1)}{D}, \quad D = \Delta^2 + 2u_1 + 1. \qquad (3)$$

In the intersection of the coordinate domains corresponding to the inversions with centers P and Q, a local isotropic vector field of S^n corresponds to local fields of H and K respectively, by the tangent stereographic maps. If we choose the field $A(u) = e_1 + ie_2$ on H, $(e_1, e_2, \ldots, e_n)$ being the canonical orthonormal basis of H, a field $B(w)$ on K is associated to A, its components being given, relying on (3), by

$$\begin{aligned}
&B_2 = 2i/D - w_2(1 - w_{n+1} + iw_2) \\
&B_k = -w_k(1 - w_{n+1} + iw_2), \quad k = 3, 4, \ldots, n, \\
&B_{n+1} = -2/D + (1 - w_{n+1})(1 - w_{n+1} + iw_2).
\end{aligned} \qquad (4)$$

The field B obviously is isotropic, everywhere defined and non-vanishing on $K \setminus \{P\}$.

Note that

$$\frac{2}{D} = \frac{w_2^2 + \ldots + w_n^2 + (1 - w_{n+1})^2}{2}.$$

In the open subset of S^n defined by $-1 \le x_{n+1} < 1 - \alpha$, $0 \le \alpha < 1$, there exists a field of non-zero isotropic vectors obtained from A by the inversion with center P, and in the open set defined by $-1 + \beta < x_1 \le 1$, $0 \le \beta < 1$, there exists a field of everywhere non-zero isotropic vectors (except at P), obtained from B by the inversion with center Q; on the intersection of both open sets, the fields coincide.

We will try to replace the image of B by a homotopic deformation which will be defined and non-vanishing on the whole of $-1 + \beta < x_1 \leq 1$.

Choose α_1, β_1 such that $0 < \alpha_1 < \beta_1 \leq 1/2$.

We can find a real C^∞ function $(\rho, w_{n+1}) \to \theta(\rho, w_{n+1})$ such that

$$0 \leq \theta(\rho, w_{n+1}) \leq \pi/2,$$
$$\theta(\rho, w_{n+1}) = \pi/2, \quad \text{if } \rho \leq \alpha_1 \text{ and } |1 - w_{n+1}| \leq \alpha_1,$$
$$\theta(\rho, w_{n+1}) = 0, \quad \text{if } \rho \geq \beta_1 \text{ or } |1 - w_{n+1}| \geq \beta_1.$$

The existence of such a (non-unique) function follows from a classical result. Writing θ for $\theta(\rho, w_{n+1})$, we define

$$\tilde{w}_2 = w_2 - (1/k)w_{n+1} \sin \theta,$$
$$1 - \tilde{w}_{n+1} = \cos \theta - w_{n+1} + k \sin \theta,$$
$$\lambda^2 \rho^2 = \sum_{j=2}^{n}(w_j)^2 + (1 - w_{n+1})^2, \quad \rho \geq 0,$$

k and λ being real constants to be fixed later on.

Let us write $C(w) = B(\tilde{w}_2, w_3, \ldots, w_n, \tilde{w}_{n+1})$, C is an isotropic field coinciding with B when $\rho \geq \beta_1$ or $|1 - w_{n+1}| \geq \beta_1$. Now we prove that for $\rho \leq \beta_1$ and $|1 - w_{n+1}| \leq \beta_1$, C does not vanish anywhere.

Indeed, if C would vanish,

$$\tilde{w}_2 = i(1 - \tilde{w}_{n+1})$$

would follow from $C_2 = 0$ and $C_{n+1} = 0$, hence

$$\begin{cases} w_2 = (1/k)w_{n+1} \sin \theta \\ w_{n+1} = \cos \theta + k \sin \theta \end{cases} \tag{5}$$

then $2/\tilde{D} = \frac{1}{2}(\sum_3^n(w_j)^2 + (\tilde{w}_2)^2 + (1 - \tilde{w}_{n+1})^2) = 0$ would lead to

$$w_3 = w_4 = \ldots = w_n = 0.$$

From (5) we deduce that

$$\frac{1}{k^2}(\cos \theta + k \sin \theta)^2 \sin^2 \theta + (1 - \cos \theta - k \sin \theta)^2 = \lambda^2 \rho^2, \tag{6}$$

θ standing for $\theta(\rho, w_{n+1})$.

The existence of a pair (w_2, w_{n+1}) for which C vanishes is equivalent to that of some θ, $0 < \theta \leq \pi/2$ for which (6) holds. Now we fix $\beta_1 = k = 1/2$.

For each solution where C vanishes we must ensure that $|1 - w_{n+1}| < 1/2$. Direct consideration of $1 - \cos \theta - k \sin \theta$ shows, however, that

$$(1 - \sqrt{1 + k^2}) \leq 1 - w_{n+1} \leq 1 - k,$$

so that $|1 - w_{n+1}| \leq 1/2$ is automatically satisfied.

If $\theta = \pi/2$, $1 - w_{n+1} = 1/2$, $w_2 = 1$ and the corresponding value of $\lambda^2 \rho^2$ is $5/4$, but since $1 - w_{n+1} = 1/2$ implies $\theta = 0$ by the definition of θ, a contradiction is reached.

Similarly, if $\theta = \pi/2 - \epsilon$ is a solution, ϵ being arbitrarily small, $(1 - w_{n+1})^2 + w_2^2 = \lambda^2 \rho^2$ is arbitrarily close to $5/4$, but since $\rho \geq \alpha_1$, the choice of a sufficiently large λ again leads to a contradiction.

Therefore a maximal value $\theta_0 < \pi/2$ exists for the solution.

Denoting the left-hand side of (6) by $f^2(\theta)$, if $\theta_0 < \theta_1 < \pi/2$, $(1/\lambda^2)f^2(\theta_1) = \rho_1^2$ can be made smaller than $1/4$ if λ is sufficiently large; and we can even make sure that $(1/\lambda^2)f^2(\theta) < 1/4$ for all $\theta \leq \theta_1$. Since $|1 - w_{n+1}^1| = |1 - \cos\theta_1 - k\sin\theta_1|$ can be assumed strictly smaller than $1/2$, a function $\theta(\rho, w_{n+1})$ can be found (in infinitely many ways) so as to satisfy the previous conditions and $\theta(\rho_1, w_{n+1}^1) = \theta_1$, and whose graph does not intersect the graph of $\theta \to (\theta, (1/\lambda)f(\theta), 1 - w_{n+1}(\theta))$ in any acceptable solution. C cannot vanish.

For $\rho = 0$, $\theta(\rho, w_{n+1}) = \pi/2$, $\tilde{w}_2 = -1/k$, $1 - \tilde{w}_{n+1} = k - 1$, $C_2(P) = ai$, $C_{n+1}(P) = -a$, $C_j(P) = 0$, $j = 3, \ldots, n$, where $2a = 1/k^2 + (k-1)^2 \neq 0$.

In this way, B has been deformed to a non-vanishing isotropic field for $\rho \leq \beta_1$; the image of this field under the inversion with center Q yields, along with A obtained previously, a global field of non-zero isotropic vectors on S^n; we have

Proposition 14.6.1 *On every sphere S^n, $n \geq 2$, there exist fields of everywhere non-zero isotropic vectors which are sections of the complexified tangent bundle of S^n.*

If $n \geq 4$, we choose an orthonormal basis $e_1, e_2, e_3, e_4, \ldots$ and consider in H the fields $A(u) = e_1 + ie_2$ and $A'(u) = e_3 + ie_4$. We construct in K the analog B' of B for A', its components $B_k'(w)$ are, relying on (3) :

$$
\begin{aligned}
B_2' &= -w_2(w_3 + iw_4), \\
B_3' &= 2/D - w_3(w_3 + iw_4), \\
B_4' &= 2i/D - w_4(w_3 + iw_4), \\
B_k' &= -w_k(w_2 + iw_4), \\
B_{n+1}' &= (1 - w_{n+1})(w_3 + iw_4).
\end{aligned}
\tag{7}
$$

We deform this field in the same way as B. The deformed field C' is everywhere non-zero since $C_3' = C_4' = 0$ would imply $iw_3 = w_4$, so $w_3 = w_4 = 0$ and $2/D$ would vanish, $w_4 = \ldots = w_n = 0$ and $\tilde{w}_2 = 1 - \tilde{w}_{n+1}$, which is impossible.

B and B' clearly are linearly independent everywhere; this property carries over to their deformations C and C' since $\alpha(w)B(\tilde{w}) + \beta(w)B'(\tilde{w}) = 0$ with $\tilde{w} = (\tilde{w}_2, w_3, \ldots, w_n, \tilde{w}_{n+1})$ would imply that B and B' are linearly dependent at $\tilde{w}$, which is impossible.

Repeating this procedure, a global field of isotropic r-vectors can be constructed on the sphere S^n, $n = 2r$ or $2r + 1$. So, S^n being orientable :

Proposition 14.6.2 *On every sphere S^n, $n = 2r$ or $2r + 1$, there exist global fields of isotropic r-vectors, and S^n therefore has a spinor structure.*

Remark. Replacing $e_1 + ie_2$ by $e_1 - ie_2$ in the proof of Proposition (14.6.1), two complex conjugate isotropic vector fields are obtained on S^n; similarly, two complex conjugate global fields of isotropic r-vectors can be constructed on S^n, giving rise to a Witt decomposition of the tangent space at every point.

The reader should not conclude from this that, for even n, S^n has an almost complex structure : this is known only to hold if $n = 2$ or 6.

The following considerations should clarify this :

We associate the field $\bar{J}$ to the field J constructed on S^n; $(J + \bar{J})/2$ is a real vector field[2] on S^n. Considering the real parts of $(B_2, \ldots B_{n+1})$ and $(C_2, \ldots, C_{n+1})$, the vanishing of its components leads to the same conditions as the vanishing of J or $\bar{J}$; $(J + \bar{J})/2$ is therefore an everywhere non-zero field of real vectors. Of course, it is a section of the complexified tangent bundle; we know that such a situation is impossible in the real bundle of S^n, $n = 2r$.

So if the bundle determined by the global field of isotropic r-vectors and the conjugate field were almost complex, the reduction to a spinoriality group would occur in the real orthogonal group and there would exist non-vanishing global sections of the real tangent bundle on the sphere, which is impossible.

Since there exists a global real non-vanishing vector field on every odd-dimensional sphere, there exist n linearly independent sections of the complexified tangent bundle of any sphere and hence :

Corollary 14.6.1 *The complexified tangent bundle of any sphere is trivial.*

Remarks. It is easy to construct many different non-vanishing complex vector fields on even-dimensional spheres.

For instance, in $\mathbf{R}^5 = (e_1, e_2, e_3, e_4, e_5)$; we can consider the sphere S^4 and the complex fields[3]

$$U = -x_2e_1 + x_1e_2 + (x_4 + ix_5)e_3 - x_3e_4 - ix_3e_5,$$
$$V = -ix_4e_1 - x_3e_2 + x_2e_3 + (x_5 + ix_1)e_4 - x_4e_5,$$

(the x_i being the usual Cartesian coordinates). U and V are clearly tangent to S^4 and do not vanish anywhere on it.

On the sphere S^2 the field

$$I = [(1-z) - x(x + iy)]e_1 + [(1-z)i - y(x + iy)]e_2 + (1-z)(x + iy)e_3$$

---

[2]Meaning that, at every point, it coincides with its conjugate.

[3]But note that these examples are too naïve to be used in the proof of Proposition (14.6.1).

is tangent and isotropic, non-zero unless $x = y = 0$ and $z = 1$; the field

$$J = [(1 - x) - y(y + iz)]e_2 + [(1 - x)i - z(y + iz)]e_3 + (1 - x)(y + iz)e_1$$

is also tangent and isotropic, it vanishes only at $x = 1$, $y = z = 0$.

It is easy to verify that they determine the same isotropic direction in every point where they are both defined, so their union defines a global field of isotropic directions on S^2, again proving that S^2 has an enlarged spinor structure.

14.6.2

Now consider the projective plane $P^n(\mathbf{R})$ which can be obtained from S^n by an identification of antipodal points under the symmetry s relative to 0.

$$S^n \xrightarrow{\pi} P^n(\mathbf{R})$$

where π is the projection from S^n to $P^n(\mathbf{R})$. The metric on $P^n(\mathbf{R})$ is naturally deduced from that on S^n and every spinor structure on $P^n(\mathbf{R})$ will give rise to a spinor structure on S^n, which will then be invariant under the isomorphism induced by s. Conversely, there corresponds a spinor structure on $P^n(\mathbf{R})$ to every spinor structure on S^n which is invariant under s.

Let us examine the action of s on a field of isotropic vectors on S^n. Locally, we can use an inversion with center P to consider the isotropic field J of $S^n \setminus \{P\}$ obtained from $A(u) = e_1 + ie_2$, whose components at $M = (x_1, x_2, \ldots, x_{n+1})$ are given by

$$\begin{aligned}
X_1 &= (1 - x_{n+1}) - x_1(x_1 + ix_2), \\
X_2 &= i(1 - x_{n+1}) - x_2(x_1 + ix_2), \\
X_k &= -x_k(x_1 + ix_2), \\
X_{n+1} &= (1 - x_{n+1})(x_1 + ix_2);
\end{aligned}$$

and $(s(J))(M) = -J(M)$.

First assume that n is even. The symmetry s determines an element $\gamma \in$ Pin$(n + 1)$ which transforms the field of isotropic r-vectors f of a spinor structure of S^n to $f' = \gamma f \gamma^{-1}$; this structure will be invariant under s if :

$$f' = N(\gamma)f,$$

by the results of section 2.3. Since $f' = (-1)^r f$ and $N(\gamma) = 1$, the metric being positive definite, a *spinor structure on $P^n(\mathbf{R})$ exists if and only if* $(-1)^r = N(\gamma) = 1$, i.e. if $r = 2r'$ $(n = 4, 8, 12, \ldots)$.

Note that if the metric was *negative definite*, γ should be replaced by γ' where $N(\gamma') = -1$, since s corresponds to an odd number of hyperplane symmetries, $N(a) = -1$ for a vector in $\mathbf{R}^{n+1}$ given this metric. *In this case, the existence condition is* $(-1)^r = -1$, or $r = 2r' + 1$ $(n = 2, 6, 10, \ldots)$. $P^n(\mathbf{R})$ is non-orientable if n is even, so the structures are Pin(Q) spinor structures.

Now assume that n is odd. By the general results on Clifford algebras, $C^+(Q)$ is, in the odd-dimensional case, isomorphic to the Clifford algebra of an n-dimensional space with a metric of opposite signature (cf. Chapter 3, 3.2). If, therefore, *the metric is positive definite*, we reobtain the previous situation with $N(\gamma') = -1$, $r = 2r' + 1$, for $n = 3, 7, 11, \ldots$, there exists a spinor structure on $P^n(\mathbf{R})$ (which is orientable).

If *the metric is negative definite*, we reobtain the previous situation with $N(\gamma) = 1$, but the determinant of the isometry determined by γ is negative, which is inadmissible since $P^n(\mathbf{R})$ is orientable. *For $n = 5, 9, 13, \ldots$ there exists no spinor structure on $P^n(\mathbf{R})$*, regardless of the definite signature.

These results could also be obtained by methods from algebraic topology, which are not of interest for our purposes.

14.6.3

The complex projective space $P^n(\mathbf{C})$ is obtained from the space $\mathbf{R}^{2n+2}$, given a linear operator J such that $J^2 = -\operatorname{Id}$ which determines a complex space structure $\mathbf{C}^{n+1}$ on it. $P_n(\mathbf{C})$ is the quotient of the sphere S^{2n+1} (the set of points $x \in \mathbf{C}^{n+1}$ for which $x\bar{x} = 1$) by the equivalence relation for which y and z are equivalent if and only if $z = J_\varphi(y)$ where $J_\varphi = \cos \varphi + J \sin \varphi$ (this corresponds to $z = (\exp i\varphi)y$ in $\mathbf{C}^{n+1}$).

J_φ is an isometry, since $(J_\varphi x)^2 = x^2$, x and $J(x)$ being orthogonal. Note that $J_0 = \operatorname{Id}$, $J_\pi = -\operatorname{Id}$.

The action of J_φ on a standard isotropic r-vector of H, written f, is expressed by :

$$f_1 = \gamma_\varphi f \gamma_\varphi^{-1} = (\cos n\varphi + i \sin n\varphi)f,$$

γ_φ lifts J_φ to $\operatorname{Spin}(2n+2)$, since

$$(\cos \varphi + J \sin \varphi)(e_1 + ie_2) = (\cos \varphi + i \sin \varphi)(e_1 + ie_2)$$

can be assumed; by Lemma (14.3.1) there exists $\delta_\varphi \in \operatorname{Spin}(2n+2)$ such that

$$(\cos \frac{n\pi}{2} + i \sin \frac{n\varphi}{2})f = \delta_\varphi f,$$

δ_φ is defined modulo $\pm \operatorname{Id}$, and the Fundamental Lemma (10.3.1) ensures that

$$\delta_\varphi f \delta_\varphi^{-1} = (\cos n\varphi + i \sin n\varphi)f,$$
$$(\delta_\varphi^{-1} \gamma_\varphi)f(\delta_\varphi^{-1} \gamma_\varphi)^{-1} = f.$$

On S^{2n+1} we reobtain the case examined in 6.2 for $P^n(\mathbf{R})$. There exists a spinor structure if and only if n is odd.

Chapter 15

SPIN DERIVATIONS.

15.1 Spin connections—covariant derivation.

15.1.1

We consider a pseudo-riemannian (possibly riemannian) manifold V with a spinor structure in one of the senses introduced in the previous chapter. The dimension of V is $n = 2r$.

The spinors are constructed using minimal left ideals by the method of isotropic r-vectors in the complexified Clifford algebra. ζ is the spinor bundle and $\Gamma(\zeta)$ the space of its sections.

We have already noted that the covariant derivation associated to an euclidean connection ∇ naturally extends to the manifold's Clifford algebra but not, in general, to its spinor fields.

Let U be an open set in V on which a local field of Witt bases $\{x_\alpha, x_{\beta^*}\}$, $\alpha, \beta = 1, 2, \ldots, r$ is defined, such that the product $x_{1^*} x_{2^*} \ldots x_{r^*} = f$ represents the pseudo-field of isotropic r-vectors which we know to exist if V has a $\mathrm{Pin}(Q)$ spinor structure. Over U we can express the local spinor fields in terms of the frame $\{x_{i_1} \ldots x_{i_h} f\}$.

The local components of the connection form ∇ will, if ∇ is euclidean, satisfy the conditions worked out in the previous chapter :

$$\omega_\beta^{\alpha^*} + \omega_\alpha^{\beta^*} = 0, \quad \omega_{\beta^*}^\alpha + \omega_{\alpha^*}^\beta = 0, \quad \omega_\beta^\alpha + \omega_{\alpha^*}^{\beta^*} = 0, \tag{1}$$

which can also be written

$$\omega_j^i + \omega_{i^*}^{j^*} = 0, \quad i, j = 1, 2, \ldots, n, \quad * = \pm r. \tag{2}$$

If we also express that these forms have values in the Lie algebra $\mathcal{L}(\mathcal{G})$ of a spinoriality group, we get

$$\sum \omega_\alpha^\alpha = 0 \tag{3}$$

and

$$\omega_{\beta^*}^\alpha = 0 \tag{4}$$

where (4), on its own, expresses that these forms belong to an enlarged spinoriality group (the condition of real-valuedness has not been expressed here).

204

15.1.2 Spin-euclidean connections.

Definition 15.1.1 *Every connection satisfying the conditions (1), (3) and (4) is called a proper spin-euclidean connection; a connection satisfying (1) and (4) is called an enlarged spin-euclidean connection.*

Proposition 15.1.1 *Let ∇ be an euclidean connection, then ∇ is proper spin-euclidean if and only if $\nabla f = 0$.*

Proof. By a computation carried out in the previous chapter,

$$\nabla f = \sum_{\alpha^*} \omega_{\alpha^*}^\lambda x_{1^*} \dots x_\lambda \dots x_{r^*} + \sum_{\lambda^*} \omega_{\lambda^*}^{\lambda^*} f.$$

∎

We have seen that such a connection naturally induces a covariant derivation of the spinor fields.

Proposition 15.1.2 *If the connection is proper spin-euclidean and if we set*

$$u = \tfrac{1}{4}\omega_j^i x_i x^j,$$

raising some of the indices by using the 'metric', then for each element

$$\psi = x_{\alpha_1} x_{\alpha_2} \dots x_{\alpha_h} f$$

of the local spin basis,

$$\nabla \psi = u\psi.$$

Proof. Since $\nabla f = 0$,

$$\nabla(x_{\alpha_1} \dots x_{\alpha_h} f) = \sum_{\alpha \in I} \omega_\alpha^i x_{\alpha_1} \dots x_i \dots x_{\alpha_h} f$$

where $I = \{\alpha_1, \alpha_2, \dots, \alpha_h\}$ and x_i replaces x_α. Now

$$\sum_{\alpha \in I} \omega_\alpha^\beta x_{\alpha_1} \dots x_\beta \dots x_{\alpha_h} f = \sum_{\alpha=1}^r \omega_\alpha^\beta x_\beta x_{\alpha^*}(x_{\alpha_1} \dots x_{\alpha_h} f).$$

Indeed, if $\alpha \notin I$, x_{α^*} anticommutes with $x_{\alpha_1}, \dots, x_{\alpha_h}$ and, passed over the product $x_{\alpha_1} \dots x_{\alpha_h}$, vanishes when multiplied by f. If $\alpha \in I$, $\alpha = \alpha_i$ for some $1 \le i \le h$ and x_{α^*} anticommutes with the elements before x_{α_i} in the product $x_{\alpha_1} \dots x_{\alpha_i} \dots x_{\alpha_h}$, since

$$x_{\alpha_i^*} x_{\alpha_i} = -x_{\alpha_i} x_{\alpha_i^*} + 1$$

we get two terms, one of which contains $x_{\alpha_i^*}$ and vanishes when multiplied by f. Then x_β is moved to the location of x_{α_i} and we note that $\omega_\alpha^\beta x_\beta x_{\alpha^*} = \tfrac{1}{2}\omega_\alpha^\beta x_\beta x^\alpha$.

By a similar computation,

$$\sum_{\alpha \in I} \omega_\alpha^{\beta^*} x_{\alpha_1} \dots x_{\beta^*} \dots x_{\alpha_h} f$$

is transformed to

$$\tfrac{1}{2} \sum_{\alpha=1}^{r} \omega_\alpha^{\beta^*} x_{\beta^*} x_{\alpha^*} (x_{\alpha_1} \dots x_{\alpha_h}) f,$$

$\omega_{\alpha^*}^{\beta^*} x_{\beta^*} x_{\alpha^*} = \tfrac{1}{4} \omega_\alpha^{\beta^*} x_{\beta^*} x^\alpha$. So $\nabla \psi = u\psi$ where $u = \tfrac{1}{2} \omega_\alpha^\beta x_\beta x^\alpha + \tfrac{1}{4} \omega_\alpha^{\beta^*} x_{\beta^*} x^\alpha$,

$$u = \tfrac{1}{4} \omega_j^i x_i x^j$$

since $\omega_\alpha^\beta x_\beta x^\alpha = \omega_{\alpha^*}^{\beta^*} x_{\beta^*} x^{\alpha^*}$ by (1) and (3). ∎

Proposition 15.1.3 *A euclidean connection naturally induces a derivation of spinor fields if and only if its forms have values in an enlarged spinoriality group.*

This was already proved (Proposition (14.1.5)). We obtained :

$$\nabla f = \sum (\omega_{\alpha^*}^{\alpha^*}) f$$

or $\omega_{\alpha^*}^{\alpha^*} f$ using the summation convention.

Proposition 15.1.4 *If ∇ is an enlarged spin-euclidean connection, for all $\psi = x_{\alpha_1} x_{\alpha_2} \dots x_{\alpha_h} f$,*

$$\nabla \psi = u\psi + \tfrac{1}{2} \omega_{\alpha^*}^{\alpha^*} \psi$$

where $u = \tfrac{1}{4} \omega_j^i x_i x^j$.

Proof. $\nabla \psi$ is the sum of $\sum_{\alpha \in I} \omega_\alpha^i x_{\alpha_1} \dots x_i \dots x_{\alpha_h} f$ (as in Proposition (15.1.2)) and

$$\sum (\omega_{\alpha^*}^{\alpha^*}) \psi = x_{\alpha_1} x_{\alpha_2} \dots x_{\alpha_h} (\sum \omega_{\alpha^*}^{\alpha^*} f),$$

since $\nabla f = \sum (\omega_{\alpha^*}^{\alpha^*}) f$. As before, the first term can be expressed as $v\psi$, where

$$v = \tfrac{1}{2} \omega_\alpha^\beta x_\beta x^\alpha + \tfrac{1}{4} \omega_\alpha^{\beta^*} x_{\beta^*} x^\alpha,$$

but here

$$\tfrac{1}{2} \omega_\alpha^\beta x_\beta x^\alpha = \tfrac{1}{4} \omega_\alpha^\beta x_\beta x^\alpha + \tfrac{1}{4} \omega_{\alpha^*}^{\beta^*} x_{\beta^*} x^{\alpha^*} + \tfrac{1}{2} \sum \omega_\alpha^\alpha$$

since :

$$\omega_{\alpha^*}^{\beta^*} x_{\beta^*} x^{\alpha^*} = -\omega_\beta^\alpha x_{\beta^*} x^{\alpha^*} = -2\omega_\beta^\alpha x_{\beta^*} x_\alpha = 2\omega_\beta^\alpha x_\alpha x_{\beta^*} - 2 \sum \omega_\alpha^\alpha$$
$$= \omega_\beta^\alpha x_\alpha x^\beta - 2 \sum \omega_\alpha^\alpha$$
$$v\psi = u\psi + \tfrac{1}{2} \sum \omega_\alpha^\alpha \psi$$
$$\nabla \psi = u\psi + \tfrac{1}{2} (\sum \omega_{\alpha^*}^{\alpha^*}) \psi = (u + \tfrac{1}{2} \sum (\omega_{\alpha^*}^{\alpha^*})) \psi.$$

The curvature tensor of a proper or enlarged spin-euclidean connection.
Let ∇ be proper spin-euclidean. ψ having the same meaning as before :

$$\nabla_X \psi = u(X)\psi, \quad u(X) \in \mathcal{L}(\mathrm{Spin}(Q))$$

if the connection is real.

$$\begin{aligned} \nabla_X \nabla_Y &= \tfrac{1}{4}\nabla_X(\omega_j^i(Y)x_i x^j \psi), \quad X, Y \in D^1(V) \\ &= \tfrac{1}{4}X(\omega_j^i(Y))x_i x^j \psi + u(X)u(Y)\psi. \end{aligned}$$

Let $\tilde{\varphi}$ be the isomorphism from $\mathcal{L}(\mathrm{Spin}(Q))$ to $\mathcal{L}(\mathrm{SO}(Q))$, then

$$4(\nabla_X \nabla_Y - \nabla_Y \nabla_X)\psi = \tilde{\varphi}^{-1}(X\omega(Y) - Y\omega(X))\psi + \tilde{\varphi}^{-1}[\omega(X), \omega(Y)]\psi$$

$$\begin{aligned} 4(\nabla_X \nabla_Y - \nabla_Y \nabla_X - \nabla_{[X,Y]})\psi &= \tilde{\varphi}^{-1}(X\omega(Y) - Y\omega(X))\psi + \tilde{\varphi}^{-1}[\omega(X), \omega(Y)]\psi \\ &\quad - \tilde{\varphi}(\omega[X, Y])\psi \\ &= \tilde{\varphi}^{-1}(R(X, Y))\psi \\ &= R_j^i(X, Y)x_i x^j \psi, \end{aligned}$$

R being the curvature tensor of ω ($R = 2\Omega$ in the standard notations). If we set $\nabla_X \nabla_Y - \nabla_Y \nabla_X - \nabla_{[X,Y]} = P(X, Y)$, where $P(X, Y)$ is supposed to act on spinors, we get

$$P(X, Y) = \tfrac{1}{4}R_j^i(X, Y)x_i x^j. \tag{5}$$

Now take for ∇ an enlarged spin-euclidean connection.

$$\nabla \psi = u\psi + \tfrac{1}{2}\omega_{\alpha\bullet}^{\alpha\bullet}\psi$$

by Proposition (15.1.4).

$$\begin{aligned} \nabla_X \nabla_Y(\psi) &= \tfrac{1}{4}X(\omega_j^i(Y))x_i x^j \psi + u(X)u(Y)\psi + \tfrac{1}{2}u(Y)\omega_{\alpha\bullet}^{\alpha\bullet}(X)\psi \\ &\quad + \tfrac{1}{2}X(\omega_{\alpha\bullet}^{\alpha\bullet}(Y))\psi + \tfrac{1}{2}\omega_{\alpha\bullet}^{\alpha\bullet}(Y)u(X)\psi + \tfrac{1}{4}\omega_{\alpha\bullet}^{\alpha\bullet}(X)\omega_{\alpha\bullet}^{\alpha\bullet}(Y)\psi. \end{aligned}$$

Forming $4(\nabla_X \nabla_Y - \nabla_Y \nabla_X - \nabla_{[X,Y]})\psi$ we find one more term than in the previous case :

$$2(X\omega_{\alpha\bullet}^{\alpha\bullet}(Y) - Y\omega_{\alpha\bullet}^{\alpha\bullet}(X) - \omega_{\alpha\bullet}^{\alpha\bullet}([X,Y])) = 4d\omega_{\alpha\bullet}^{\alpha\bullet}(X, Y),$$

by a classical formula, and finally

$$P(X, Y) = \tfrac{1}{4}R_j^i(X, Y)x_i x^j + d\omega_{\alpha\bullet}^{\alpha\bullet}(X, Y).$$

15.1.3 General spin connections.

Let ω be the euclidean connection form, then the isomorphism between the Lie algebras associates to it a form u with values in the Lie algebra of $\mathrm{Spin}(Q)$.

By the method used in Chapter 6, 1.4, noticing that every element of the Lie algebra of $\mathrm{Pin}(Q)$ can be written as $a_j^i x_i x^j$, the summation being from 1 to n, where

$\{x_\alpha, x_{\beta^*}\}$ is a special Witt basis of general element x_i and $x_{i^*} = x_j$ with $j = i \pm r$, the indices being raised and lowered using the standard 'metric',

$$a^i_j + a^{j^*}_{i^*} = 0,$$
$$(a^i_j x_i x^j) x_k - x_k (a^i_j x_i x^j) = 4 a^i_j x_i$$

so that the matrix $\|a^l_k\|$ of the Lie algebra of $O(Q)$ corresponds to $\frac{1}{4} \sum a^i_j x_i x^j$ in the Lie algebra of $\mathrm{Pin}(Q)$:

$$\tilde{\varphi}(a^i_j x_i x^j) = 4\|a^l_k\|.$$

In terms of the euclidean connection form ω, the x_i standing for local sections of special Witt bases, we have

$$\tilde{\varphi}(\omega^i_j x_i x^j) = 4\|\omega^l_k\|.$$

To the euclidean connection with form ω, there corresponds a spin connection with $\mathcal{L}(\mathrm{Pin}(Q))$-valued form

$$u = \tfrac{1}{4}\omega^i_j x_i x^j,$$

under the principal morphism from the bundle η to the bundle ξ (cf. Definition (14.2.2)), which is a connection on the spinor bundle ζ associated to the principal bundle η.

Definition 15.1.2 *For any element $\psi = x_{\alpha_1} \ldots x_{\alpha_h} f$ of the local spin basis we define, if $u = \tfrac{1}{4}\omega^i_j x_i x^j$,*

$$D\psi = u\psi \quad \text{and} \quad D_X \psi = u(X)\psi, \quad X \in D^1(V). \tag{6}$$

D is the spin derivation 'over' the euclidean derivation ∇; it is said to be *canonically associated* to ∇.

By Proposition (15.1.2), D is the natural extension of the proper spin-euclidean connection when the spinoriality structural group of the bundle is replaced by $\mathrm{Pin}(Q)$.

Note that if $\psi = z^{\alpha_1 \cdots \alpha_k} x_{\alpha_1} \ldots x_{\alpha_k} f = z^A \psi_A$,

$$D_X \psi = X(z^A)\psi_A + z^A u(X)\psi_A. \tag{7}$$

Remark. By a classical result from connection theory, an infinite number of connections acting on the spinor bundle ζ can be obtained by adding to u any 1-form field θ with values in the linear endomorphism algebra of the spin sections, i.e. by the properties highlighted before, a Clifford algebra valued 1-form field.

The resulting forms are no longer $\mathcal{L}(\mathrm{Pin}(Q))$-valued. θ is called the *Clifford excess* of the connection.

By Proposition (15.1.4), an enlarged spin-euclidean connection can be considered as a particular spin connection with non-vanishing Clifford excess (it equals $\tfrac{1}{2}\omega^{\alpha^*}_{\alpha^*}$).

The covariant derivative of cospinors.

By a classical procedure the connection extends to the dual ζ^* of the bundle ζ, and the connection form on the dual is $-u^T$.

Assuming that the structural group of the pseudo-riemannian bundle can be reduced to its subgroup of elements $\varphi(G_0)$ (such that $N(g_{\alpha\beta}(x)) = 1$), then a field $x \to f(x)$ of isotropic r-vectors can be defined and, by the definition of the bilinear form $\mathcal{B}$ (Chapter 8, 2.1), there exists a field of bilinear forms, still denoted $\mathcal{B}$ such that :

$$\mathcal{B}(vf)wf = \mathcal{B}(vf, wf)f.$$

$\mathcal{B}$ is non-degenerate and $\mathcal{B}_x$ defines a duality between $\mathrm{Clif}'_x(V)f$ and the minimal right ideal $f\mathrm{Clif}'_x(V)$.

The dual bundle of ζ, considered as $f\mathrm{Clif}'_x(V)$, can therefore be identified with the bundle $f\mathrm{Clif}'_x(V)$, which we will call the cospinor bundle.

If vf and wf are canonical basis sections $vf = x_{\alpha_1} \ldots x_{\alpha_h} f$, $wf = x_{\beta_1} \ldots x_{\beta_k} f$, the definition of the dual connection (still denoted by D) leads to the relations :

$$\langle fv, Dwf \rangle + \langle D(fv), wf \rangle = d\mathcal{B}(\beta(fv), wf) = 0,$$
$$\mathcal{B}(\beta(fv), uwf) + \mathcal{B}(\beta(Dfv), wf) = 0,$$
$$fvuwf = -(Dfv)wf, \quad \forall wf.$$

$$D(fv) = -fvu \tag{8}$$

a formula that solves completely this problem and, if the connection has no Clifford excess, expresses that if the cospinor φ equals $\beta(\psi)$:

$$D\varphi = \beta(D\varphi)$$

i.e.

$$D(\beta\psi) = \beta(D\psi).$$

From this, covariant derivatives can be obtained on all the fields considered before, using the duality and the tensor product.

Proposition 15.1.5 *If D is the spin derivative canonically associated to ∇, $D\mathcal{B} = 0$.*

Proof. D extends to tensor products of spinors by a classical procedure and commutes with contractions, so by the previous computations and the definition of the dual connection, using the same notations :

$$\langle -\beta(vf)u, wf \rangle + \langle \beta(vf), uwf \rangle = d\mathcal{B}(vf, wf) = 0$$
$$= (D\mathcal{B})(vf, wf) + \mathcal{B}(Dvf, wf) + \mathcal{B}(vf, Dwf)$$

and the result follows from

$$\mathcal{B}(Dvf, wf)f + \mathcal{B}(vf, Dwf)f = \beta(uvf)wf + \beta(vf)uwf = \beta(vf)(u + \beta(u))wf = 0$$

since $u + \beta(u) = 0$.

The physicist's fundamental tensor-spinor. Let us still take $\psi = x_{\alpha_1} \ldots x_{\alpha_h} f$ and $\frac{1}{2} x^i = x_{i^*}$. The connection extending D to $\text{Hom}(\zeta, \zeta) = \zeta^* \otimes \zeta$ is defined as usual :

$$(D(\Theta))(s) = D(\Theta(s)) - \Theta(Ds), \quad \Theta \in \text{Hom}(\zeta, \zeta),$$

$s \in \Gamma(\zeta)$ of the same form as ψ.

Consider the linear operator $\tilde{x}^i$, the left multiplication by x^i on the spin sections :

$$(D\tilde{x}^i)(s) = D(x^i s) - x^i u s = (u x^i - x^i u)s,$$

$D(\tilde{x}^i) = u x^i - x^i u$.

But $u x_i - x_i u = \omega_i^k x_k$,

$$u x^i - x^i u = \omega^{ki} x_k = -\omega^{ik} x_k = -\omega_k^i x^k = \nabla x^i,$$

so that $D(\tilde{x}^i) - \nabla x^i = 0$, or $D(\tilde{x}^i) + \omega_k^i x^k = 0$.

In terms of components we would write this as

$$\sigma_C^A \gamma_B^{Ci} - \gamma_C^{Ai} \sigma_B^C + \omega_k^i \gamma_B^{Ak} = 0, \quad A, B = 1, \ldots, 2^r, \quad i = 1, 2, \ldots, n. \tag{9}$$

Imagine that we define the connection associated to $\delta = D \otimes I_n + I_{2^r} \otimes \nabla$ on the tensor-spinor fields (i.e. the tensor products of spin and tensor sections).

If B is the sum of the tensor products of sections of $\text{Hom}(\zeta, \zeta)$ and sections of the tangent bundle, $B = B^k \otimes x_k$, we get :

$$\begin{aligned}
\delta_X B &= \delta_X B^k \otimes x_k + B^k \otimes \delta_X x_k, \quad X \in D^1(V), \\
&= [\delta_X, B^i] \otimes x_i + B^k \otimes \omega_k^i(X) x_i
\end{aligned}$$

$\delta_X B = 0$ being expressed by $[\delta_X, B^i] + \omega_k^i(X) B^k = 0$, which is formally equivalent to (9) for a 'fundamental tensor-spinor' denoted by γ rather than B, γ^i being the operator $\tilde{x}^i$ and $\gamma = \tilde{x}^i \otimes x_i = \tilde{x}_i \otimes x^i$.

In fact, physicists use an orthonormal basis e_i, but these e_i being linear combinations with constant coefficients of the x_i, the reasonings and the results are identical.

Our approach does not require the notion of a fundamental tensor-spinor, but to ease the comparison with some results in the literature, we will use

$$D\gamma^i = D(\tilde{x}^i) + \omega_k^i x^k = u x^i - x^i u + \omega_k^i x^k = 0$$

and :

Proposition 15.1.6 *The spin connections $\mathcal{D}$ such that $\mathcal{D}\gamma^i = 0$ have scalar Clifford excess.*

Proof. If u' is the form of $\mathcal{D}$,

$$u'x^i - x^i u' = ux^i - x^i u,$$
$$(u' - u)x^i = x^i(u' - u)$$

and since the x^i generate the Clifford algebra, $(u' - u)$ is scalar-valued. ∎

Remark. Instead of the field of Witt coframes (x^i), we can consider an arbitrary coframe field (E^i) such as the field $E^i = g^{ij}(\partial/\partial u^j)$ constructed from a natural frame of local coordinates $(u^1, u^2, \ldots, u^n)$, the g^{ij} forming the inverse matrix of the $g_{ij} = B(\partial/\partial u^i, \partial/\partial u^j)$.

We still have that $D(\tilde{E}_i) - \nabla E^i = 0$, since if $E^i = A^i_k x^k$, it equals

$$dA^i_k(x^k - x^k) + A^i_k(D\tilde{x}^k - \nabla x^k).$$

We can still say that the 'fundamental tensor-spinor' corresponding to the E^i has zero covariant derivative, in the same sense as γ. But note that

$$E^i E^j + E^j E^i = 2g^{ij}.$$

Proposition 15.1.7 *D being the canonical spin connection, setting*

$$D_X D_Y - D_Y D_X - D_{[X,Y]} = P(X,Y), \tag{10}$$

$P(X,Y)$ acts from the left on the spinor fields and

$$P(X,Y) = \tfrac{1}{4} R^i_j(X,Y)x_i x^j, \tag{11}$$

R being the curvature tensor of the euclidean connection.

Proof. The computation used for a proper spin-euclidean connection can be repeated here. ∎

P is said to define the spin curvature.

15.1.4 The covariant derivative of $\mathcal{H}$.

Just as we did when we introduces the field $\mathcal{B}$, we will assume that the structural group of the pseudo-riemannian bundle reduces to the subgroup of elements $\varphi(G_0)$, such that $N(g_{\alpha\beta}(x)) = 1$. Under these conditions :

Lemma 15.1.1 *A global field $\mathcal{H}$ of hermitian sesquilinear forms exists on V.*

Proof. Using the notations of Proposition (14.2.1), if $g_{\alpha\beta}(x) \in H$, $g_{\alpha\beta}(x)f = \epsilon f$, $\epsilon = \pm 1$, and if γf is the pure spinor defining $\bar{f}$, $\bar{f} = \gamma f \gamma^{-1}$,

$$\gamma^{-1} g_{\alpha\beta} \gamma f = \epsilon f = g_{\alpha\beta} f$$
$$g_{\alpha\beta} \gamma g_{\alpha\beta}^{-1} f = \gamma f,$$

we can set $g_{\alpha\beta}fg_{\alpha\beta}^{-1} = \gamma f$ where $hf = f$. γf lies in the standard Clifford algebra.

$$\varphi_\alpha^x(g_{\alpha\beta}(x)\gamma g_{\alpha\beta}^{-1}(x)) = \varphi_\alpha^x(\gamma h(x)), \quad x \in U_\alpha \cap U_\beta,$$
$$\varphi_\beta^x(\gamma)\varphi_\beta^x(f) = \varphi_\alpha^x(\gamma)\varphi_\alpha^x(h(x))\varphi_\alpha^x(f),$$
$$\varphi_\beta^x(\gamma f) = \varphi_\alpha^x(\gamma f).$$

We could introduce a field $x \to (\gamma f)_x$ over V.

Therefore a field of pure spinors can be associated to the complex conjugation. If $x \to (vf)_x$, $x \to (wf)_x$ are spinor fields, setting (Chapter 10, 3.2) :

$$\beta(\overline{vf}(x), wf(x)) = a\mathcal{H}(vf(x), wf(x))(\gamma f)_x, \quad a^2 = \epsilon\epsilon',$$

a field $x \to \mathcal{H}(x)$ of hermitian sesquilinear forms is uniquely defined.   ∎

Lemma 15.1.2 *Under the assumptions of Lemma (15.1.1) there exists a charge conjugation*

$$\mathcal{C} : x \to \mathcal{C}_x$$

on V.

Proof. This follows at once from the existence of the field $x \to (\gamma f)_x$ and from the definitions in Chapter 10, 3.3 : $\mathcal{C}_x(vf)_x = (\bar{v}\gamma f)_x$.   ∎

Proposition 15.1.8 *If D is a spin connection with zero Clifford excess,*

$$D\mathcal{H} = 0. \tag{12}$$

Proof. Consider the tensor product

$$\mathcal{H} \otimes \psi \otimes \varphi \otimes \gamma f$$

where ψ and φ are frame sections as in Definition (15.1.2). Note that by Chapter 10, γf can be chosen as a frame section of the same type. D commutes with the contractions so that, contracting $\mathcal{H}$ with ψ and φ, we get

$$D(\mathcal{H}(\psi, \varphi)\gamma f) = u\mathcal{H}(\psi, \varphi)\gamma f$$

on the one hand,

$$(D\mathcal{H})(\psi, \varphi)\gamma f + \mathcal{H}(D\psi, \varphi)\gamma f + \mathcal{H}(\psi, D\varphi)\gamma f + \mathcal{H}(\psi, \varphi)u\gamma f$$

on the other. But, by the definition of D, $\mathcal{H}(D\psi, \varphi)\gamma f + \mathcal{H}(\psi, D\varphi)\gamma f = 0$ and

$$u + \beta(u) = 0;$$

ψ and φ being arbitrary, $D\mathcal{H} = 0$.   ∎

Proposition 15.1.9
$$D \circ \mathcal{C} = \mathcal{C} \circ D. \tag{13}$$

Proof. Immediate, since u is real.   ∎

Note that a 'Dirac adjoint' can be defined by $\mathcal{A} \circ \mathcal{C} = \beta$.

From $\mathcal{A}(vf) = \epsilon'\beta(\mathcal{C}(uf))$, proved in Chapter 10, 3.3, we get

$$D \circ \mathcal{A} = \mathcal{A} \circ D. \tag{14}$$

15.1.5 Some results from 'spinor analysis'.

a) The laplacian on spinor fields. ψ being an arbitrary spinor field, $\psi \in \Gamma(\zeta)$, we define

$$\Delta\psi = E^\alpha E^\beta D_\alpha D_\beta \psi \qquad (15)$$

where the summation is over α, β from 1 to n, the E_α are a local field of frames and E^α their dual coframe;

$$D_\alpha = D_{E_\alpha}$$

and D is the spin connection with zero Clifford excess, corresponding naturally to the (torsion-free) pseudo-riemannian connection.

$$
\begin{aligned}
\Delta\psi &= \tfrac{1}{2}(E^\alpha E^\beta + E^\beta E^\alpha)D_\alpha D_\beta \psi + \tfrac{1}{2}(E^\alpha E^\beta - E^\beta E^\alpha)D_\alpha D_\beta \psi \\
&= g^{\alpha\beta}D_\alpha D_\beta \psi + \tfrac{1}{2}(E^\alpha E^\beta - E^\beta E^\alpha)D_\alpha D_\beta \psi \\
&= D^\alpha(D_\alpha \psi) + \tfrac{1}{2}E^\alpha E^\beta(D_\alpha D_\beta - D_\beta D_\alpha)\psi.
\end{aligned}
$$

For the computation, it is more convenient to assume that the (E_α) form a natural frame of local coordinates, so that by (10),

$$(D_\alpha D_\beta - D_\beta D_\alpha)\psi = \tfrac{1}{4}R^\lambda{}_{\mu,\alpha\beta}E_\lambda E^\mu \psi, \quad \text{(the 'Ricci identity')}$$
$$\Delta\psi = D^\alpha(D_\alpha \psi) + \tfrac{1}{8}R_{\lambda\mu,\alpha\beta}E^\alpha E^\beta E^\lambda E^\mu \psi.$$

Relying on the Bianchi identities

$$
\begin{aligned}
R_{\lambda\mu,\alpha\beta} &= R_{\alpha\beta,\lambda\mu}, \\
R_{\alpha\beta,\lambda\mu} + R_{\alpha\lambda,\mu\beta} + R_{\alpha\mu,\beta\lambda} &= 0,
\end{aligned}
$$

we get

$$R_{\alpha\beta,\lambda\mu}E^\alpha(E^\beta E^\lambda E^\mu + E^\lambda E^\mu E^\beta + E^\mu E^\beta E^\lambda) = 0$$
$$R_{\alpha\beta,\lambda\mu}E^\alpha(3E^\beta E^\lambda E^\mu + 2g^{\mu\beta}E^\lambda + 2g^{\beta\mu}E^\lambda - 2g^{\lambda\beta}E^\mu) = 0$$

where the factors in the last two terms have been changed to the order of the first one.

$$R_{\alpha\beta,\lambda\mu}E^\alpha E^\beta E^\lambda E^\mu = -2\mathcal{R}_{\alpha\beta}E^\alpha E^\mu$$

where $\mathcal{R}_{\alpha\beta} = R^\lambda{}_{\alpha\lambda\beta}$ is the Ricci tensor, $\mathcal{R}_{\alpha\beta} = \mathcal{R}_{\beta\alpha}$.

$$R_{\alpha\beta,\lambda\mu}E^\alpha E^\beta E^\lambda E^\mu = -2\mathcal{R}_{\alpha\beta}g^{\alpha\beta} = -2\mathcal{R},$$

where $\mathcal{R}$ is the scalar Riemann curvature.

Finally we obtain :

$$\Delta\psi = D^\alpha(D_\alpha \psi) - \tfrac{1}{4}\mathcal{R}\psi. \qquad (16)$$

Let φ be a cospinor, $\psi = \beta(\varphi)$, then

$$\Delta\beta(\varphi) = E^\lambda E^\mu \beta(D_\lambda D_\mu \varphi)$$

by the remark following (8),

$$\Delta\beta(\varphi) = \beta(D_\lambda D_\mu\varphi)E^\mu E^\lambda$$

leading us to the definition

$$\Delta\varphi = (D_\lambda D_\mu\varphi)E^\mu E^\lambda \tag{17}$$

so that $\Delta \circ \beta(\varphi) = \beta \circ \Delta(\varphi)$.

From $\Delta\varphi = \beta(\Delta\psi)$ and (16) it follows that

$$\Delta\varphi = D^\alpha D_\alpha - \tfrac{1}{4}\mathcal{R}\varphi, \tag{18}$$

which is formally identical to (16).

The definition of Δ does not depend on the choice of the local frame E_α. If these are chosen real, from

$$D \circ C = C \circ D$$

it follows, by the very definition of C, that

$$C \circ \Delta\psi = \Delta \circ C\psi. \tag{19}$$

b) The Dirac operators. The first order Dirac operators are defined on the fields of 'contravariant' spinors ψ by

$$L\psi = E^\alpha D_\alpha\psi - m\psi, \quad L'\psi = E^\alpha D_\alpha\psi + m\psi,$$

m being a constant and $E^\alpha = g^{\alpha\beta}E_\beta$.

The E^α can be considered as the components of a 'fundamental tensor-spinor' $\boldsymbol{\gamma}$ such that

$$D\boldsymbol{\gamma} = 0$$

(cf. 1.3). Using well-known notations from theoretical physics,

$$L\psi = \boldsymbol{\gamma}^\alpha D_\alpha\psi - m\psi = D_\alpha(\boldsymbol{\gamma}^\alpha\psi) - m\psi,$$

so that

$$L \circ L'(\psi) = (\boldsymbol{\gamma}^\alpha D_\alpha - m)(\boldsymbol{\gamma}^\beta D_\beta + m)\psi = (\boldsymbol{\gamma}^\alpha\boldsymbol{\gamma}^\beta D_\alpha D_\beta - m^2)\psi = (\Delta - m^2)\psi.$$

If dual (or 'covariant') spinors are considered, it follows from the results on the derivative of cospinors that for

$$\hat{L}\varphi = -(D_\alpha\varphi\boldsymbol{\gamma}^\alpha + m\varphi), \quad \hat{L}'\varphi = -(D_\alpha\varphi\boldsymbol{\gamma}^\alpha - m\varphi),$$

$$\hat{L} \circ \hat{L}'\varphi = D_\alpha D_\beta\varphi\boldsymbol{\gamma}^\alpha\boldsymbol{\gamma}^\beta - m^2\varphi$$

will hold, and by (17), this is equal to $(\Delta - m^2)\varphi$. Under the assumption that a global charge conjugation $\mathcal{C}$ exists (cf. Lemma (15.1.2)), choosing the E_α real and relying on (14), we obtain for the Dirac adjoint $\mathcal{A}(\psi) = \hat{\psi}$ that

$$L\psi = \boldsymbol{\gamma}^\alpha D_\alpha - m\psi,$$
$$\widehat{L\psi} = -\widehat{D_\alpha\psi}\boldsymbol{\gamma}^\alpha - m\hat{\psi} = -D_\alpha\hat{\psi}\boldsymbol{\gamma}^\alpha - m\hat{\psi}$$

or $\hat{L}\hat{\psi} = \widehat{L\psi}$ and similarly $\hat{L}'\hat{\psi} = \widehat{L'\psi}$.

c) The tensor product of spinor and cospinor fields (1-1 type spinors). We saw in Chapter 8, 3 that $S \otimes \beta(S)$ and $C(Q')$ can be identified by $vf \otimes fw \to vfw$ and that

$$gvf \otimes fwg^{-1} \to gvfwg^{-1}, \quad g \in G'.$$

Consequently, there exists an isomorphism from the tensor product of the spinor bundle ζ with the cospinor bundle ζ^* introduced before to the complexified Clifford bundle; hence also an isomorphism from this tensor product to the exterior algebra bundle.

If we set $P\psi = \boldsymbol{\gamma}^\alpha D_\alpha\psi$ and $\hat{P}\psi = -D_\alpha\psi\boldsymbol{\gamma}^\alpha$,

$$P^2\psi = \boldsymbol{\gamma}^\alpha\boldsymbol{\gamma}^\beta D_\alpha D_\beta\psi$$

ψ being a field of 1-1 type spinors.

A similar computation to a) yields :

$$P^2\psi = D^\alpha(D_\alpha\psi) + \tfrac{1}{2}\boldsymbol{\gamma}^\alpha\boldsymbol{\gamma}^\beta(D_\alpha D_\beta - D_\beta D_\alpha)\psi.$$

A 'Ricci identity' for 1-1 type spinors is easy to establish : if the E_α form a natural frame,

$$(D_\alpha D_\beta - D_\beta D_\alpha)\psi = \tfrac{1}{4}R^\lambda_{\mu,\alpha\beta}E_\lambda E^\mu\psi - \tfrac{1}{4}R^\lambda_{\mu,\alpha\beta}\psi E_\lambda E^\mu.$$

By a computation carried out in a) this gives

$$P^2\psi = D^\alpha(D_\alpha\psi) - \tfrac{1}{4}\mathcal{R}\psi - \tfrac{1}{8}R_{\alpha\beta,\gamma\mu}\boldsymbol{\gamma}^\alpha\boldsymbol{\gamma}^\beta\psi\boldsymbol{\gamma}^\gamma\boldsymbol{\gamma}^\mu.$$

$\hat{P}^2\psi$ can be expressed in a similar way;

$$P^2\psi = \Delta\psi = \boldsymbol{\gamma}^\alpha\boldsymbol{\gamma}^\beta D_\alpha D_\beta\psi = D_\alpha D_\beta\psi\boldsymbol{\gamma}^\beta\boldsymbol{\gamma}^\alpha$$

and these formulas hold for arbitrary E^α and $\boldsymbol{\gamma}^\alpha$.

By the bundle isomorphism just mentioned the spin connection D which is canonically associated with the euclidean connection ∇, corresponds to the extension of ∇ to Clifford fields.

The parallel transport along ∇ transforms an orthonormal frame in another orthonormal frame, so the linear identification of Clifford algebras and euclidean algebra fields is preserved by the covariant derivative.

Take a local field (E_α) of orthonormal frames with dual frame (E^α), and identify the bundle $\zeta \otimes \zeta^*$ with the Clifford bundle on the cotangent spaces to V.

The linear isomorphism associates to the form Θ with components $\Theta_{\alpha_1 \alpha_2 \ldots \alpha_p}$ the Clifford field $\dot\psi = E^{\alpha_1} E^{\alpha_2} \ldots E^{\alpha_p} \Theta_{\alpha_1 \alpha_2 \ldots \alpha_p}$, which is also a type 1-1 spin field : $\dot\psi = s(\Theta)$.

$$Ps(\Theta) = E^\alpha D_\alpha \dot\psi = E^\alpha \nabla_\alpha \dot\psi,$$

(∇ is assumed torsion-free here), or by formula (9) in Chapter 3, 1.5,

$$Ps(\Theta) = \sum_\alpha (E^\alpha \wedge \nabla_\alpha \dot\psi + i_{E^\alpha}(\nabla_\alpha \dot\psi)) = d\dot\psi - \delta\dot\psi = (d - \delta)(\dot\psi),$$

using classical notations from differential geometry.

$$P^2 s(\Theta) = (d - \delta)^2(\dot\psi) = -(d\delta + \delta d)(\dot\psi)$$

since $d^2 = \delta^2 = 0$. $\Delta\dot\psi = \Delta_{DR}\dot\psi$ where $\Delta_{DR} = -(d\delta + \delta d)$ is De Rham's laplacian of $\dot\psi$.

15.2   The Lie derivative of spinors.

15.2.1

Let X be a Killing field of V (i.e. a field which induces a local group of isometries, hence such that $L_X(Q) = 0$, by a well-known result). We have mentioned in Chapter 14, 1.2 that the Lie derivative L_X defined by X can be naturally extended to the Clifford algebra of V.

Consider a connection ∇ with form ω on V, assumed euclidean and torsion-free. By a classical result from differential geometry,

$$L_X(Y) = [X, Y] = X^k \nabla_k Y - Y^k \nabla_k X, \quad \forall Y \in D^1(V). \tag{1}$$

$\{x_\alpha, x_{\beta^*}\}$ being a local Witt basis, we set

$$L_X(x_j) = a_j^i x_i. \tag{2}$$

If X is a Killing field,

$$L_X(f) = L_X(x_1 \cdot x_2 \cdot \ldots x_{r^*}) = \sum_{\alpha^* = 1^*}^{r^*} x_{1^*} \ldots L_X(x_{\alpha^*}) \ldots x_{r^*},$$

and we immediately obtain :

Proposition 15.2.1 *If X is a Killing field, the Lie derivative extends naturally from the Clifford algebra to the spinors if and only if*

$$L_X(x_{\alpha\bullet}) = a^i_{\alpha\bullet}.x_i, \quad a^\beta_{\alpha\bullet} = 0$$

and in that case

$$L_X(f) = \sum_{\alpha^\bullet=1^\bullet}^{r^\bullet} a^{\alpha^\bullet}_{\alpha\bullet} f$$

or $a^{\alpha^\bullet}_{\alpha\bullet} f$ for short.

For $x \in V$, the matrix of L_X belongs to an enlarged spinoriality group.

Let us assume in particular that this matrix corresponds to an element of $\mathcal{L}(\mathcal{G})$ and that ∇ is torsion-free and spin-euclidean (the existence of such a connection may cause problems for the topology of V; we have proved in *Schémas d'Einstein-Dirac en spin 1/2*, Ann. I.H.P. Vol. 23, n. 3, 1975, that the Ricci tensor of the connection must vanish identically in this case). Let ψ be a local Witt section as in section 1 of this chapter.

$$L_X(\psi) = \sum_{\alpha \in I} x_{\alpha_1} \ldots X^k(\nabla_k x_\alpha) \ldots x_{\alpha_h} f - \sum_{\alpha \in I} x_{\alpha_1} \ldots (\nabla_\alpha X) \ldots x_{\alpha_h} f$$

where $I = (\alpha_1, \alpha_2, \ldots, \alpha_h)$ and the factor x_{α_1} does not occur in the first term of both sums, x_{α_2} does not occur in the second etc. up to x_{α_h}.

The first set of terms can be transformed as the analogous expression found in the proof of Proposition (15.1.2) and yields

$$\tfrac{1}{4}(\Gamma^i_{jk} X^k x_i x^j)\psi = X^k \nabla_k \psi,$$

the Γ^i_{jk} being the coefficients of the connection ∇. As to $(\nabla_\alpha X)$, we can write

$$\nabla_\alpha X = (\nabla_\alpha X)^i x_i = \nabla_X x_\alpha - L_X(x_\alpha),$$

relying on (1).

Transforming the second set of terms yields

$$-\tfrac{1}{4}(\nabla_j X)^i x_i x^j \psi$$

and finally

$$L_X(\psi) = X^k(\nabla_k \psi) - \tfrac{1}{4}(\nabla_j X)^i x_i x^j \psi \tag{3}$$

the right-hand side of which can again be transformed, relying on (1), to

$$\tfrac{1}{4} L_X(x_j) x^i \psi,$$

so

$$L_X(\psi) = \tfrac{1}{4}(a^i_j x_i x^j)\psi, \tag{4}$$

which is formally identical to the result obtained for the covariant derivative.

Now, if we make no assumptions on the Killing field X nor on the matrix field (a_j^i), we set, if $\psi = x_{\alpha_1} x_{\alpha_2} \ldots x_{\alpha_h} f$,

$$L_X(\psi) = \tfrac{1}{4}(a_j^i x_i x^j)\psi, \tag{5}$$

with $L_X(x_j) = a_j^i x_i$.

If $\psi = z^{\alpha_1 \ldots \alpha_h} x_{\alpha_1} \ldots x_{\alpha_h} f = z^A \psi_A$,

$$L_X(\psi) = X(z^A)\psi_A + z^A L_X(\psi_A). \tag{6}$$

In general, this derivation will no longer be associated to the Lie derivative in the Clifford algebra. The Lie derivative is then extended to the dual spinors and to tensor products by requiring the usual properties and, particularly, the commutation with contractions. A similar computation to that for the covariant derivative proves that if $\Theta = f x_{\alpha_1} \ldots x_{\alpha_h}$ is a cospinor,

$$L_X(\Theta) = -\tfrac{1}{4}\Theta a_j^i x_i x^j,$$

and that

$$L_X(\Theta) = X^k \nabla_k \Theta + \tfrac{1}{4}(\nabla_j X)^i \Theta x_i x^j.$$

It is then a routine task to compute the Lie derivatives of all types of spinors. Finally, we may define

$$\hat{L}_X = L_X \otimes I_n + I_{2^r} \otimes L_X,$$

the Lie derivative allowing the extension of this operator to 'tensor-spinor' fields, just as we did for the covariant derivative.

15.2.2 Generalization.

Let us examine under what conditions a derivation D of the tensor algebra of V, preserving the type and commuting with the contractions, will 'naturally' extend to spinor fields. First, it must extend to the Clifford algebra, so it should preserve the ideal J generated by the elements $X \otimes X - Q(X)$. The same computation as for the covariant derivative and the Lie derivative shows that $D(Q) = 0$ is the condition for D to extend to the Clifford algebra. The restriction of D to vector fields defines an infinitesimal isometry. In the canonical Witt bases, the matrix $||D_j^i||$ of D has to satisfy the condition $D_{\beta^{\bullet}}^{\alpha} = 0$, which is clearly a sufficient condition. The action of D on the spinor ψ belonging to a local Witt basis, is then expressed by

$$\psi \to \tfrac{1}{4}(D_j^i x_i x^j)\psi.$$

D can then be extended using the same procedure as before, to the cospinor fields and to all tensor products of tensors and spinors, under the sole condition that $||D_j^i||$ be the matrix of an infinitesimal isometry.

Chapter 16

THE DIRAC EQUATION.

16.1

We will express the main results concerning the Dirac equation, using the concepts and the notations of this book.

In this chapter, we use a standard Minkowski space E with complexification E'; its signature is $(+ - - -)$, an orthonormal basis is given by e_1, e_2, e_3, e_4 where $(e_1)^2 = 1$ and a Witt basis

$$x_1 = \frac{e_1 + e_4}{2}, \qquad x_2 = \frac{ie_2 + e_3}{2},$$
$$y_1 = \frac{e_1 - e_4}{2}, \qquad y_2 = \frac{ie_2 - e_3}{2}, \qquad f = y_1 y_2.$$
$$B(x_i, y_i) = \tfrac{1}{2}\delta_{ij}, \qquad i, j = 1, 2, 3, 4.$$

On the spacetime manifold, the local fields of orthonormal frames will also be denoted by e_i and the field of Witt bases by (x_i, y_j).

D is the spin connection which corresponds naturally to the torsion-free pseudo-riemannian connection.

The structural group of the pseudo-riemannian bundle will be assumed reducible at least to the image group of G_0, the elements of G_0 having spin norm 1. Then there will exist a field of bilinear forms $\mathcal{B}$ which are invariant under the action of G_0 and a field of hermitian sesquilinear forms $\mathcal{H}$ naturally associated to $\mathcal{B}$, in the sense of Chapter 10, 4.

There exist Majorana spinors which are compatible with the action of $G^+(X)$.

If the manifold is also oriented, a spinor field is said to have a well-determined chirality if $ie_N\psi = \psi$ or $ie_N\psi = -\psi$, $e_N = e_1 e_2 e_3 e_4$; since $e_N f = -if$:

$$ie_N(\alpha_1 f + \alpha_2 x_1 x_2 f + \alpha_3 x_1 f + \alpha_4 x_2 f) = \alpha_1 f + \alpha_2 x_1 x_2 f - \alpha_3 x_1 f - \alpha_4 x_2 f,$$

and $ie_N\psi = \psi$ is equivalent to $\alpha_3 = \alpha_4 = 0$ whereas $ie_N\psi = -\psi$ is equivalent to $\alpha_1 = \alpha_2 = 0$. The chirality property is therefore equivalent to the existence of a well-determined parity.

The conjugation spinor will be $\gamma f = -ix_2 f = e_2 f$ and C will be defined by

$$C(uf) = e_N \bar{u} \gamma f,$$

then C commutes with the action of G^+, $C^2 = \mathrm{Id}$ and C is globally defined since the manifold is oriented. One easily verifies that

$$C(f) = x_2 f, \quad C(x_1 f) = -x_1 x_2 f.$$

For the chosen conjugation, a spinor is of Majorana type if $\alpha_1 = -\bar{\alpha}_4$ and $\alpha_2 = -\bar{\alpha}_3$.
A basis for the Majorana spinors is given by :

$$f + x_2 f = s_1, \quad i x_1 f + i x_1 x_2 f = s_2, \quad if - i x_2 f = s_3, \quad x_1 f - x_1 x_2 f = s_4. \quad (1)$$

16.2   The Dirac equation and the associated quantities.

16.2.1

We consider the equation

$$e^k D_k \psi = mi\psi, \tag{1}$$

m being a real constant; the factor i on the right-hand side can be justified as follows :
if $C\psi = \psi$,

$$C(e^k D_k \psi) = C\varphi = -\varphi$$

and φ belongs to the Majorana spinor space defined by this property; this space can be obtained from the space of spinors such that $C\psi = \psi$ using a multiplication by i (cf. Chapter 10, 3.7).
  Indeed, $C(e^k D_k \psi) = -e^k C(D_k \psi)$ since $e_N e^k = -e^k e_N$; the conjugation $uf \to \bar{u}\gamma f$ commutes with D_k (Proposition (15.1.9)) and $D_k(\gamma_1 \gamma_2 \gamma_3 \gamma_4) = 0$, since $\gamma_1 \gamma_2 \gamma_3 \gamma_4$ can be obtained from the fundamental tensor-spinors γ_1, γ_2, γ_3, γ_4 by a product which corresponds to a contraction, and D commutes with contractions; $D\gamma_i = 0$ then leads to

$$C(e^k D_k \psi) = -e^k D_k(C\psi) = -e^k D_k \psi.$$

Note that we have proved $e^k D_k(C\psi) = miC(\psi)$, so

Proposition 16.2.1 *The Dirac equation is invariant under the conjugation* C.

  Let $\psi = a^1 s_1 + a^2 s_2 + a^3 s_3 + a^4 s_4$. Assume for the moment that spacetime is the Minkowski vector space E. A routine computation expresses (1) as the system :

$$\begin{aligned}
\partial_1 a^4 + \partial_2 a^3 + \partial_3 a^1 + \partial_4 a^4 &= ma^3, \\
\partial_1 a^3 + \partial_2 a^4 - \partial_3 a^2 - \partial_4 a^3 &= -ma^4, \\
\partial_1 a^2 + \partial_2 a^1 - \partial_3 a^3 + \partial_4 a^2 &= -ma^1, \\
\partial_1 a^1 + \partial_2 a^2 + \partial_3 a^4 - \partial_4 a^1 &= ma^2,
\end{aligned} \tag{2}$$

where ∂_k stands for $\partial/\partial u^k$, u^1, u^2, u^3, u^4 being the coordinates on E, u^1 the time coordinate.

Consider, for instance, the case where the spinor ψ does not depend on the space coordinates; (2) then splits in two independent systems :

$$\begin{cases} \partial_1 a^3 = -ma^4, \\ \partial_1 a^4 = ma^3 \end{cases} \tag{3}$$

and

$$\begin{cases} \partial_1 a^1 = ma^2, \\ \partial_1 a^2 = -ma^1. \end{cases} \tag{4}$$

If we set

$$\psi_1 = a^1 s_1 + a^2 s_2, \quad \psi_2 = a^3 s_3 + a^4 s_4, \quad A = \begin{pmatrix} 0 & 1 \\ -1 & 0 \end{pmatrix},$$

$$\psi_1 = \exp(mu^1 A)\psi_1(0), \quad \psi_2 = \exp(-mu^1 A)\psi_2(0)$$

are the solutions of (3) and (4) which equal $\psi_1(0)$ and $\psi_2(0)$ for $u^1 = 0$.

Now return to the general case of a spacetime manifold.

We will construct several tensor fields which physicists associate to the spinor fields. In doing so, we will rely on our results from Chapter 10, 4 and use the antisymmetric bilinear forms $\mathcal{B}$ and $\tilde{\mathcal{B}}$ which are invariant under G_0^+.

16.2.2 Bilinear vector-valued interaction operations.

The generalized triality principle (cf. Chapter 13) yields for $\mathcal{B}$, where

$$\mathcal{B}(xuf, vf) = \mathcal{B}(uf \circ vf, x), \quad \forall x \in E', \quad uf \in S^+, \quad vf \in S^-,$$

and $guf \circ gvf = g(uf \circ vf)g^{-1}$, $g \in G_0$ (equivariance), the commutative operation $\circ$:

$$\begin{aligned} f \circ x_1 f &= -2y_2, \\ f \circ x_2 f &= 2y_1, \\ x_1 x_2 f \circ x_1 f &= -2x_1, \\ x_1 x_2 f \circ x_2 f &= -2x_2, \end{aligned} \tag{5}$$

which, when applied to the Majorana basis, gives the table :

$\circ$	s_1	s_2	s_3	s_4
s_1	$2(e_1 - e_4)$	$2e_2$	0	$2e_3$
s_2	$2e_2$	$2(e_1 + e_4)$	$-2e_3$	0
s_3	0	$-2e_3$	$2(e_1 - e_4)$	$2e_2$
s_4	$2e_3$	0	$2e_2$	$2(e_1 + e_4)$

$$\tag{6}$$

If $\psi = a^i s_i$ and $\varphi = b^i s_i$, $a^i, b^i \in C$, the components of $\psi \circ \psi$ are given by :

$$
\begin{aligned}
2(a^1b^1 + a^2b^2 + a^3b^3 + a^4b^4)e_1, \\
2(a^2b^1 + a^1b^2 + a^3b^4 + a^4b^3)e_2, \\
2(a^1b^4 - a^2b^3 - a^3b^2 + a^4b^1)e_3, \\
2(-a^1b^1 + a^2b^2 - a^3b^3 + a^4b^4)e_4,
\end{aligned}
\tag{7}
$$

or, in a Witt basis, if

$$
\psi = \alpha_1 f + \alpha_2 x_1 x_2 f + \alpha_3 x_1 f + \alpha_4 x_2 f
$$

and a similar expression for φ,

$$
\psi \circ \varphi = -2(\alpha_2\beta_3 + \alpha_3\beta_2)x_1 - 2(\alpha_2\beta_4 + \alpha_4\beta_2)x_2 + 2(\alpha_1\beta_4 + \alpha_4\beta_1)y_1 - 2(\alpha_1\beta_3 + \alpha_3\beta_1)y_2
\tag{8}
$$

and the vector $\psi \circ \varphi$ is isotropic if and only if

$$
B(\psi \circ \varphi, \psi \circ \varphi) = (\alpha_2\beta_1 - \alpha_1\beta_2)(\alpha_3\beta_4 - \alpha_4\beta_3) = 0
$$

as can easily be verified using (8), i.e. if $\alpha_1 = \lambda\beta_1$ and $\alpha_2 = \lambda\beta_2$ or $\alpha_3 = \mu\beta_3$ and $\alpha_4 = \mu\beta_4$. In particular, this happens when $\varphi = \psi$, but the necessary and sufficient condition is that $\psi^+ = \lambda\varphi^+$ or $\psi^- = \lambda\varphi^-$.

Proposition 16.2.2 $\psi \circ \varphi$ is isotropic if and only one pair of chiral components is identical up to a scalar factor. In particular, $\psi = \varphi$ is sufficient.

Remark. We have that

$$
B(\psi \circ \varphi, \psi \circ \varphi) = B(\psi^-, \varphi^-)B(\psi^+, \varphi^+).
$$

Replacing B by $\tilde{B}$, another operation $\tilde{\circ}$ is obtained for which $uf\tilde{\circ}vf = vf\tilde{\circ}uf$ and Proposition (16.2.2) still holds.

Any bilinear form invariant under G_0^+ is known to be of the form $B_1 = aB + b\tilde{B}$ so, if X^B stands for the vector $\psi \circ \varphi$,

$$
X^{aB+b\tilde{B}} = aX^B + bX^{\tilde{B}} = (a - b)X^B = (b - a)X^{\tilde{B}}.
$$

We have proved the

Proposition 16.2.3 For any bilinear form $B_1 = aB + b\tilde{B}$ with $a \neq b$, the time component of X^{B_1} is a (positive or negative) definite bilinear form of the component spinors.

Definition 16.2.1 $\frac{1}{2}X^{B_1}$ is called a 'current' associated to the pair of spinors ψ and φ; it is denoted by $J_1(\psi, \varphi)$. $J(\psi, \varphi)$ and $\tilde{J}(\psi, \varphi)$ correspond to B and $\tilde{B}$.

From the Dirac equation we deduce that if ψ and ψ' are solutions for the coefficients m and m' :

$$\mathcal{B}(\psi', e^\alpha D_\alpha \psi) = mi\mathcal{B}(\psi', \psi),$$
$$\mathcal{B}(e^\alpha D_\alpha \psi', \psi) = m'i\mathcal{B}(\psi', \psi),$$

and adding both sides :

$$\mathcal{B}(e^\alpha D_\alpha \psi', \psi) + \mathcal{B}(\psi', e^\alpha D_\alpha \psi) = (m + m')i\mathcal{B}(\psi', \psi).$$

Introducing the fundamental tensor-spinor γ^α associated to e^α, and relying on $\beta(e^\alpha) = e^\alpha$,

$$\mathcal{B}(D_\alpha \gamma^\alpha \psi', \psi) + \mathcal{B}(\gamma^\alpha \psi', D_\alpha \psi) = (m + m')i\mathcal{B}(\psi', \psi);$$

now since $D\mathcal{B} = 0$,

$$D_\alpha(\mathcal{B}(e^\alpha \psi', \psi)) = (m + m')i\mathcal{B}(\psi', \psi)$$

and the definition itself leading to $\mathcal{B}(e^\alpha uf, u'f) = (uf \circ u'f)^\alpha$,

$$D_\alpha J^\alpha(\psi', \psi) = \frac{m + m'}{2} i\mathcal{B}(\psi', \psi). \tag{9}$$

In particular, if $\psi = \psi'$ or even if $\mathcal{B}(\psi', \psi) = 0$, for all choices of m, m' :

$$D_\alpha J^\alpha = 0, \tag{10}$$

holding also if $m = m' = 0$.

Considering $\tilde{\mathcal{B}}$ instead of $\mathcal{B}$, one can prove in a similar way that

$$D_\alpha \hat{J}^\alpha(\psi', \psi) = \frac{m - m'}{2} i\tilde{\mathcal{B}}(\psi', \psi) \tag{11}$$

so that if $\tilde{\mathcal{B}}(\psi', \psi) = 0$ (which, as a matter of fact, is easily verified to be equivalent to $\mathcal{B}(\psi', \psi) = 0$), (10) will hold for all currents $J_1(\psi', \psi)$.

Proposition 16.2.4 For all currents $J_1(\psi', \psi)$ such that $\mathcal{B}_1(\psi, \varphi) = 0$, the conservation law $D_\alpha J_1^\alpha = 0$ holds if ψ and φ are solutions of the Dirac equation.

16.2.3 Sesquilinear vector-valued interaction operations.

We now want to replace the bilinear forms by hermitian sesquilinear ones naturally associated to them by the procedure of Chapter 10, 4; the conjugation $\mathcal{C}$ will be the product of the one used in Chapter 10, 4, by $e_N = e_1 e_2 e_3 e_4$.

Let us define

$$\mathcal{H}_1(\mathcal{C}(uf), vf) = \lambda \mathcal{B}_1(uf, vf), \quad \lambda \in \mathbf{C}^* \tag{12}$$

or $\mathcal{H}_1(uf, vf) = \lambda \mathcal{B}_1(\mathcal{B}(uf), vf)$.

Considering $\mathcal{B}$:

$$\begin{aligned}
\mathcal{H}(uf,vf)f &= \lambda\beta(\mathcal{C}uf)vf \\
&= \lambda\beta(\gamma)\beta(\overline{uf})e_Nvf, \\
\mathcal{H}(uf,vf)\gamma f &= -\lambda\beta(e_N\overline{uf})vf,
\end{aligned}$$

since $\gamma = e_2$.

$\mathcal{H}$ will then be hermitian if :

$$\bar\lambda\beta(\overline{vf})e_N u_f = \epsilon\overline{\mathcal{H}(uf,vf)}\bar{f}\gamma^{-1} = -\epsilon\overline{\mathcal{H}(uf,vf)}\gamma f$$

or

$$(\bar\lambda/\lambda)\mathcal{H}(vf,uf)\gamma f = \epsilon\overline{\mathcal{H}(uf,vf)}\gamma f$$

and since $\epsilon = -1$, $\lambda = -\bar\lambda$ so that λ can be chosen equal to i.

Then $\mathcal{H}$ is defined by

$$\beta(e_N\overline{uf})vf = \mathcal{H}(uf,vf)x_2 f. \tag{13}$$

For $\tilde\beta$, on the other hand, we see that $\lambda = 1$ since $\tilde\beta(\gamma) = \gamma$ and

$$\tilde\beta(e_N\overline{uf})vf = -i\tilde{\mathcal{H}}(uf,vf)x_2 f. \tag{14}$$

Remarks. In the first case, $\mathcal{H}(uf,vf) = i\mathcal{B}((\mathcal{C}uf),vf)$ and if uf and vf are Majorana spinors, the remark in Chapter 10, 4 applies so that $\mathcal{B}(uf,vf)$ must be real, which can easily be verified for s_1, s_2, s_3, s_4.

In the second case, $\tilde{\mathcal{H}}(uf,vf) = \tilde{\mathcal{B}}(\mathcal{C}uf,vf)$ and $\tilde{\mathcal{H}}$ is pure imaginary for Majorana spinors, which can again be verified for the (explicit) s_1, s_2, s_3, s_4.

Also note that

$$\begin{aligned}
\mathcal{H}(uf,xvf) &= \mathcal{H}(-\bar{x}uf,vf), \\
\tilde{\mathcal{H}}(uf,xvf) &= \tilde{\mathcal{H}}(\bar{x}uf,vf)
\end{aligned}$$

and that[1]

$$\begin{aligned}
\tilde{\mathcal{H}}(\mathcal{C}uf,\mathcal{C}vf) &= \tilde{\mathcal{B}}(uf,\mathcal{C}vf) = -\tilde{\mathcal{B}}(\mathcal{C}vf,uf), \\
\tilde{\mathcal{H}}(\mathcal{C}uf,\mathcal{C}vf) &= -\overline{\tilde{\mathcal{H}}(uf,vf)}.
\end{aligned}$$

Our aim is to construct, using $\mathcal{H}_1$, a sesquilinear extension of the operation $\circ$, denoted by $\hat\circ$, such that

$$\mathcal{H}_1(uf,xvf) = \lambda\mathcal{B}_1(\mathcal{C}uf,xvf) = B(uf\hat\circ vf,x) \tag{15}$$

for all pairs $(uf,vf) \in S \times S$ and all $x \in E'$. We will prove the

Proposition 16.2.5 *The only sesquilinear extensions of the operation $\circ$ (in the sense of the preceding remark) are constructed using $\mathcal{H}_1 = k\tilde{\mathcal{H}}$, $k \in \mathbf{R}$, $\tilde{\mathcal{H}}$ defined by (14) and $uf\hat\circ vf = \overline{vf\hat\circ uf}$.*

---

[1]These $\mathcal{H}$ and $\tilde{\mathcal{H}}$ differ from those in Chapter 10, 4.

Proof. Setting $uf = u^+f + u^-f$, $vf = v^+f + v^-f$, we have defined

$$B(u^+f, xv^-f) = B(u^+f \circ v^-f, x);$$

but

$$\begin{aligned} B(uf, xvf) &= B(u^+f, xv^-f) + B(u^-f, xv^+f) \\ &= B(u^+f, xv^-f) - B(v^+f, xu^-f) \\ &= B(u^+f \circ v^-f - v^+f \circ u^-f, x), \end{aligned}$$

whereas

$$\begin{aligned} \tilde{B}(uf, xvf) &= \tilde{B}(u^+f, xv^-f) + B(u^-f, xv^+f) \\ &= \tilde{B}(u^+f, xv^-f) + \tilde{B}(v^+f, xu^-f) \\ &= B(u^+f \circ v^-f + v^+f \circ u^-f, x) \end{aligned}$$

and $B(uf \circ vf, x)$ is only obtained for $\tilde{B}$.

$\mathcal{H}(uf, vf) = B(uf\hat{o}vf, x)$ for arbitrary uf and vf therefore does not make sense. Now write

$$\begin{aligned} \tilde{\mathcal{H}}(uf, xvf) = \tilde{\mathcal{H}}(\bar{x}uf, vf) &= B(uf\hat{o}vf, x) = \overline{\tilde{\mathcal{H}}(vf, \bar{x}uf)} \\ &= \overline{B(vf\hat{o}uf, \bar{x})} = B(\overline{vf\hat{o}uf}, x) \end{aligned}$$

since B is the extension of a real bilinear form. This gives

$$uf\hat{o}vf = \overline{vf\hat{o}uf}. \tag{16}$$

Indeed, for $\tilde{\mathcal{H}}$ the following table is obtained :

ô	f	x_1x_2f	x_1f	x_2f
f	$2y_1$	$-2x_2$	0	0
x_1x_2f	$2y_2$	$2x_1$	0	0
x_1f	0	0	$2x_1$	$2x_2$
x_2f	0	0	$-2y_2$	$2y_1$

(17)

We have computed $\tilde{\mathcal{H}}(x_2f, x_1x_2f) = \tilde{\mathcal{H}}(f, x_1f) = 1$, from which a table for ô can be determined; *for Majorana spinors*, it coincides completely with (6).

For general spinors ψ and φ, (7) is replaced by

$$\begin{aligned} &2(\bar{a}^1b^1 + \bar{a}^2b^2 + \bar{a}^3b^3 + \bar{a}^4b^4)e_1, \\ &2(\bar{a}^2b^1 + \bar{a}^1b^2 + \bar{a}^3b^4 + \bar{a}^4b^3)e_2, \\ &2(\bar{a}^1b^4 - \bar{a}^2b^3 - \bar{a}^3b^2 + \bar{a}^4b^1)e_3, \\ &2(-\bar{a}^1b^1 + \bar{a}^2b^2 - \bar{a}^3b^3 + \bar{a}^4b^4)e_4. \end{aligned} \tag{18}$$

$\psi\hat{o}\psi$ is real. Clearly $\exp(i\theta)\psi\hat{o}\exp(i\theta)\varphi = \psi\hat{o}\varphi$, a fundamental result for quantum mechanics, which did not hold for $\psi \circ \varphi$.

If $2U$ is the vector defined by (18), it is the current associated to $\tilde{\mathcal{H}}$. Computing in a Majorana basis, we see that

$$2U = \mathcal{C}(\psi) \circ \varphi = J(\mathcal{C}(\psi), \varphi),$$

hence

$$u f \hat{o} v f = \mathcal{C}(u f) \circ v f \tag{19}$$

and since $\tilde{\mathcal{H}}(\psi, \varphi) = \tilde{B}(\mathcal{C}\psi, \varphi)$ we get, relying on previous computations,

$$B(\psi \hat{o} \varphi, \psi \hat{o} \varphi) = \tilde{B}((\mathcal{C}\psi)^-, \varphi^-)\tilde{B}((\mathcal{C}\psi)^+, \varphi^+) = \tilde{B}(\mathcal{C}\psi^+, \varphi^-)\tilde{B}(\mathcal{C}\psi^-, \varphi^+).$$

$$B(\psi \hat{o} \varphi, \psi \hat{o} \varphi) = \tilde{\mathcal{H}}(\psi^+, \varphi^-)\tilde{\mathcal{H}}(\psi^-, \varphi^+).$$

Hence

Proposition 16.2.6 *In general, $\psi \hat{o} \psi$ lies in the interior of the isotropic cone. $\psi \hat{o} \varphi$ is isotropic if ψ and φ are of the same chirality. $\psi \hat{o} \psi$ is isotropic if and only if ψ^+ and ψ^- are orthogonal for $\mathcal{H}$.*

Note that the *time component of $\psi \hat{o} \varphi$ is a positive definite* sesquilinear form of the components.

The same proof as before will yield

$$D_\alpha U^\alpha(\psi', \psi) = \frac{m - m'}{2} i \tilde{\mathcal{H}}(\psi', \psi)$$

for solutions ψ and ψ' of the Dirac equation, relying on $D\tilde{\mathcal{H}} = 0$. Therefore if $\tilde{\mathcal{H}}(\psi', \psi) = 0$ or if $\psi = \psi'$ (and $m = m'$), or if $m = m'$:

$$D_\alpha U^\alpha(\psi', \psi) = 0. \tag{20}$$

Finally : modulo an unimportant non-zero real factor, only one hermitian sesquilinear form, invariant under G_0^+ allows the extension of the commutative operation ○ to a (non-commutative) sesquilinear operation : $\tilde{\mathcal{H}}$.

(14) can be used to easily verify that $\tilde{\mathcal{H}}$ is actually invariant under G_0 and that the composition is equivariant.

Recall that G_0^+ is the covering of the restricted Lorentz group and that G_0 is the covering of the orthochronic Lorentz group.

This form $\tilde{\mathcal{H}}$ allows the construction of the positive definite form $(\psi \circ \psi)^1$, which justifies its use by physicists in the Dirac theory.

16.2.4   The interaction operation for a positive definite sesquilinear form.

If we consider hermitian sesquilinear form which are invariant under a subgroup of G_0^+, other sesquilinear interaction operations can be found. For example, let $\mathcal{H}^+$ be invariant under the covering of the subgroup Γ which preserves a split of E into a positive and a negative part. This construction was mentioned in Chapter 10.

Introduce a conjugation $\mathcal{C}$ such that

$$\mathcal{C}(u f) = e_2 e_3 e_4 u^c x_1 x_2 f,$$

where u^c stands for the conjugation which takes into account the signature ($\gamma' f = x_1 x_2 f$ is this conjugation's pure spinor). The product $e_2 e_3 e_4 = e_P$ is globally defined since we work in G_0, $(e_P)^2 = -1$.

$$C(f) = -ix_2 f, \quad C(x_1 f) = -ix_1 x_2 f, \quad C^2 = \text{Id}.$$

We have the Majorana basis

$$\sigma_1 = f - ix_2 f, \quad \sigma_2 = ix_1 f - x_1 x_2 f, \quad \sigma_3 = -if + x_2 f, \quad \sigma_4 = x_1 f - ix_1 x_2 f.$$

$uf\hat{o}vf$ will be defined so that

$$\mathcal{H}^+(uf, xvf) = B(uf\hat{o}vf, x)$$

where

$$\mathcal{H}^+(uf, vf)\gamma' f = \epsilon\beta(uf)^c vf,$$
$$\mathcal{H}^+(Cuf, Cvf) = \mathcal{H}^+(vf, uf).$$

It is easily seen, as for $\tilde{\mathcal{H}}$, that

$$uf\hat{o}vf = (vf\hat{o}uf)^c.$$

Using

$$\mathcal{H}^+(f, f) = \mathcal{H}^+(ix_2 f, ix_2 f) = \mathcal{H}^+(ix_1 f, ix_1 f) = \mathcal{H}^+(x_1 x_2 f, x_1 x_2 f) = 1,$$

so that by easy computations,

$$x_1 f\hat{o}f = 2y_1, \quad x_2 f\hat{o}f = 2y_2, \quad x_1 x_2 f\hat{o}x_1 f = -2y_2,$$
$$f\hat{o}x_1 f = 2x_1, \quad f\hat{o}x_2 f = 2x_2, \quad x_1 f\hat{o}x_1 x_2 f = -2x_2,$$
$$x_1 x_2 f\hat{o}x_2 f = 2y_1$$
$$x_2 f\hat{o}x_1 x_2 f = 2x_1,$$

and in a Majorana basis, the following table is obtained; the first factor standing in the left column :

$\hat{o}$	σ_1	σ_2	σ_3	σ_4
σ_1	$-2ie_3$	0	$2ie_2$	$2(e_1 + e_4)$
σ_2	0	$-2ie_3$	$-2(e_1 - e_4)$	$2ie_2$
σ_3	$2ie_2$	$-2(e_1 + e_4)$	$2ie_3$	0
σ_4	$2(e_1 - e_4)$	$2ie_2$	0	$2ie_3$

(21)

in which $uf\hat{o}vf = (vf\hat{o}uf)^c$ is verified.

Finally for $\psi\hat{o}\varphi = (\sum a^i \sigma_i)\hat{o}(\sum b^i \sigma_i)$, we get the components

$$2(\bar{a}^1 b^4 - \bar{a}^2 b^3 - \bar{a}^3 b^2 + \bar{a}^4 b^1)e_1,$$
$$2i(\bar{a}^1 b^3 + \bar{a}^2 b^4 + \bar{a}^3 b^1 + \bar{a}^4 b^2)e_2,$$
$$2i(-\bar{a}^1 b^1 - \bar{a}^2 b^2 + \bar{a}^3 b^3 + \bar{a}^4 b^4)e_3,$$
$$2(-\bar{a}^4 b^1 - \bar{a}^3 b^2 + \bar{a}^2 b^3 + \bar{a}^4 b^4)e_4.$$

In a Witt basis we see that

$$B(\psi\hat{o}\varphi, \psi\hat{o}\varphi) = (\bar{\alpha}_3\beta_3 + \bar{\alpha}_4\beta_4)(\bar{\alpha}_1\beta_1 + \bar{\alpha}_2\beta_2),$$

$$B(\psi\hat{o}\varphi, \psi\hat{o}\varphi) = \mathcal{H}^+(\psi^-, \varphi^-)\mathcal{H}^+(\psi^+, \varphi^+) \tag{22}$$

an we have the

Proposition 16.2.7 *In general, $\psi\hat{o}\psi$ lies in the interior of the isotropic cone. $\psi\hat{o}\varphi$ is isotropic if ψ and φ have different chirality.*

Lemma 16.2.1 *The conjugation $C : C(uf) = e_P u^c x_1 x_2 f$ commutes with the covariant derivative relative to a connection $\mathcal{D}$ whose form has values in the Lie algebra of the reduced group Γ.*

Proof. We know that the spin connection has the form

$$\tfrac{1}{4}\sum_1^4 w^i_j e_i e^j = \tfrac{1}{4}\sum_1^4 w^{ij} e_i e_j,$$

but for a real basis adapted to the chosen decomposition,

$$\tfrac{1}{4}w^{11}(e_1)^2 + \sum_2^4 w^{ij} e_i e_j$$

is obtained. Since $e_P(e_k)^c = -e_k e_P$, C commutes with the connection form.

Note that if $X^k D_{e_k}\psi = X^k D_k \psi$, C commutes with D_k.

From the Dirac equation for $\mathcal{D}$, it follows that if ψ is a solution, $C\psi$ is also a solution and that

$$\begin{aligned}
(e^k)^c \mathcal{D}_k e_P \psi &= -m i e_P \psi, \\
\mathcal{H}^+((e^k)^c \mathcal{D}_k e_P \psi, \varphi) &= m i \mathcal{H}^+(e_P \psi, \varphi), \\
\mathcal{H}^+(\mathcal{D}_k e_P \psi, e^k \varphi) &= m i \mathcal{H}^+(e_P \psi, \varphi);
\end{aligned}$$

then if ψ' is a solution with coefficient m',

$$\mathcal{H}^+(e_P \psi, e^k \mathcal{D}_k \psi') = m' i \mathcal{H}^+(e_P \psi, \psi'),$$

from which it follows that

$$\mathcal{D}_k \mathcal{H}^+(e_P \psi, e^k \psi') = (m + m') i \mathcal{H}^+(e_P \psi, \psi'),$$

and if we set $\psi\hat{o}\psi' = 2V(\psi, \psi')$,

$$\mathcal{D}_k V^k(e_P \psi, \psi') = \frac{m + m'}{2} i \mathcal{H}^+(e_P \psi, \psi') \tag{23}$$

so that $\mathcal{D}_k V^k(e_P \psi, \psi')$ vanishes if $m + m' = 0$ or if $\mathcal{H}^+(e_P \psi, \psi') = 0$.

This construction is equivariant under the action of Γ.

16.2.5 Bivector-valued sesquilinear interaction operations.

Using the hermitian sesquilinear form $\tilde{\mathcal{H}}$, we will now construct $uf\,\hat{\circ}\,vf$, which will have values in the exterior algebra $\wedge^2(E')$, identified with a subspace of the complexified Clifford algebra once an orthonormal basis is fixed.

Let us call B_2 the extension of B to $\wedge^2(E')$, a space which is linearly isomorphic to the Lie algebra of $O'(1,3)$. The choice of B_2 is arbitrary up to a non-zero constant coefficient.

We will take

$$B_2(e_re_s, e_ie_j) = g_{rj}g_{is} - g_{sj}g_{ir} = -2(\wedge^2 B)(e_r \wedge e_s, e_i \wedge e_j), \quad r < s, \quad i < j,$$

with $g_{ij} = B(e_1, e_j)$.

Define

$$\tilde{\mathcal{H}}(uf, zvf) = B_2(uf\,\hat{\circ}\,vf, z), \quad \forall z \in \wedge^2(E'),$$

the components of $uf\,\hat{\circ}\,vf$ are given by

$$\tilde{\mathcal{H}}(uf, e_ke_lvf) = B_2(uf\,\hat{\circ}\,vf, e_ke_l).$$

Note that

$$\tilde{\mathcal{H}}(uf, e_ke_lvf) = \tilde{\mathcal{H}}(e_le_kuf, vf) = \overline{\tilde{\mathcal{H}}(vf, e_le_kuf)}$$

from which we immediately deduce that

$$uf\,\hat{\circ}\,vf = -\overline{vf\,\hat{\circ}\,uf}.$$

This composition is computed by the substitution of uf and vf in $\tilde{\mathcal{H}}(uf, e_ke_lvf)$ by

$$\psi = \alpha_1 f + \alpha_2 x_1 x_2 f + \alpha_3 x_1 f + \alpha_4 x_2 f,$$
$$\varphi = \beta_1 f + \beta_2 x_1 x_2 f + \beta_3 x_1 f + \beta_4 x_2 f$$

and, explicitly, relying on

$$B_2(e_\alpha e_\beta, e_\alpha e_\beta) = -1,$$
$$B_2(e_1 e_\alpha, e_1 e_\alpha) = 1, \quad \alpha, \beta = 2, 3, 4,$$

we obtain the components :

$$
\begin{aligned}
&i(-\bar{\alpha}_1\beta_4 + \bar{\alpha}_3\beta_2 - \bar{\alpha}_4\beta_1 + \bar{\alpha}_2\beta_3)e_1e_2, \\
&(-\bar{\alpha}_1\beta_4 + \bar{\alpha}_3\beta_2 + \bar{\alpha}_4\beta_1 - \bar{\alpha}_2\beta_3)e_1e_3, \\
&(-\bar{\alpha}_1\beta_3 + \bar{\alpha}_3\beta_1 - \bar{\alpha}_4\beta_2 + \bar{\alpha}_2\beta_4)e_1e_4, \\
&i(\bar{\alpha}_1\beta_3 + \bar{\alpha}_3\beta_1 - \bar{\alpha}_4\beta_2 - \bar{\alpha}_2\beta_4)e_2e_3, \\
&-i(\bar{\alpha}_1\beta_4 + \bar{\alpha}_3\beta_2 + \bar{\alpha}_4\beta_1 + \bar{\alpha}_2\beta_3)e_2e_4, \\
&(-\bar{\alpha}_1\beta_4 - \bar{\alpha}_3\beta_2 + \bar{\alpha}_4\beta_1 + \bar{\alpha}_2\beta_3)e_3e_4.
\end{aligned}
\tag{24}
$$

This antisymmetric tensor is said to be the *spin associated to the pair* $(\psi, \varphi)^2$. It is only defined up ta a constant factor.

This spin can also be expressed in term of the pairwise orthogonal bases $e_3e_4 + ie_1e_2 = \epsilon_1$, $e_2e_4 + ie_1e_3 = \epsilon_2$ and $e_2e_3 + ie_1e_4 = \epsilon_3$ related to the complex formalism (cf. Chapter 8, 3.2), and their conjugate bases :

$$
\begin{aligned}
&(\bar{\alpha}_2\beta_3 - \bar{\alpha}_1\beta_4)\epsilon_1 - i(\bar{\alpha}_1\beta_4 + \bar{\alpha}_2\beta_3)\epsilon_2 + i(\bar{\alpha}_3\beta_1 - \bar{\alpha}_4\beta_2)\epsilon_3 \\
&-(\bar{\alpha}_3\beta_2 - \bar{\alpha}_4\beta_1)\bar{\epsilon}_1 - i(\bar{\alpha}_4\beta_1 + \bar{\alpha}_3\beta_2)\bar{\epsilon}_2 + i(\bar{\alpha}_1\beta_3 - \bar{\alpha}_2\beta_4)\bar{\epsilon}_3.
\end{aligned}
\tag{25}
$$

As we expected, this tensor is pure imaginary when $\psi = \varphi$.

It vanishes identically if $\psi = \varphi$ and if the spinor has a well-defined chirality.

Remark. A density tensor clearly can be constructed using the hermitian sesquilinear form $\mathcal{H}^+$; the computations are the same and the components are, in the same order :

$$
\begin{aligned}
&i(\bar{\alpha}_1\beta_2 - \bar{\alpha}_2\beta_1 - \bar{\alpha}_3\beta_4 + \bar{\alpha}_4\beta_3)e_1e_2, \\
&(\bar{\alpha}_1\beta_2 + \bar{\alpha}_2\beta_1 - \bar{\alpha}_3\beta_4 - \bar{\alpha}_4\beta_3)e_1e_2, \\
&(\bar{\alpha}_1\beta_1 - \bar{\alpha}_2\beta_2 - \bar{\alpha}_3\beta_3 + \bar{\alpha}_4\beta_4)e_1e_4, \\
&i(\bar{\alpha}_1\beta_1 - \bar{\alpha}_2\beta_2 + \bar{\alpha}_3\beta_3 - \bar{\alpha}_4\beta_4)e_2e_3, \\
&-i(\bar{\alpha}_1\beta_2 - \bar{\alpha}_2\beta_1 - \bar{\alpha}_3\beta_4 - \bar{\alpha}_4\beta_3)e_2e_4, \\
&(-\bar{\alpha}_1\beta_2 + \bar{\alpha}_2\beta_1 - \bar{\alpha}_3\beta_4 + \bar{\alpha}_4\beta_3)e_3e_4.
\end{aligned}
\tag{26}
$$

16.2.6   Trivector-valued sesquilinear interaction operations.

The bilinear form B_3 is defined on the space of antisymmetric tensors of order 3 (identified, in an orthonormal basis, with the Clifford numbers of order 3) by

$$
B_3(e_ie_je_k, e_ie_je_k) = (e_ie_je_k)^2
$$

(a similar property was true for B_2).

This definition implies that

$$
\begin{aligned}
B_3(e_Ne_k, e_Ne_k) &= (e_Ne_k)^2 = (e_k)^2, \\
B_3(e_Nx, e_Ny) &= B_1(x, y), \quad x, y \in E'.
\end{aligned}
$$

We construct the operation $uf\,\hat{\hat{\circ}}\,vf$ such that if $i < j < k$,

$$
\tilde{\mathcal{H}}(uf, e_ie_je_kvf) = B_3(uf\,\hat{\hat{\circ}}\,vf, e_ie_je_k) = B_1(e_N(uf\,\hat{\hat{\circ}}\,vf), e_Ne_ie_je_k)
$$

but by formula (14), $\tilde{\mathcal{H}}(e_Nuf, e_Nvf) = -\tilde{\mathcal{H}}(uf, vf)$, so

$$
\tilde{\mathcal{H}}(e_Nuf, e_Ne_ie_je_kvf) = B_1(e_Nuf\hat{\circ}vf, e_Ne_ie_je_k),
$$

---

[2]It is also called 'electromagnetic moment density'.

$e_N(uf\,\hat{\hat{o}}\,vf) = -(e_Nuf\hat{o}vf),$

$$uf\,\hat{\hat{o}}\,vf = -(e_Nuf\hat{o}vf)e_N, \qquad (27)$$

and $uf\,\hat{o}\,vf$ defines the spin density tensor.

Obviously all these constructions have the correct equivariance property relative to the action of the structural group.

16.2.7 An energy-momentum tensor.

Up to some normalization coefficient, we will take

$$2T_l^k(\psi,\varphi) = \tilde{\mathcal{H}}(\psi, e^k D_l\varphi) + \tilde{\mathcal{H}}(e^k D_l\varphi, \psi).$$

Then

$$2T_l^k(\psi,\varphi) = (\psi\hat{o}D_l\varphi)^k + (D_l\varphi\hat{o}\psi)^k \qquad (28)$$

and the T_l^k defined by this are real valued, relying on (16). T_l^k also equals

$$U^k(\psi, D_l\varphi) + U^k(D_l\psi, \varphi).$$

Proposition 16.2.8 *If the space is flat,*

$$D_k T_l^k(\psi,\varphi) = 0. \qquad (29)$$

Proof. This conservation property will be verified if D_l satisfies the same Dirac equation as ψ, i.e. with the same coefficient m. But $e^k D_k(D_l\varphi) = e^k D_l D_k\varphi$, if the curvature vanishes, this also equals $D_l(\gamma^k D_k\varphi) = mi(D_l\varphi)$. ∎

If the space is curved, (29) does not hold, since

$$e^k D_k D_l\varphi = e^k D_l D_k\varphi + \frac{e^k}{4} R_{ijkl} e^i e^j,$$

where R is the pseudo-riemannian curvature tensor.

$\tilde{\mathcal{H}}(\psi, e^k D_k D_l\varphi) = \tilde{\mathcal{H}}(e^k\psi, D_k D_l\varphi) = mi\tilde{\mathcal{H}}(\psi, D_l\varphi) + \frac{1}{4}R_{ijkl}\tilde{\mathcal{H}}(\psi, e^k e^i e^j\varphi)$
$\tilde{\mathcal{H}}(e^k D_k\psi, D_l\varphi) = -mi\tilde{\mathcal{H}}(\psi, D_l\varphi)$

if ψ satisfies the same Dirac equation as φ. Hence

$D_k\tilde{\mathcal{H}}(e^k\psi, D_l\varphi) = \frac{1}{4}R_{ijkl}\tilde{\mathcal{H}}(\psi, e^k e^i e^j\varphi)$
$2D_k T_l^k(\psi,\varphi) = \frac{1}{4}R_{ijkl}[\tilde{\mathcal{H}}(\psi, e^k e^i e^j\varphi) + \overline{\tilde{\mathcal{H}}(\psi, e^k e^i e^j\varphi)}]$
$2D_k T_l^k(\psi,\varphi) = \frac{1}{4}R_{ijkl}(\tilde{\mathcal{H}}(\psi, e^k e^i e^j\varphi) + \tilde{\mathcal{H}}(\varphi, e^j e^i e^k\psi)).$

If $\varphi = \psi$ we obtain :

$$R_{ijkl}(e^k e^i e^j + e^j e^i e^k) = 2R_{ijkl}(g^{ki}e^j - g^{jk}e^i) = 4R_{lj}e^j,$$

where $\mathcal{R}$ is the Ricci tensor, so that

$$2D_k T_l^k(\psi, \psi) = \mathcal{R}_{lj}\tilde{\mathcal{H}}(\psi, e^j\psi)$$

and

$$D_k T_l^k(\psi, \psi) = \mathcal{R}_{lj}U^j(\psi, \psi). \tag{30}$$

We have proved :

Proposition 16.2.9 *If ψ is a solution of the Dirac equation, if the space is curved, has a torsion-free pseudo-euclidean connection and if D is the associated connection,*

$$D_k T_l^k(\psi, \psi) = \mathcal{R}_{lj}U^j(\psi, \psi)$$

where $\mathcal{R}$ is the Ricci tensor.

16.3   Selected references.

- N. N. Bogolioubov, D. V. Chirkov, *Introduction à la théorie quantique des champs*, Dunod, Paris.

- E. Durand, *Mécanique quantique. Tome III—Spin et relativité*, Masson, Paris.

- D. Kastler, *Introduction à l'électrodynamique quantique*, Dunod, Paris.

Chapter 17

SYMPLECTIC CLIFFORD ALGEBRAS AND ASSOCIATED GROUPS.

17.1 Common symplectic Clifford algebras or Weyl algebras.

E is an n-dimensional vector space over $\mathbf{K}$ ($\mathbf{R}$ or $\mathbf{C}$), F is an antisymmetric bilinear form.

17.1.1

Definition 17.1.1 *The quotient of the tensor algebra $\otimes(E)$ by the two-sided ideal $\mathcal{N}(F)$ generated by the elements*

$$x \otimes y - y \otimes x - F(x,y), \quad x,y \in E$$

is called a common symplectic Clifford algebra (or a Weyl algebra).

If $F = 0$, we reobtain $\vee(E)$ (cf. Chapter 2, 3). The quotient algebra will be denoted by $C_S(E,F)$, often shortened to $C_S(F)$.

Let ρ_F be the canonical mapping $\otimes(E) \to \otimes(E)/\mathcal{N}(F) = C_S(F)$; $\otimes(E)$ being generated by E, $C_S(F)$ will be generated by $\rho_F(E)$ and we will see later that $\rho_F(E)$ can be identified with E. By its definition,

$$\rho_F(x)\rho_F(y) - \rho_F(y)\rho_F(x) = F(x,y). \tag{1}$$

$x \otimes y - y \otimes x$ and $F(x,y)$ being even elements of $\otimes(E)$, C_S is graded over $\mathbf{Z}_2$, just as in the orthogonal case :

$$C_S(F) = C_S^+(F) \oplus C_S^-(F)$$

and $C_S^+(F)$ is a sub-algebra of $C_S(F)$.

Remark. If we define the $(2r+1)$-dimensional Lie algebra H over the space $E \oplus \mathbf{K}$, with bracket

$$[x,y] = F(x,y), \quad [x,1] = 0, \quad x,y \in E,$$

i.e. a Heisenberg algebra, $C_S(F)$ is isomorphic to the enveloping algebra of H.

233

17.1.2   The universal property of $C_S(F)$.

Theorem 17.1.1 *(Fundamental Theorem.) Let A be an associative algebra over K, $*$ its multiplication and u a linear mapping from E to A such that*

$$u(x) * u(y) - u(y) * u(x) = F(x,y), \quad \forall x, y \in E; \tag{2}$$

then there exists a unique homomorphism $\hat{u}$ from $C_S(F)$ to A such that $u = \hat{u} \circ \rho_F$.

Proof. If it exits, $\hat{u}$ is unique since $\rho_F(E)$ generates $C_S(F)$. u extends to a unique homomorphism v from $\otimes(E)$ to A such that

$$v(x_1 \otimes x_2 \otimes \ldots \otimes x_h) = u(x_1) * u(x_2) * \ldots * u(x_h), \quad x_i \in E.$$

From (2) it follows that $v(x \otimes y - y \otimes x - F(x,y)) = 0$, hence v vanishes on $\mathcal{N}(F)$ and $\hat{u}$ is obtained by taking the quotient. ∎

Consequences :

1. Take $A = C_S(F)$, $u = -\rho_F$, then there exists a principal automorphism α of $C_S(F)$, such that $\alpha \circ \rho_F = -\rho_F$ on E.

2. Take $K = C$, for A the algebra associated to $C_S(F)$ and $u = i\rho_F$; then there exists a main anti-isomorphism β of $C_S(F)$ such that $\beta \circ \rho_F = i\rho_F$ on E.

3. Consider a symplectic basis $\{e_\alpha, e_\beta\}$, $\alpha, \beta = 1, \ldots, r$, F being of rank $n = 2r = \dim E$. (cf. Proposition (1.2.3)).

$$F(e_\alpha, e_\beta) = F(e_{\alpha^*}, e_{\beta^*}) = 0, \quad F(e_\alpha, e_{\beta^*}) = \delta_{\alpha\beta}. \tag{3}$$

   Take A as in 2, but K arbitrary :

$$u(e_\alpha) = \rho_F(e_\alpha), \quad u(e_{\alpha^*}) = \rho_F(e_\alpha).$$

   From this we deduce that there exists a main anti-automorphism β of $C_S(F)$ such that

$$\beta(\rho_F e_\alpha) = \rho_F(e_{\alpha^*}), \quad \beta(\rho_F e_{\alpha^*}) = \rho_F(e_\alpha).$$

Note that the definition of β in 3 depends on the chosen symplectic basis.

17.1.3   The linear isomorphism with the symmetric algebra $\vee(E)$ and the rôle of the 'bosonic' creation and annihilation operators.

Using the notations and the results of Chapter 1, 1.2 and of Chapter 2, 4.2, we define

$$\gamma_y : x \to \tfrac{1}{2}F(x,y),$$

γ_y is an element of the dual E^* of E, so $\gamma_y = y^*$ where $y^*(x) = \langle x, y^* \rangle = \frac{1}{2}F(y, x)$. Since $\varphi \in E^*$, our result

$$j_\varphi \circ e_x - e_x \circ j_\varphi = \langle x, \varphi \rangle, \tag{4}$$

means in this setting that

$$j_{y^*} \circ e_x - e_x \circ j_{y^*} = \frac{1}{2}F(y, x). \tag{5}$$

(We will sometimes write j_{y^*} as j_y^F.)

Keeping the analogy with the orthogonal case, we write

$$L_x = j_{x^*} + e_x,$$

L_x is an endomorphism of $\vee(E)$ such that

$$L_x \circ L_y - L_y \circ L_x = F(x, y);$$

the universal property of $C_S(F)$ now allows us to consider the unique homomorphism $\hat{u}$ from $C_S(F)$ to $\mathrm{End}(\vee(E))$ such that

$$\hat{u} \circ \rho(x) = L_x, \quad \forall x \in E.$$

This yields a representation of $C_S(F)$ in the symmetric algebra of E.

Lemma 17.1.1 *If j_φ is the endomorphism of $\otimes(E)$ satisfying (4) (with $j_\varphi(1) = 0$), given a bilinear form G on E, there exists a unique linear endomorphism λ_G of $\otimes(E)$ such that*

$$\lambda_G(1) = 1 \tag{6}$$

$$\lambda_G \circ e_x = (e_x + j_x^G) \circ \lambda_G, \quad x \in E. \tag{7}$$

Proof. (6) determines λ_G on $\mathbf{K}$. If $(x \otimes a) \in \otimes^h(E)$,

$$\lambda_G(x \otimes a) = x \otimes \lambda_G(a) + j_x^G(\lambda_G(a))$$

and the lemma follows as in Chapter 3, 1.5. ∎

Lemma 17.1.2 *For all $\varphi \in E^*$, λ_G and j_φ commute.*

Proof. We use induction; a simple computation involving

$$j_\varphi(x \otimes a) = x \otimes j_\varphi(a) + \varphi(x)a$$

(cf. Chapter 2, 4.2),

$$j_\varphi \circ e_x - e_x \circ j_\varphi = \varphi(x),$$

$$j_\varphi \circ j_x^F - j_x^F \circ j_\varphi = 0$$

gives the result if $\lambda_F \circ j_\varphi$ is applied to $(x \otimes a)$. ∎

Lemma 17.1.3 *If G and G' are two bilinear forms on E,*

$$\lambda_G \circ \lambda_{G'} = \lambda_{G+G'},$$

and λ_G is bijective on $\otimes(E)$.

Proof. Again using induction, we only need to verify that

$$\lambda_G \circ \lambda_{G'} \circ e_x = \lambda_{G+G'} \circ e_x$$

applied to $\otimes^{h-1}(E)$.

If $G = 0$, $\lambda_G = \mathrm{Id}$ by induction. The computations are similar to those in Chapter 3, 1.5. ∎

Let $\mathcal{J}$ be the two-sided ideal generated by the $(x \otimes y - y \otimes x)$.

Proposition 17.1.1 *λ_F maps the ideal $\mathcal{N}(F)$ into $\mathcal{J}$ and defines an vector space isomorphism from $C_S(F)$ to $\vee(E)$.*

Proof. Just as in Chapter 3, it is sufficient to prove that

$$\lambda_F(\mathcal{N}(F)) \subseteq \mathcal{J}.$$

The reasoning is similar to the orthogonal case. Using the previous lemmas and definition, one verifies that

$$\lambda_F(x \otimes y - y \otimes x - F(x,y)) \in \mathcal{J}.$$

Then for $c = (x \otimes y - y \otimes x - F(x,y)) \otimes b$, $\lambda_F(c) \in \mathcal{J}$, and $\lambda_F(a \otimes c)$ is computed, assuming that $\lambda_F(c) \in \mathcal{J}$, starting with $a = y \in E$.

All of these are routine computations. ∎

Proposition (17.1.1) allows us to identify $\rho_F(\alpha)$ with $\alpha \in \mathbf{K}$ and $\rho_F(x)$ with $x \in E$.

If (e_i) is an ordered basis of E, $i = 1, 2, \ldots, n$, the products $(e_1)^{\alpha_1}(e_2)^{\alpha_2} \ldots (e_n)^{\alpha_n}$ with $\alpha_i \in \mathbf{N}$ form a basis of $C_S(F)$. Note that every element of $C_S(F)$ is a finite linear combination of elements of this basis.

Consequence. Formula 7 applied to $u \in C_S(F)$ gives

$$\lambda_F(xu) = x \vee \lambda_F(u) + j_{x\bullet}(\lambda_F(u)) \tag{8}$$

where the product in $C_S(F)$ is written as the juxtaposition of the factors.

Just as for Clifford algebras, there are two algebraic operations on $\vee(E)$, identifying $\lambda_F(v)$ with $v \in C_S(F)$, the second one is defined by

$$xu = x \vee u + j_{x\bullet}(u), \quad x \in E, \quad u \in \vee(E). \tag{9}$$

But just as in Chapter 3, the identifications are only linear and (9) gives a computational algorithm rather than a definition. Note also that the creation and annihilation operators generate a symplectic Clifford algebra which is isomorphic to an endomorphism algebra of $\vee(E)$ so that the situation is analogous to the orthogonal case (cf. Chapter 3, 1.5).

Definition 17.1.2 *If G_1 and G_2 are two associative algebras over **K**, the tensor product of G_1 and G_2 is the space $G_1 \otimes G_2$ with the multiplication*

$$(a_1 \otimes a_2)(b_1 \otimes b_2) = a_1 b_1 \otimes a_2 b_2,$$

extended linearly to the whole of $G_1 \otimes G_2$.

Proposition 17.1.2 *If F and F' are two antisymmetric bilinear forms on the n resp. n'-dimensional vector spaces E and E', there exists an isomorphism from $C_S(F \oplus F')$ to $C_S(F) \otimes C_S(F')$.*

Proof. The space $E \oplus E'$ is given the bilinear form $F \oplus F'$, extending the similar definition in the orthogonal case (Chapter 8, 1.3). We set

$$u(x, x') = (x \otimes 1) + (1 \otimes x'), \quad x \in E, \quad x' \in E',$$

then u maps $E \oplus E'$ into $C_S(F) \otimes C_S(F')$ and it is easily verified that u satisfies condition (2). Applying the universal property, there exists a homomorphism $\hat{u}$ from $C_S(F \oplus F')$ to $C_S(F) \otimes C_S(F')$ which is an algebra homomorphism; u is surjective since the $x \otimes 1$ and $1 \otimes x'$ generate $C_S(F) \otimes C_S(F')$. But upon consideration of the bases for these algebras, $\hat{u}$ is seen to be injective. ∎

Remark.

1. The linear isomorphism λ_F from $\vee(E)$ to $C_S(F)$ commutes with the action of the symplectic group; this follows easily from formula (8), using induction.

2. The isomorphism λ_F does not associate the elements $(e_1)^{\alpha_1} \vee (e_2)^{\alpha_2} \vee \ldots \vee (e_n)^{\alpha_n}$ with $(e_1)^{\alpha_1}(e_2)^{\alpha_2}\ldots(e_n)^{\alpha_n}$; for instance $\lambda_F(e_1 e_2) = e_1 \vee e_2 - \frac{1}{2}$, where $n = 2$ and $F(e_1, e_2) = 1$ is assumed. The situation is more complicated than the orthogonal case where $\lambda_B(e_{i_1} e_{i_2} \ldots e_{i_p}) = e_{i_1} \wedge e_{i_2} \wedge \ldots \wedge e_{i_p}$ if the (e_i) form an orthonormal basis (cf. formula 8 in Chapter 3, 1.5), λ_B commutes with the action of the orthogonal group.

3. To add a symmetric bilinear form to $\frac{1}{2}F$ does not invalidate Proposition (17.1.2) or formula (8).

17.1.4

Proposition 17.1.3 *If F is non-degenerate, the center of $C_S(F)$ is **K**.*

Proof. Choose a symplectic basis $\{e_\alpha, e_{\beta^*}\}$, the elements of $C_S(F)$ which can be expressed as linear combinations of products not involving e_{α^*} commute with e_α. Let u be an element of the center, $u = u_1 + v$, v not containing any power of e_{α^*}; since $u_1 e_\alpha - e_\alpha u_1 = 0$, if u_1 contains a term in $(e_{\alpha^*})^k$ and since

$$(e_{\alpha^*})^k e_\alpha - e_\alpha (e_{\alpha^*})^k = -k(e_{\alpha^*})^{k-1}, \tag{10}$$

u_1 must vanish and u cannot contain any power of e_{α^*}; the same reasoning holds for all e_i, $i = 1, 2, \ldots, n$, so the center must be trivial. ∎

Formula (10) allows the stepwise expression of $(e_{\alpha^*})^k (e_\alpha)^l$ as

$$
\begin{aligned}
(e_{\alpha^*})^k (e_\alpha)^l &= (e_\alpha)^l (e_{\alpha^*})^k - lk(e_\alpha)^{l-1}(e_{\alpha^*})^{k-1} + \ldots \\
&\quad + (-1)^p C_l^p C_k^p p! (e_\alpha)^{l-p}(e_{\alpha^*})^{k-p} + \ldots, \quad p \le l, \quad p \le k,
\end{aligned}
\tag{11}
$$

and in an arbitrary basis e_i, $i = 1, 2, \ldots, n$,

$$
\begin{aligned}
(e_i)^k (e_j)^l &= (e_j)^l (e_i)^k + lk F(e_i, e_j)(e_j)^{l-1}(e_i)^{k-1} + \ldots \\
&\quad + (F(e_i, e_j))^p C_l^p C_k^p p! (e_j)^{l-p}(e_i)^{k-p} + \ldots
\end{aligned}
\tag{12}
$$

17.1.5 Degenerate algebras.

Assume that the rank $2r$ of F is strictly smaller than n. We set $E = E_1 \oplus E_2$, $\dim E_1 = 2r$, E_2 is the radical of (E, F), F induces F_1 on E_1, a non-degenerate bilinear form of rank $2r$. Then from Proposition (17.1.2) :

Proposition 17.1.4 *The symplectic Clifford algebra $C_S(F)$ is isomorphic to the tensor product of $C_S(F_1)$ and $\vee(E_2)$.*

Proposition 17.1.5 *The center of $C_S(F)$ is $\vee(E_2)$.*

Proof. Obvious. ∎

17.2 Enlarged symplectic Clifford algebras.

17.2.1

The Weyl algebras constructed in this way do not allow the construction of Clifford groups and spin groups having properties similar to the orthogonal case. This caused them to be treated as mathematical curiosities until a very recent date. In fact, they were commonly defined using differential operators with constant coefficients. Now we will construct new algebras that will fully generalize the Clifford algebras to the symplectic case. These algebras are called 'enlarged symplectic Clifford algebras' and their elements are infinite linear combinations of the products $(e_1)^{\alpha_1}(e_2)^{\alpha_2} \ldots (e_n)^{\alpha_n}$ used for $C_S(F)$. Such sums will have a formal or a non-formal meaning, depending on the introduction of convergence conditions.

It will be convenient to introduce an antisymmetric bilinear form hF, which is a non-degenerate symplectic form, h will either be an indeterminate or a constant in $\mathbf{K}$ which we will call *Planck's constant*.

Definition 17.2.1 *If $H = \mathbf{K}[[h]]$ is the algebra of formal power series in the indeterminate h, with coefficients in $\mathbf{K}$, let*

$$
E_H = E \otimes \mathbf{K}[[h]].
$$

E_H is a free module over $\mathbf{K}[[h]]$ (and an infinite-dimensional vector space over $\mathbf{K}$) called the *Planck module*.

17.2.2 The formal symplectic Clifford algebras.

Henceforth we assume that G has maximal rank $n = 2r$. E_H is given the symplectic form hF. We define the algebra $C_S(hF)$ in the same way as $C_S(F)$. Nothing essential must be changed; we reobtain the same results (such as the universal property, the linear isomorphism with $\vee(E_H)$ and the trivial center).

It will also be useful to define an enlarged algebra, rather than to consider each element as a finite linear combination of generators $(e_1)^{\alpha_1}(e_2)^{\alpha_2}\ldots(e_n)^{\alpha_n}$, $\alpha_i \in \mathcal{N}$. We consider formal power series with coefficients in $\mathbf{K}[[h]]$ and, since formula (12) is now expressed by

$$
\begin{aligned}
(e_i)^k(e_j)^l &= (e_j)^l(e_i)^k + hlkF(e_i,e_j)(e_j)^{l-1}(e_i)^{k-1} + \ldots \\
&\quad + (h)^p(F(e_i,e_j))^p C_l^p C_k^p p!(e_j)^{l-p}(e_i)^{k-p} + \ldots
\end{aligned}
\tag{1}
$$

it will be convenient to consider h as a indeterminate of degree 2.

Let us call this module over $\mathbf{K}[[h]]$, $\check{C}_S(hF)$.

Proposition 17.2.1 $\check{C}_S(hF)$ is an associative algebra.

Proof. It is sufficient to verify that the product of two elements in $\check{C}_S(hF)$ is again an element of $\check{C}_S(hF)$. The rest is obvious.

Proposition (17.1.2) still holds for $\check{C}_S(hF)$ and allows us to consider a two-dimensional space and elements

$$
\check{u} = \sum_k \frac{a_k}{k!}(e_{1*})^k, \quad \check{v} = \sum_i \frac{b_l}{l!}(e_1)^l,
$$

where (e_1, e_{1*}) is a symplectic basis and the a_k and b_k are formal power series in h (the other cases either are trivial or can be reduced to this case).

The coefficient in $\mathbf{K}$ of $h^p(e_1)^l(e_{1*})^k$ in the expansion of $\check{u}\check{v}$ comes, by (1), from :

$$
\frac{1}{k!}\frac{1}{l!}(a_k b_l - h a_{k+1}b_{l+1} + \ldots + (-h)^s \frac{a^{k+s}b^{l+s}}{s!} + \ldots)
$$

and, in fact, from the coefficient of h^p in $a_k b_l$, the coefficient of h^{p-1} in $a_{k+1}b_{l+1}$, ..., of h^{p-s} in $a_{k+s}b_{k+s}$; so from a finite number of terms, since $s \le p$. ∎

This also ensures that the definition of $\check{C}_S(hF)$ does not depend on the basis.

The universal property, the isomorphism with an enlarged symmetric algebra and the center are still as before and can be proved in the same way.

Definition 17.2.2 *The elements of $\check{C}_S(hF)$ are called symplectic formal series (or s.f.s. for short).*

h being considered as a indeterminate of degree 2, the total degree of a monomial in $h, e_1, e_2, \ldots, e_n$ is invariant for any change of basis, as can be verified using (1). The formal series can therefore be considered as series with coefficients in $\mathbf{K}$ constructed on $2r + 1$ indeterminates (e_i) and h. This convention allows us to formulate :

Definition 17.2.3 *The total order $\omega(\check{u})$ of $\check{u} \neq 0$ is the smallest integer $q \geq 0$ such that the homogeneous part of degree q does not vanish.*

Then, just as in a commutative algebra :

$$\omega(\check{u} + \check{v}) \geq \min(\omega(\check{u}), \omega(\check{v})), \quad (\check{u} + \check{v} \neq 0),$$
$$\omega(\check{u}\check{v}) \geq \omega(\check{u}) + \omega(\check{v}).$$

Proposition 17.2.2 *(Substitution principle.) If the X_i are symplectic formal series with vanishing constant term and $f(e_1, e_2, \ldots, e_n, h)$ is a symplectic formal series, then $f(X_1, X_2, \ldots, X_n, h)$ is also a symplectic formal series.*

Proof. If $k + l$ is fixed, only a finite number of terms make up the right-hand side of (1); when the products are ordered, each coefficient can be obtained from a finite number of terms. ∎

Proposition 17.2.3 *A symplectic formal series is invertible if and only if its constant term in $\mathbf{K}$ is non-zero.*

Proof. Apply the substitution principle to the inverse of $1 - X_i$. ∎

17.2.3 The truncated formal symplectic Clifford algebras.

Consider the quotient $\bar{H}$ of the algebra $\mathbf{K}[h]$ of polynomials in H by the ideal of multiples of h^p, p being a fixed integer ($p \geq 2$). $\bar{H}$ is a p-dimensional algebra over $\mathbf{K}$ and its general element is given by

$$\sum_{i=0}^{p-1} a_i \bar{h}^i, \quad a_i \in \mathbf{K}, \quad \bar{h} = \mathrm{Cl}(h) \,(\mathrm{mod}\, h^p).$$

The $\bar{H}$-module $E \otimes \bar{H}$, also denoted by $E_{\bar{H}}$, is free and has an (np)-dimensional vector space structure over $\mathbf{K}$. It is called a truncated Planck module. $\overset{\triangledown}{C}_S(hF)_p$ (also written $\check{C}_S(\bar{h}F)$) is the Clifford algebra obtained by the same procedure as before, it still has the properties proved in 1.3 : given an ordered basis of E, its elements are truncated symplectic formal series (t.s.f.s.). They can also be considered as formal series with coefficients in $\mathbf{K}$ in the indeterminates $e_1, e_2, \ldots, e_n, \bar{h}$. Nothing needs to be changed in the definition of the total order ω, and both Propositions (17.2.2) and (17.2.3) still hold. Note that in the right-hand side of (1) at most p terms can occur in any case.

Proposition 17.2.4 *If $p_2 > p_1$, there exists a natural homomorphism $m_{p_2 p_1}$ from $\overset{\triangledown}{C}_S(hF)_{p_2}$ to $\overset{\triangledown}{C}_S(hF)_{p_1}$.*

Proof. More precisely, if $\bar{H}_{(p)}$ is the quotient $\bar{H}$ modulo h^p, there exists a homomorphism from the algebra $\bar{H}_{(p_2)}$ to the algebra $\bar{H}_{(p_1)}$. A linear homomorphism from $E_{\bar{H}_{(p_2)}}$ to $E_{\bar{H}_{(p_1)}}$ can be derived from it, this is a linear homomorphism u from $E_{\bar{H}_{(p_2)}}$ to $\overset{\triangledown}{C}_S(hf)_{p_1}$ such that :

$$u(x) *_{p_1} u(y) - u(y) *_{p_1} u(x) = (\bar{h})_{p_1} F(x,y), \quad x,y \in E_{\bar{H}_{(p_2)}};$$

indeed, as in the proof of the Fundamental Theorem (17.1.2), u is extended to v from $\otimes(E_{\bar{H}_{(p_2)}})$ to $\overset{\triangledown}{C}_S(hF)_{p_1}$, where

$$v(x_1 \otimes x_2 \otimes \ldots \otimes x_h) = u(x_1) *_{p_1} u(x_2) *_{p_1} \ldots *_{p_1} u(x_h),$$
$$v(x \otimes y - y \otimes x - (\bar{h})_{p_2} F(x,y)) = u(x) *_{p_1} u(y) - u(y) *_{p_1} u(x) - (\bar{h})_{p_1} F(x,y) = 0.$$

from which we find $\hat{u}$, taking the quotient, and $m_{p_2 p_1} = \hat{u}$. ∎

Corollary 17.2.1 *The set of homomorphisms*

$$m_{p_2 p_1}, \ (p_2 > p_1) \quad \text{and algebras} \quad \overset{\triangledown}{C}_S(hF)_{p_1}$$

defines a projective system of mappings and associative algebras with limit $\overset{\triangledown}{C}_S(hF)$.

17.2.4 The formal symplectic Clifford algebras over $\mathbf{K}((h))$.

Now the ring $\mathbf{K}[[h]]$ is replaced by the field consisting of its rational fractions, meaning that in the expansions of Clifford elements a finite number of negative exponents of h are allowed. The elements of $\mathbf{K}((h))$ will also be called formal series in h and the new algebra will still be denoted by $\overset{\triangledown}{C}_S(hF)$ (abusively).

An arbitrary symplectic basis $\{e_\alpha, e_{\beta^*}\}$ being chosen, the general element of this algebra is called a symplectic formal series and its general term $\check{u}$ is of the form

$$\lambda_{h_1 h_2 \ldots h_r k_1 k_2 \ldots k_r}(e_1)^{h_1}(e_2)^{h_2} \ldots (e_r)^{h_r}(e_{1^*})^{k_1}(e_{2^*})^{k_2} \ldots (e_{r^*})^{k_r},$$

where the h_i and k_i are positive or zero integers. Symbolically, we will write

$$\check{u} = \sum_{H,K^*} \lambda_{HK^*} e^H e^{K^*}, \quad e_\emptyset = 1, \tag{2}$$

where the H, K^* stand for multi-indices and the λ_{HK^*} are formal power series in $\mathbf{K}((h))$.

Symplectic spinors for $\overset{\triangledown}{C}_S(hF)$. Consider the quotient of the algebra $\overset{\triangledown}{C}_S(hF)$ by the left ideal J_m generated by the (e_{α^*}), when a symplectic basis $\{e_\alpha, e_{\alpha^*}\}$ has been chosen. The elements of J_m are of the form $\sum \lambda_{HK^*} e^J e^{K^*}$, where K^* is non-empty. The representation of $\overset{\triangledown}{C}_S(hF)$ in this quotient space is defined by left multiplication;

the space of this representation will be written $\check{C}_S(hF)\Phi^*$ (by analogy with the orthogonal case), Φ^* being just a symbol resembling the isotropic r-vector of the orthogonal case. *The space $\check{C}_S(hF)\Phi^*$ is called the space of symplectic spinors.* A change of symplectic basis gives rise to an equivalent representation.

Note that the space $\check{C}_S(hF)\Phi^*$ is linearly isomorphic to the symmetric algebra on the vectors $(e_1, e_2, \ldots, e_r)$.

The following proposition justifies this terminology :

Proposition 17.2.5 *The representation of $\check{C}_S(hF)$ in $\check{C}_S(hF)\Phi^*$ by left multiplication is irreducible.*

Proof. Indeed, the left ideal J_m is maximal, for if an element $u = \sum \lambda_H e^H$ is added to it, the multiplication table (11) of 1.4 shows that, h being invertible, 1 will be an element of the resulting ideal, which must therefore coincide with the whole algebra. J_m being maximal, the representation is irreducible by the results in Chapter 5. ■

Proposition 17.2.6 *In a sense that will specified later, the algebra $\check{C}_S(hF)$ with coefficients in $\mathbf{K}((h))$ is simple.*

Proof. We use induction on the dimension of E to prove the existence of a set of maximal left ideals with zero intersection.

Assume that $\dim E = 2$ and choose the ordered symplectic bases (e_1, e_{1^*}) and $(-e_{1^*}, e_1)$. The corresponding minimal left ideals are

$$J_1 = \{u = \textstyle\sum_{k,l} \lambda_{kl}(e_1)^k(e_{1^*})^l, \quad l \geq 1, \quad \lambda_{k,l} \in \mathbf{K}((h))\},$$
$$J_2 = \{u = \textstyle\sum_{k,l} \mu_{kl}(e_{1^*})^k(e_1)^l, \quad l \geq 1, \quad \mu_{k,l} \in \mathbf{K}((h))\}.$$

If $v = \sum_{k,l} \theta_{kl}(e_1)^k(e_{1^*})^l$, $l \geq 1$ is an element of $J_1 \cap J_2$, $\theta_{0l} = 0$ since if $v = \sum \theta_{0l}(e_{1^*})^l$, we see, expressing it in the basis $(-e_{1^*}, e_1)$, that it cannot be an element of J_2 unless it vanishes. But for $k = a \neq 0$ given, $\theta_{al} = 0$ since, multiplying v from the left with $(e_{1^*})^a$ the multiplication table of the first basis (e_1, e_{1^*}) creates in $(e_{1^*})^a v \in J_1 \cap J_2$ terms in $\theta_{al}(e_{1^*})^l$, up to a non-zero coefficient, and this implies $\theta_{al} = 0$.

If $\dim E = n = 2r$ and $(e_1, \ldots, e_r, e_{1^*}, \ldots, e_{r^*})$ is a symplectic basis for E, we consider the maximal left ideals :

$$J_1' = \sum_{\alpha=1}^{r} \check{C}_S(hf)e_{\alpha^*}, \quad J_2' = \check{C}_S(hF)e_1 + \sum_{\alpha=2}^{r} \check{C}_S(hF)e_{\alpha^*}.$$

Ordering the basis as follows :

$$e_2, \ldots, e_r, e_1, e_{1^*}, e_{2^*}, \ldots, e_{r^*},$$

an element of $J_1' \cap J_2'$ is either of the form

$$u = \sum_{k,l} \lambda_{kl}(e_1)^k(e_{1^*})^l + u', \quad l \geq 1$$

or of the form

$$u = \sum_{k,l} \mu_{kl}(e_{1\cdot})^k (e_1)^l + u'', \quad l \geq 1,$$

where λ_{kl} and μ_{kl} are coefficients which contain no powers of e_1 or $e_{1\cdot}$ and commute with e_1 and $e_{1\cdot}$; u' and u'' contain no factor in e_1 or $e_{1\cdot}$. The same reasoning as when $n = 2$ proves that $\lambda_{kl} = \mu_{kl} = 0$. The problem is then reduced to $n - 2$ dimensions and the induction hypothesis can be applied.

So there exists a set $J_{m_1}, J_{m_2}, \ldots, J_{m_k}$ of maximal left ideals with zero intersection. Let φ_{m_i} be the canonical homomorphism :

$$\check{C}_S(hF) \rightarrow \check{C}_S(hF)/J_{m_i} = \hat{J}_{m_i},$$

$\prod(\varphi_{m_i})$ is an injective homomorphism from $\check{C}_S(hF)$ to $\prod(\hat{J}_{m_i})$ since

$$\prod(\varphi_{m_i}(u)) = \prod(\varphi_{m_i})(u')$$

implies that $u' - u \in J_{m_i}$ for all J_{m_i}.

J_{m_i} being maximal, $\hat{J}_{m_i}$ is simple and $\prod(\hat{J}_{m_i})$ is said to be semisimple, just as its inverse image under $\prod(\varphi_{m_i})$.

The J_{m_i} are left ideals which are naturally associated to symplectic bases; there exists a symplectic transformation sending one of these bases to another one, and this symplectic transformation defines a representation equivalence. All simple modules obtained in this fashion are isomorphic, and such an algebra is said to be simple[1]. ∎

Remark. *All two-sided ideals of $\check{C}_S(hF)$ are trivial.* If J is a two-sided ideal and if $u \in J$, consider a minimal term for the lexicographical ordering of the e_α and $e_{\alpha\cdot}$. If it is $\lambda_{HK\cdot} e^H e^{K\cdot}$ (where there is no summation), a left multiplication by $e^{H\cdot}$ followed by a right multiplication by e^K produces an element of J with a non-zero term of degree 0. This element is invertible and J must be trivial.

17.2.5   Large non-formal symplectic Clifford algebras.

h is considered here as a real or complex parameter. Using the notations of formula (2), we choose, after a certain rank,

$$|\lambda_{HK\cdot}| \leq \frac{\sigma(\hat{u})\rho(\hat{u})^{|H|+|K^*|}}{(H!K^*!)^{\alpha+1/2}}, \tag{3}$$

where $H! = h_1! h_2! \ldots h_r!$, $K^*! = k_1! k_2! \ldots k_r!$, $|H| = \sum h_i$, $|K^*| = \sum k_i$, α is a real number, $\alpha > 0$, and does not depend on $\hat{u}$; $\sigma(\hat{u})$ and $\rho(\hat{u})$ are positive constants.

Condition (3) is independent of the chosen symplectic basis. By elementary convergence rules, it can be verified that an algebra over **K** is obtained; we will

---

[1]A priori, this definition is not equivalent to the one given in Chapter 5 above.

denote it by $\check{C}_S(hF)_\alpha$. The only non-obvious fact to check concerns the product, which can be handled as before in 2.2.

$$\check{u} = \sum_k \frac{a_k}{k!}(e_{1*})^k, \quad \check{v} = \sum_l \frac{b_l}{l!}(e_1)^l,$$

with the conditions

$$|a_k| \le a(k!)^{1/2-\alpha}t^k, \quad |b_l| \le b(l!)^{1/2-\alpha}(t')^l,$$

for k and l sufficiently large and some positive constants a, b, t, t'. By the binomial formula :

$$(k+s)! \le 2^{k+s}k!s!,$$

so that

$$\frac{|a_{k+s}b_{l+s}|}{s!k!l!}|h|^s \le \frac{ab|h|^s 2^{(2s+k+l)(1/2-\alpha)}(k!l!)^{1/2-\alpha}t^k t'^l (tt')^s}{k!l!(s!)^{2\alpha}}$$

and if $M(\alpha)$ is the sum of the convergent series

$$\sum_{s=0}^{\infty} \frac{2^{2s(1/2-\alpha)}(tt')^s|h|^s}{(s!)^{2\alpha}}, \tag{4}$$

we get

$$\sum_{s=0}^{\infty} \frac{|a_{k+s}b_{l+s}||h|^s}{s!k!l!} \le \frac{abM(\alpha)}{(k!l!)^{1/2+\alpha}}(At)^k(At')^l \tag{5}$$

where $A = 2^{1/2-\alpha}$.

The general element of $C_S(hF)_\alpha$ can be written as

$$\check{u} = \sum_{H,K^*} \frac{M(H,K^*)e^H e^{K^*}}{(H!K^*!)^{1/2+\alpha}}, \quad \alpha > 0 \tag{6}$$

where $|M(H,K^*)| \le \sigma(\check{u})\rho(\check{u})^{|H|+|K^*|}$ for sufficiently large $|H|$ and $|K^*|$.

17.2.6 The exponential function.

The element $\exp ta^2 \in \check{C}_S(hF)$, where $a \in E$, $t \in \mathbf{K}$ is defined by the formal power series

$$\exp ta^2 = 1 + ta^2 + \frac{t^2 a^4}{2!} + \dots + \frac{t^k a^{2k}}{k!} + \dots \tag{7}$$

with inverse $\exp(-ta^2)$.

It is easily verified that $\exp t(e_\alpha)^2$ and $\exp t(e_{\alpha*})^2$ satisfy condition (3) for $\alpha = 0$; indeed, if

$$\check{u} = \exp t(e_{\alpha*})^2 = \sum \frac{t^k}{k!}(e_{\alpha*})^{2k},$$

the coefficients $a_{2k}, a_{2k+1}, \ldots$ satisfy

$$\frac{a_{2k}}{(2k)!} = \frac{t^k}{k!}, \quad a_{2k+1} = 0,$$

but from $((2k)!)^{1/2} \leq 2^k k!$ it follows that

$$\frac{|a_{2k}|}{(2k)!} = \frac{|t|^k}{k!} \leq \frac{(2|t|)^k}{((2k)!)^{1/2}}.$$

If $\alpha = 0$, the convergence of (4) requires an upper bound for $\rho(\check{u})$. The series in (4) converges if $\sup(\rho(\check{u}), \rho(\check{v})) < 1/\sqrt{2h}$; if h is Planck's constant, this upper bound is of the order of 10^{13} in C.G.S. units, hence it is very large.

17.2.7 Clifford, toroplectic and spinoplectic groups.

Definition 17.2.4 *The subgroup of invertible elements $\gamma \in \check{C}_S(hF)$ such that for all $x \in E_H$, $\gamma x \gamma^{-1} \in E_H$ is called the* formal symplectic Clifford group G_S^H.

We immediately see that G_S^H is a group, that $\gamma \to p(\gamma)$ where $p(\gamma)(x) = \gamma x \gamma^{-1}$ is a homomorphism from G_S^H to $\mathrm{Sp}(2r, \mathbf{K}[[h]])$, the group of $\mathbf{K}[[h]]$-linear isomorphisms σ from E_H such that $F(\sigma x, \sigma y) = F(x, y)$, $x, y \in E_H$, and that the kernel of p is the set $\mathbf{K}[[h]]^*$ of invertible elements in $\mathbf{K}[[h]]$.

Proposition 17.2.7 *The exponential elements $\exp(ta^2)$, $t \in \mathbf{K}[[h]]$, $a \in E_H$ are elements of G_S^H and satisfy*

$$p(\exp ta^2)x = x + 2hF(a, x)ta. \tag{8}$$

Proof. Obvious. ∎

When $\mathbf{K} = \mathbf{R}$ or $\mathbf{C}$, we consider h either as an indeterminate or as an element in $\mathbf{K}$. h can be replaced by $\hat{h} \in \mathbf{K}$ if the indeterminate h is given a numerical value in $\mathbf{K}$, but the notation $\hat{h}$ will be avoided when this will not cause confusion. Similarly, any expression in which h is replaced by $\hat{h}$ will be written with a hat.

Set $\hat{p}(\exp ta^2)(x) = x + 2\hat{h}F(a, x)ta$, if $t \in \mathbf{K}$, $a \in E$, a symplectic transvection is obtained on the right-hand side (Chapter 1, 4.1); we know that the symplectic group $\mathrm{Sp}(2r, \mathbf{K})$ is generated by the set of symplectic transvections. Note that a formal algebraic identity allows us to write

$$\exp sa^2 \exp tb^2 \exp(-sa^2) = \exp(-2h^2\alpha^2 st) \exp(tb^2 + 4h\alpha stab + 4h^2s^2t\alpha^2a^2)$$

where $\alpha = F(a, b)$, giving rise to a convergent scalar factor if h is replaced by $\hat{h}$. This leads us to the following :

Definition 17.2.5 *The quotient of the subgroup of elements*

$$\prod_i \lambda_i(h) \exp(t_i(a_i)^2), \quad \lambda_i(h) \in \mathbf{K}[[h]], \quad a_i \in E, \quad t_i \in \mathbf{K}$$

in G_S^H *by the subgroup of elements* $1 + u$, $u \in \mathbf{K}[[h]]$ *is called the symplectic Clifford group* G_S.

The following sequence is exact :

$$1 \to \mathbf{K}^* \to G_S \xrightarrow{\hat{p}} \text{Sp}(2r, \mathbf{K}) \to 1. \tag{9}$$

Since $\text{Sp}(2r, \mathbf{K})$ is connected, it is generated by an arbitrarily small neighborhood of the identity, hence every element of $\text{Sp}(2r, \mathbf{K})$ is a finite product of elements $\hat{u}$ such that $|\rho(\hat{u})| < 1/\sqrt{2\hat{h}}$ from 2.6 will be satisfied in the non-formal case.

Definition 17.2.6 *(Symplectic spin norm.)* *Let* $\mathbf{K} = \mathbf{C}$ *and write* $G_S'^H$ *and* G_S' *for the corresponding groups. We use the main antiautomorphism* β *of the symplectic Clifford algebra, for which* $\beta|_{E'} = i \, \text{Id}$, *(E' being the vector space under consideration) and define the spin norm* $N(\gamma)$ *by* $N(\gamma) = \beta(\gamma)\gamma$.

Obviously if $\gamma \in G_S'^H$,

$$N(\gamma) = \beta(\gamma)\gamma \in \mathbf{C}[[h]], \quad N(\gamma) \neq 0,$$
$$N(\gamma\gamma') = N(\gamma)N(\gamma'),$$

if $\gamma \in G_S'$, taking the quotient ensures that $N(\gamma) \in \mathbf{C}^*$ (replacing h by $\hat{h}$ where necessary in the computation of $\beta(\gamma)\gamma$).

The toroplectic (or metaplectic) group. *If* E' *is the complexification of a real symplectic space* E, *of real dimension* $n = 2r$, *and if* G_S' *is its symplectic Clifford group, the subgroup of elements* $\gamma \in G_S'$ *such that* $\gamma x \gamma^{-1} \in E$, $\forall x \in E$, *with* $|N(\gamma)| = 1$ *(or, if one prefers,* $|\hat{N}(\gamma)| = 1$*) is called the toroplectic (or metaplectic) group, denoted by* $\text{Mp}(r)$.

The following sequence is exact :

$$1 \to S^1 \to \text{Mp}(r) \to \text{Sp}(2r, \mathbf{R}) \to 1. \tag{10}$$

The spinoplectic group. Under the same conditions as before, if $N(\gamma) = 1$ is required, a group Sp_2 is obtained for which the following sequence is exact :

$$1 \to \mathbf{Z}_2 \to \text{Sp}_2(r) \to \text{Sp}(2r, \mathbf{R}) \to 1. \tag{11}$$

The clear analogy would motivate the name of 'symplectic spin group' for this group; but we will see later on that there exist groups for which $\mathbf{Z}_2$ in the exact sequence is replaced by $\mathbf{Z}_q$ and $\mathbf{Z}$, so we should specify this and call it a spin group of order 2; we will use the name 'spinoplectic' for it.

17.2.8

In this paragraph, we will prove that G_S and the derived groups are Lie groups. We will initially assume that $\mathbf{K} = \mathbf{R}$, that h is a real indeterminate and that E is a real $(n = 2r)$-dimensional space with complexification E'. It is well-known that an elliptic metric $(\cdot, \cdot)$ can be defined on E, along with an $\mathbf{R}$-linear orthogonal endomorphism for this metric, denoted by J, such that $J^2 = -\operatorname{Id}$ and $(Jx, y) = F(x, y)$. Then $F(Jx, Jy) = F(x, y)$ and J is simultaneously symplectic and orthogonal.

Let us decompose E' as the direct sum of two complex conjugate subspaces $\mathcal{E}_1$ and $\mathcal{E}_2$ with bases (ϵ_α) and (ϵ_{α^*}), $\alpha = 1, 2, \ldots, r$, satisfying

$$(\epsilon_\alpha, \epsilon_\beta) = (\epsilon_{\alpha^*}, \epsilon_{\beta^*}) = 0, \qquad (\epsilon_\alpha, \epsilon_{\beta^*}) = \delta_{\alpha\beta}.$$
$$F(\epsilon_\alpha, \epsilon_\beta) = F(\epsilon_{\alpha^*}, \epsilon_{\beta^*}) = 0, \qquad F(\epsilon_\alpha, \epsilon_{\beta^*}) = i\delta_{\alpha\beta}.$$

The ϵ_α and ϵ_{β^*} form a 'real' Witt basis for the metric of E'.

$$J|_{\mathcal{E}_1} = i\operatorname{Id}, \quad J|_{\mathcal{E}_2} = -i\operatorname{Id},$$
$$J\epsilon_\alpha = i\epsilon_\alpha,$$
$$J\epsilon_{\alpha^*} = -i\epsilon_{\alpha^*}$$
$$Je_\alpha = e_{\alpha^*},$$
$$Je_{\alpha^*} = -e_\alpha.$$

If we define $e_\alpha = (\epsilon_{\alpha^*} + \epsilon_\alpha)/\sqrt{2}$, $e_{\alpha^*} = (\epsilon_{\alpha^*} - \epsilon_\alpha)/(i\sqrt{2})$, the (e_α, e_{β^*}) form a symplectic and orthogonal basis of E. We will write $E = E_1 \oplus E_2$, $e_\alpha \in E_1$, $e_{\alpha^*} \in E_2$. Every element of $\operatorname{Sp}(2r, \mathbf{R})$ which commutes with J will be called unitary; the group $\operatorname{U}(r)$ is the set of all elements in $\operatorname{SO}(2r)$ which commute with J; $\operatorname{U}(r)$ is the pairwise intersection of the groups $\operatorname{Sp}(2r, \mathbf{R})$, $\operatorname{SO}(2r)$ and $\operatorname{GL}(r, \mathbf{C})$.

If G_S contains a local Lie group Γ, Γ is endowed with a smooth manifold structure and contains the neutral element 1; $g \in \Gamma \to g^{-1} \in \Gamma$ will be a C^∞ map and $(g, g') \to gg'$ will be a C^∞ mapping from an open set $U \times U$ in $\Gamma \times \Gamma$ to γ. The Lie algebra $\underline{\Gamma}$ of this local group will contain the elements a^2, $\forall a \in E$, this being the tangent to $t \to \exp ta^2$ at $t = 0$; Γ will contain $\exp(-sb^2)\exp(ta^2)\exp(sb^2)$ for sufficiently small s and t, hence $\underline{\Gamma}$ will contain $\exp(-sb^2)a^2\exp(sb^2)$ obtained when t varies; now letting s vary, $\underline{\Gamma}$ must contain $a^2b^2 - b^2a^2 = 4\hat{h}F(a, b)ab - 2h^2(F(a, b))^2$, and since $\underline{\Gamma}$ must also contain the tangent to $t \to \exp \lambda t$, $\lambda \in \mathbf{R}$, it must be the Lie subalgebra of terms of degree 0 and 2, in the symplectic Clifford algebra.

Lemma 17.2.1 *For all $a \in E$, $\exp(ta^2) \in \check{C}_S(hF)_0$ if t is small enough.*

Proof. If $\{e_\alpha, e_{\beta^*}\}$ is a symplectic basis, we write $a = \sum_1^r (\lambda_\alpha e_\alpha + \mu_\alpha e_{\alpha^*})$, $\lambda_\alpha, \mu_\alpha \in \mathbf{R}$ and

$$\frac{a^{2k}}{k!} = \frac{1}{k!} \sum \frac{(2k)!}{n_1! n_2! \ldots n_r!} (\lambda_1 e_1 + \mu_1 e_{1^*})^{2n_1} (\lambda_2 e_2 + \mu_2 e_{2^*})^{2n_2} \ldots (\lambda_r e_r + \mu_r e_{r^*})^{2n_r},$$

the sum extending over all positive or zero integers $n_1, n_2, \ldots, n_r$ such that $n_1 + n_2 + \ldots + n_r = k$; since $(\lambda_i e_i + \mu_i e_{i*})^{2n_i}$ commutes with $(\lambda_j e_j + \mu_j e_{j*})^{2n_j}$, $i \neq j$, this leads us to consider the case $n = 2$ first. Let us determine the coefficient of $(e_1)^i (e_{1*})^j$ in the expansion of $(\lambda e_1 + \mu e_{1*})^{2k+2p}/(k+p)!$, $p = 0, 1, 2, \ldots$, $i+j = 2k$, i, j, k fixed, first without taking into account the powers of λ and μ. Every monomial contributing $(e_1)^i (e_{1*})^j$ will, after reduction, be of the form

$$(e_{1*})^{j+p-\sum \beta_\rho}(e_1)^{\alpha_1}(e_{1*})^{\beta_1} \ldots (e_1)^{\alpha_\rho}(e_{1*})^{\beta_\rho} \ldots (e_1)^{\alpha_\nu}(e_{1*})^{\beta_\nu}(e_1)^{i+p-\sum \alpha_\rho}$$

where $\sum \alpha_\rho \leq i + p$ and $\sum \beta_\rho \leq j + p$.

a) First consider $(e_{1*})^{j+p-\beta}(e_1)^{\alpha}(e_{1*})^{\beta}(e_1)^{i+p-\alpha}$, where $1 \leq \alpha \leq i + p$, $1 \leq \beta \leq j + p$; it gives rise to coefficients whose absolute value equals

$$\frac{\alpha!}{\nu!(\alpha - \nu)!} \frac{\beta!}{\nu!(\beta - \nu)!} \frac{(j+p-\nu)!(i+p-\nu)!}{j!(p-\nu)!i!(p-\nu)!}\nu!(p - \nu)!|\hat{h}|^p, \quad 0 \leq \nu \leq \alpha, \; 0 \leq \nu \leq \beta,$$

from which, adding for fixed α and β, the following upper bound for the coefficient of $(e_1)^i(e_{1*})^j$ is obtained :

$$\sum_\nu 2^{i+j+2p}p!2^{\alpha+\beta}|\hat{h}|^p 2^{-2\nu} \leq \tfrac{4}{3}2^{i+j+2p}p!|\hat{h}|^p 2^{\alpha+\beta}.$$

Since there are at most $(i+p)(j+p)$ pairs α, β and since $2^{\alpha+\beta}$ is smaller that 2^{2k+2p}, the coefficient is smaller, in absolute value, than

$$\tfrac{4}{3}2^{2k+2p}(i + p)(j + p)p!|\hat{h}|^p.$$

b) Now take

$$(e_{1*})^{j+p-\beta_1-\beta_2}(e_1)^{\alpha_1}(e_{1*})^{\beta_1}(e_1)^{\alpha_2}(e_{1*})^{\beta_2}(e_1)^{i+p-\alpha_1-\alpha_2}.$$

Changing the order of the product $(e_{1*})^{\beta_1}(e_1)^{\alpha_1}$ gives monomials of the type considered in a) :

$$(e_{1*})^{j+p-\beta_1-\beta_2}(e_1)^{\alpha_1+\alpha_2-\nu}(e_{1*})^{\beta_1+\beta_2-\nu}(e_1)^{i+p-\alpha_1-\alpha_2}\frac{\alpha_2!\beta_1!}{\nu!(\alpha_2 - \nu)!(\beta_1 - \nu)!},$$

$0 \leq \nu \leq \alpha_2$, $0 \leq \nu \leq \beta_1$, and the line ν yields a coefficient of absolute value

$$\frac{\alpha_2!\beta_1!}{\nu!(\alpha_2 - \nu)!\nu!(\beta_1 - \nu)!} \frac{(j+p-\nu)!(i+p-\nu)!}{j!(p-\nu)!i!(p-\nu)!}\nu!(p - \nu)!|\hat{h}|^p,$$

bounded by $2^{i+j+2p}p!2^{\alpha_2+\beta_1}|\hat{h}|^p 2^{-2\nu}$, and the whole sum over all pairs α_2, β_1 is bounded by

$$\tfrac{4}{3}2^{4k+4p}(i + p)(j + p)p!|\hat{h}|^p,$$

as in a). Continuing, since $\sum \alpha_\rho \leq j + p$ and $\sum \beta_\rho \leq i + p$, there are at most $2k + 2p$ pairs $(e_1)^{\alpha_i}(e_1{}_*)^{\beta_i}$ to consider, so the coefficient under consideration will be bounded by

$$\tfrac{4}{3}(2k + 2p)^3 2^{4k+2p} p! |\hat{h}|^p$$

and if p and k are sufficiently large, by

$$2^{6k+6p} p! |\hat{h}|^p.$$

Finally, to obtain the expansion of $\exp a^2$ we will get

$$\sum_{i,j} \frac{\sigma_{ij}}{k!} \lambda^i \mu^j (e_1)^i (e_1{}_*)^j, \quad i + j = 2k,$$

where σ_{ij} is the sum of a numerical series whose general term is of the order $2^{6k} 2^{6p} |\hat{h}|^p |\lambda \mu|^p$, $((k + p)! \leq 2^{k+p} k! p!)$, smaller than $2^{6p} \frac{1}{2}^p$, if $|\lambda|$ and $|\mu|$ are smaller that $1/(8\sqrt{2|\hat{h}|})$; so if $n = 2$,

$$\exp a^2 = \sum \frac{A_{ij}}{k!} (8\lambda)^i (8\mu)^j (e_1)^i (e_1{}_*)^j, \quad i + j = 2k,$$

the coefficients A_{ij} being bounded in absolute value by a constant A, for each pair i, j and every choice of λ and μ for which $|\lambda|, |\mu| < 1/(8\sqrt{2|\hat{h}|})$.

In the general case, consider $\prod_s \exp(a_s)^2$, $s = 1, 2, \ldots, r$; the expansion of $\exp(a_s)^2$ being of the same form as before, the coefficient of $e^H e^{K^*}$ is the sum of terms

$$\frac{A_{i_1 j_1} A_{i_2 j_2} \ldots A_{i_r j_r}}{k_1! k_2! \ldots k_r!} (8\lambda_1)^{i_1} \ldots (8\lambda_r)^{i_r} (8\mu_1)^{j_1} \ldots (8\mu_r)^{j_r}$$

where $k_p = \frac{1}{2}(i_p + j_p)$, $i_1 + i_2 + \ldots + i_r = |H|$, $j_1 + j_2 + \ldots + j_r = |K^*|$, $|A_{i_\rho j_\rho}| < A_\rho$, bounded by

$$\frac{A_1 A_2 \ldots A_r}{(H! K^*!)^{1/2}} (16\lambda)^{|H|} (16\mu)^{|K^*|}$$

if $\lambda = \sup |\lambda_\rho|$, $\mu = \sup |\mu_\rho|$. This proves Lemma (17.2.1). ∎

By the results of section 2.5, when $\alpha = 0$, $\exp ta^2$ and $\exp tb^2$ also belong to $\check{C}_S(hF)_0$ for sufficiently small t and s. From the formal algebraic identity :

$$\exp(-sb^2)\exp(ta^2)\exp(sb^2) = \exp(\sum_0^\infty \frac{1}{n!}(\text{ad}(-sb^2))^n(ta^2)) \qquad (12)$$

we formally deduce that

$$\exp(-b^2)a^2\exp(b^2) = \exp(a^2 b^2 - b^2 a^2) = \exp(4\hat{h}F(a,b)ab)\exp(-2\hat{h}^2(F(a,b))^2),$$

so that $\exp(tab)$ also belongs to $\check{C}_S(hF)_0$ if t is sufficiently small.

As we recalled at the beginning of this paragraph, there exist bases of E which are simultaneously orthogonal and symplectic and the isomorphism of $C_S(F)$ and the symmetric algebra induces on $\check{C}_S(F)_0$ the euclidean structure of E, independently of the basis choice; so this defines the topological structure of a normed space on the vector space $\check{C}_S(hF)_0$.

Lemma 17.2.2 *There exists a neighborhood of 0 which is injectively mapped into $\check{C}_S(hF)_0$ by the exponential map.*

Proof. This follows from Lemma (17.2.1), from its consequences and from a standard result concerning the exponential map. ∎

Lemma 17.2.3 *G_S contains a local Lie group Γ with Lie algebra L.*

Proof. By a general result, there exists a local Lie group generated by L; this group can be identified, using Lemma (17.2.2), to the group generated by the $\exp ta^2$ for sufficiently small t, since this Lie group is obtained by the so-called Hausdorff construction. ∎

Proposition 17.2.8 *G_S has a Lie group structure.*

Proof. It is sufficient to apply Proposition 118, page 112 of Bourbaki's *Groupes et Algèbres de Lie, chapitres 2 et 3* (Hermann, Paris, 1972) which, in our notations, says :
If G_S is a group and if, in the local Lie group Γ contained in G_S, an open set $\mathcal{U}$ containing the identity e can be found, such that it has an analytic manifold structure, and if V is an open set in $\mathcal{U}$ such that $V = V^{-1}$, $V^2 \subseteq \mathcal{U}$ and

1. *$(x,y) \to xy^{-1}$ is analytic from $V \times V$ to $\mathcal{U}$*

2. *$\forall g \in G_S$ there exists an open neighborhood V' of e, $V' \subseteq V$, with $gV'g^{-1} \subseteq \mathcal{U}$ and such that $x \to gxg^{-1}$ is analytic from V' to $\mathcal{U}$*

then G_S has a Lie group structure.
     Formula (12) yields for $X, Y \in L$ that

$$\exp Y \exp X \exp(-Y) = \exp(\sum \frac{1}{n!}(\operatorname{ad} Y)^n(X)) \qquad (13)$$

so ad being continuous in L, $\|\operatorname{ad}(Y)(X)\| \leq M\|x\|\|y\|$, M being a constant, if Y is fixed, setting $\alpha = M\|Y\|$, the norm of $\sum(1/n!)(\operatorname{ad} Y)^n(X)$ is bounded by $\|X\|\exp\alpha$ and X can be chosen so small ($X \in V \subseteq \mathcal{U}$) as to ensure that the right-hand side is in Γ and arbitrarily close to the identity. Since $g \in G_S$ is a finite product of exponential elements, if $X \in L$ is close enough to 0, $g\exp Xg^{-1}$ will be arbitrarily close to e. ∎

Lemma 17.2.4 *The projection $\hat{p}$ from G_S to $\mathrm{Sp}(n, \mathbf{R})$ and $\gamma \in G_S^* \to N(\gamma) \in \mathbf{C}^*$ are C^∞.*

Proof. Invoking a classical result on homomorphisms it is sufficient to prove that $\hat{p}$ is continuous.

By a remark made previously, we can note that every element of G_S can be expressed as a finite product of elements in the local group Γ, which can even be assumed of the form $\exp ta^2$, $a \in E$, but :

$$\hat{p}(\exp ta^2)(x) = x + 2\hat{h}F(a, x)ta$$

and the result follows at once.

The property of $N(\gamma)$ is trivial since $N(\exp ta^2) = 1$.   ∎

Corollary 17.2.2 *The groups $\mathrm{Mp}(r)$ and $\mathrm{Sp}_2(r)$ are topological Lie subgroups of G_S' and G_S respectively.*

Using the notations of the beginning of this paragraph, we have the

Proposition 17.2.9 $\hat{p}\exp(t\sum_\alpha \epsilon_\alpha \epsilon_{\alpha^*})$ *induces* $x \to \exp(-i\hat{h}t)x$ *on* $\mathcal{E}_1$ *and* $x \to \exp(i\hat{h}t)x$ *on* $\mathcal{E}_2$.

Proof. Since $\epsilon_\alpha \epsilon_{\alpha^*}$ commutes with $\epsilon_\beta \epsilon_{\beta^*}$ when $\alpha \neq \beta$, we have that

$$\exp(t\sum_\alpha \epsilon_\alpha \epsilon_{\alpha^*}) = \prod_\alpha \exp(t\epsilon_\alpha \epsilon_{\alpha^*})$$

and this factorization reduces the problem to the two-dimensional case in a space determined by two vectors a, b such that $ab - ba = ih$, but a direct computation then shows that

$$\exp(tab)a \exp(-tab) = \exp(-i\hat{h}t)a.$$

Corollary 17.2.3

$$\hat{p}\exp(\tfrac{\pi}{h} \sum \epsilon_\alpha \epsilon_{\alpha^*}) = -\,\mathrm{Id},$$

$$\hat{p}\exp(\tfrac{2\pi}{h} \sum \epsilon_\alpha \epsilon_{\alpha^*}) = \mathrm{Id}, \tag{14}$$

$$\hat{p}\exp(\tfrac{\pi}{2h} \sum \epsilon_\alpha \epsilon_{\alpha^*}) = J.$$

Proposition 17.2.10 *The unitary group is part of the image under $\hat{p}$ of the set of products of elements in $\mathrm{Mp}(r)$ of the form $\exp \lambda \exp(a^{\alpha\beta^*} \epsilon_\alpha \epsilon_{\beta^*})$ where there is no summation in α and β,*

$$a^{\alpha\beta^*} \in \mathbf{C}, \quad a^{\alpha\beta^*} = \bar{a}^{\beta\alpha^*}, \quad \lambda = -\frac{i\sum a^{\alpha\alpha^*}}{2}. \tag{15}$$

Proof. It is sufficient to explicit a condition for real-valuedness, to note that $\epsilon_\alpha \epsilon_{\beta^*}$ commutes with $\sum_\gamma \epsilon_\gamma \epsilon_{\gamma^*}$ and to invoke Corollary (17.2.2).   ∎

We will prove later on that this image is equal to $U(r)$.

Proposition 17.2.11 *The spinoplectic group* $\mathrm{Sp}_2(r)$ *is a connected group and a non-trivial double covering of* $\mathrm{Sp}(2r, \mathbf{R})$.

Proof. We must only prove the existence of a continuous path from (-1) to $(+1)$ in $\mathrm{Sp}_2(r)$. Consider $t \to \exp(-rit/2)\exp((t/h)\sum \epsilon_\alpha \epsilon_{\alpha^*}) = \alpha(t)$; this is a continuous path of spin norm 1, $\alpha(0) = 1$, $\alpha(2\pi) = \exp(-i\pi)\exp((2\pi/h)\sum \epsilon_\alpha \epsilon_{\alpha^*})$.

If r is odd, $\hat{p} \circ \alpha(2\pi) = \mathrm{Id}$ and by Corollary (17.2.3), $\alpha(2\pi) = \pm 1$, but by continuity,

$$\exp(\frac{2\pi}{h}\sum \epsilon_\alpha \epsilon_{\alpha^*}) = 1 \tag{16}$$

(this also follows from an easy direct computation), so that $\alpha(2\pi) = -1$.

If r is even, we use induction on the dimension of E, $E = E_1 \oplus E_2$ being a direct symplectic sum, we take $\alpha_1(t)$ inducing the identity on E_1 and $\alpha_2(t)$ as $\alpha(t)$ above; then the induction hypothesis reduces the problem to the odd-dimensional case.   ∎

Remark. From this it follows easily that $\mathrm{Mp}(r)$ is connected (this is also a consequence of a general result : $\mathrm{Sp}(2r, \mathbf{R})$ and S^1 are connected, hence so is their quotient $\mathrm{Mp}(r)$).

Proposition 17.2.12 *The Lie algebra* $\underline{\mathrm{Mp}}(r)$ *of* $\mathrm{Mp}(r)$ *has a basis formed by the* $(e_1)^2$ *and* $e_i e_j$, $i < j$, *where the* e_i, $i = 1, \ldots, 2r$ *are an arbitrary basis of* E, *and it is a Lie sub-algebra of* $C_S(hF)$ *with the bracket* $[u, v] = uv - vu$.

Proof. Note that if $\mathrm{ad}\, u(x) = ux - xu$, $u \in C_S(hF)$, $x \in E$, $\mathrm{ad}\, u(x) \in E$ if and only if u is the sum of terms of degree 0 and 2, $\mathrm{ad}\, u$ lies in the Lie algebra of the symplectic group,

$$F(\mathrm{ad}\, ux, y) + F(x, \mathrm{ad}\, uy) = 0,$$

$ux - xu \in E$, $\forall x \in E$ is equivalent to $u \in \mathbf{R} \oplus V^2(E)$.

The elements $\mathrm{ad}_{(e_i)^2}$, $\mathrm{ad}_{(e_i e_j)}$, $i < j$, are linearly independent over $\mathbf{R}$, so because of the dimension of $\mathrm{Sp}(2r, \mathbf{R})$, they form a basis for the Lie algebra of the symplectic group, the kernel of $\hat{p}$ is $\mathbf{R}^*$ in G_S and S^1 in $\mathrm{Mp}(r)$. The result follows.   ∎

Remark. The Lie algebra of $\mathrm{Sp}_2(r)$ is the quotient Lie algebra of $\underline{\mathrm{Mp}}(r)$ by $\mathbf{R}$: the $r(2r + 1)$ elements $(e_i)^2$, $e_i e_j$, $i < j$, do not form a basis for that algebra (since a bracket gives rise to elements of $\mathbf{R}$).

Writing $E' = E_\mathbf{C}$, we see that the $\epsilon_\alpha \epsilon_\beta$, $\epsilon_{\alpha^*} \epsilon_{\beta^*}$, $\alpha \le \beta$, $\epsilon_\alpha \epsilon_{\beta^*}$, 1, form a basis for the complexified Lie algebra of $\mathrm{Mp}(r)$. The elements of $\underline{\mathrm{Mp}}(r)$ $a^{ij}\epsilon_i \epsilon_j$ are characterized by :

$$a^{\alpha\beta} = \bar{a}^{\alpha^*\beta^*}, \quad a^{\alpha\beta^*} = \bar{a}^{\beta\alpha^*}, \quad \lambda - \bar{\lambda} = -i\sum a^{\alpha\alpha^*}, \tag{17}$$

if λ is their component in 1.

Proposition 17.2.13 *The elements of* $U(r)$ *can be identified with the images under* $\hat{p}$ *of* $\exp \lambda \exp(a^{\alpha\beta^*} \epsilon_\alpha \epsilon_\beta)$, *where* $a^{\alpha\beta^*} \in \mathbf{C}$ *and*

$$a^{\alpha\beta^*} = \bar{a}^{\beta\alpha^*}, \quad \lambda = -i \frac{\sum a^{\alpha\alpha^*}}{2}$$

(cf. also Proposition (17.2.10)).

Proof. The $\epsilon_\alpha \epsilon_{\beta^*}$ generate a subspace of $\underline{Mp}(r)$ which has real dimension $r^2 = \dim U(r)$. The group $U(r)$ is connected and compact, so the exponential function is surjective; finally, the spin norm is taken into account. ∎

Corollary 17.2.4 $Sp_2(r)$ *is generated by the elements of the form*

$$\exp(-(\alpha^{ij}/2)F(e_i, e_j)) \exp(\alpha^{ij} e_i e_j)$$

where the α^{ij} *are real,* e_i *is a basis for* E, $i \leq j$ *and no summation occurs on* i *and* j. $Mp(r)$ *is generated by the elements of the form* $\exp(\alpha^{ij} e_i e_j)$, *under the same conditions.*

Proof. This follows from the connectedness of these groups and from their Lie algebras. ∎

Lemma 17.2.5 $Mp'(r)$ *being obtained from* $E' = E_{\mathbf{C}}$ *in a similar way as* $Mp(r)$ *from* E, *the subgroup* G_i' *of elements in* $Mp'(r)$ *which fix all points of* $\mathcal{E}_i$, $i = 1, 2$, *is generated by the* $\exp(u)$, $u \in \mathbf{C} \oplus \vee^2 \mathcal{E}_i$.

Proof. The set of these u gives rise to a commutative Lie subalgebra of $\underline{Mp}'(r)$ and $\exp(u)$ clearly fixes all points of $\mathcal{E}_i$; conversely each element of $\check{C}_S'(hF)$ which commutes with all elements of $\mathcal{E}_i$ is a formal series in the ϵ_α if $i = 1$ or in the ϵ_{α^*} if $i = 2$. ∎

It is obvious that

$$\underline{Mp}'(r) = \underline{GL}(r, \mathbf{C}) \oplus \underline{G}_1' + \underline{G}_2',$$

where G_i' is the connected group generated by the $\exp(u)$ for $u \in \mathbf{C} \oplus \vee^2 \mathcal{E}_i$; we note that $\underline{GL}(r, \mathbf{C}) \oplus \underline{G}_i'$ is a Lie subalgebra of $\underline{Mp}'(r)$ and that $\underline{G}_i'$ is one of its ideals. Therefore G_i' is normal in the subgroup generated by the exponentials of $\underline{GL}(r, \mathbf{C}) \oplus \underline{G}_i'$.

The elements of $Mp(r)$ are products of exponentials of elements in $\underline{U}(r)$, $\underline{G}_1'$, $\underline{G}_2'$, so the consequence of Lemma (17.2.5) allow us to consider products where the exponentials of elements of $U(r)$ come first.

If $\gamma \in Sp_2(r)$, it can be factored as

$$\gamma = \sigma \tau_2' \tau_1', \quad \tau_i' \in \exp(\vee^2(\mathcal{E}_i)), \quad i = 1, 2, \quad \hat{p}(\sigma) \in U(r),$$

since the exponential is surjective on compact or on abelian Lie groups.

An analogous reasoning shows that

$$\gamma = \sigma\tau_1\tau_2, \quad \tau_i \in \exp(\vee^2(E_i)), \quad i = 1, 2.$$

This decomposition which, in this form, is necessarily unique, is a Cartan decomposition, $U(r)$ being a maximal compact subgroup of $Sp_2(r)$. Therefore, summarizing :

$$\gamma = \sigma\tau_1\tau_2, \quad \begin{cases} \sigma = \exp(-ih\sum a^{\alpha\alpha^*}/2)\exp(a^{\alpha\beta^*}\epsilon_\alpha\epsilon_{\beta^*}), \quad a^{\alpha\beta^*} = \bar{a}^{\beta\alpha^*}, \\ \tau_1 = \exp(a^{\alpha\beta}e_\alpha e_\beta), \quad \tau_2 = \exp(a^{\alpha^*\beta^*}e_{\alpha^*}e_{\beta^*}), \end{cases} \tag{18}$$

where $a^{\alpha\beta}$ and $a^{\alpha^*\beta^*}$ are real and summations should be carried out.

From $[a^{\alpha\beta^*}\epsilon_\alpha\epsilon_{\beta^*}, \epsilon_\gamma] = -iha^{\alpha\gamma^*}\epsilon_\alpha$ we deduce that $\det(\hat{p}(\sigma)) = \exp(ih\sum a^{\alpha\alpha^*})$, if $\hat{p}$ operates in the space $\mathcal{E}_2$ of the ϵ_{α^*}.

The scalar coefficient in the decomposition of σ is therefore a square root of the determinant of $\hat{p}(\sigma)$ (when h is replaced by $\hat{h}$) and has a geometric meaning. This remark should be compared with an often used method for the construction of a double covering group of the symplectic group.

17.2.9   The q-fold covering of the symplectic group and its universal covering.

For $\gamma \in Sp_2(r)$, $\gamma = \exp(i\theta)\mu$ where $\widehat{\exp}(i\theta) = (N(\mu))^{-1/2}$, $\hat{}$ meaning that h is replaced by $\hat{h}$ in its argument. This suggests the more general definition : consider the set of elements of $Mp(r)$ of the form

$$\exp(-\frac{ih}{q}\sum a^{\alpha\alpha^*})\exp(a^{\alpha\beta^*}\epsilon_\alpha\epsilon_{\beta^*})\tau_1\tau_2, \quad \tau_i \in \exp(\vee^2(E_i)), \tag{19}$$

then using the consequences of Lemma (17.2.5) it is easy to see that these elements form a closed subgroup of $Mp(r)$, with general term $\gamma = \exp(i\theta)\mu$ satisfying

$$\widehat{\exp}(i\theta) = (N(\mu))^{-1/q}.$$

This subgroup is called a symplectic spin group of order q (or spinoplectic of order q) and is denoted by $Sp_q(r)$.

Proposition 17.2.14 $Sp_q(r)$ *is a non-trivial q-fold covering of the symplectic group.*

Proof. If $\hat{p}(\mu\exp(i\theta)) = \text{Id}$, necessarily $\hat{\mu}\widehat{\exp}(i\theta) \in S^1$, taking $\sum a^{\alpha\alpha^*} = -2k\pi/h$ and relying on (16), we see that the q elements $\exp(2\pi ki/q)$ are in the kernel of the restriction of $\hat{p}$ to $Sp_q(r)$. It is easily seen that these are the only elements in the kernel (which, therefore, is isomorphic to $\mathbf{Z}_q$). The connectedness of these groups is proved as for $Sp_2(r)$.   ∎

Finally, the *universal covering* of $\mathrm{Sp}(2r, \mathbf{R})$ is defined by considering the subgroup of elements of $\mathrm{Mp}(r)$ of the form

$$\exp(i\theta)\exp(a^{\alpha\beta^*}\epsilon_\alpha\epsilon_{\beta^*})\tau_1\tau_2$$

for which $\sum a^{\alpha\alpha^*}h/\theta$ is rational. It is denoted by $\mathrm{Sp}_\infty(r)$. Details are left to the reader. The kernel is isomorphic to $\mathbf{Z}$. The notation $\mathrm{Sp}_q(r)$ will stand for any symplectic spin group, $q = 2, 3, \ldots, \infty$.

17.2.10 The Heisenberg group.

The elements of E and $\mathbf{R}$ form a $(2r+1)$-dimensional Lie algebra H. Using the exponential, a connected Lie group with multiplication

$$\exp(x+t)\exp(x'+t') = \exp(x + x' + t + t' + \tfrac{1}{2}\hat{h}F(x, x'))$$

is defined. This is the Heisenberg group. Usually, $\hat{h} = 1$.

Chapter 18

SYMPLECTIC SPINOR BUNDLES—THE MASLOV INDEX.

18.1 The symplectic Clifford bundle of an almost symplectic manifold.

18.1.1

In this chapter, V is a smooth real manifold, of dimension $n = 2r$, paracompact and endowed with an almost symplectic structure defined by a field F of alternating bilinear forms of maximal rank $n = 2r$ in every point. If $dF = 0$, V is a symplectic manifold. It is equivalent to state that V has an almost complex structure (cf. Chapter 14, 3.2); to an almost complex structure there always corresponds an almost hermitian structure. Such a manifold is known to be orientable.

F will also stand for the 2-form of the standard real symplectic space E.

The Weyl algebra of V can be defined by an adaptation of Definition (14.1.1), but in each point $x \in V$ an infinite-dimensional vector space is obtained, which is linearly isomorphic to the symmetric algebra of the space $T_x(V)$. Propositions (14.1.1) and (14.1.2) can immediately be adapted.

The definition of the Weyl algebra bundle of V is obvious; this bundle is linearly isomorphic to the symmetric algebra bundle $\vee(T(V))$.

Note that two frames which are related by a symplectic transformation define, by λ_F, the same linear isomorphism between the Weyl algebra and the symmetric algebra; this should be compared with the crucial analogous remark in the orthogonal case.

18.1.2

We generalize the notion of Weyl algebra bundle to that of symplectic Clifford bundle, both in the formal and in the non-formal sense. In the first case the fibers are isomorphic to $\check{C}_S(hF)$, in the second case to some $C_S(hF)_\alpha$—we noted that formula (3) in Chapter 17, 2.5 has a meaning which is independent of any choice of basis. It is very important to note that the fundamental group of the principal bundle of frames in $T(V)$ is reducible to the unitary (compact) group U(r), a subgroup of the orthogonal group.

Any subbundle of the symplectic Clifford bundle will be called an *amorphic spin subbundle* if for all $x \in V$ its fiber is a minimal left ideal of the formal symplectic Clifford fiber.

Sufficient conditions for the existence of such subbundles are easily found :

1. There exists a global field of maximal totally isotropic subspaces for F : such a subbundle is called *lagrangian* for $T(V)$.

2. The complex subbundle determined by $T(V)$, which, as we know, defines an amorphic orthogonal spinor bundle over V with an almost hermitian structure, also determines an amorphic symplectic spinor bundle, by the properties recalled at the beginning of 2.8 in Chapter 17. This bundle lies in the complexified Clifford bundle and is determined by a complex lagrangian subbundle.

Definition 18.1.1 *A* $\text{Sp}_2(r)$ *spinor structure (called 'spinoplectic') is said to exist on V if a principal bundle η with group $\text{Sp}(r)$ and a principal morphism h from η to the principal bundle ξ of symplectic frames can be found.*

Since the group $\text{Sp}(r)$ is not compact, the structural group reduces to the double covering of U(r), i.e. the group H_e of elements σ of the form

$$\sigma = \exp(-ih \sum a^{\alpha\alpha^*}/2) \exp(a^{\alpha\beta^*}) \epsilon_\alpha \epsilon_{\beta^*},$$

$a^{\alpha\beta^*} = \bar{a}^{\beta\alpha^*}$ (Proposition (17.1.13)). (H_e is isomorphic to the subgroup of elements $g \in \text{Spin}(Q)$ such that $gf = \exp(i\theta)f$, using the notations for orthogonal spinors).

U(r) being a subgroup of SO($2r, \mathbf{R}$), we see that if there exists a spinoplectic structure, there must exist an orthogonal spinor structure, in the strict sense of Definition (14.2.2), but with group $\text{Spin}(Q)$ since V is orientable.

An existence condition for such a structure is therefore given by Proposition (14.3.2).

Because V is orientable and its orthogonal spinor structure is of group $\text{Spin}(Q)$, there exists a field of isotropic r-vectors which completely determines the complex structure of the tangent bundle. Take as standard spinor space the quotient of the (complexified) algebra $\check{C}'_S(hF)$ by the left ideal J'_m generated by the (ϵ_{α^*}); this space will be denoted by $\check{C}'_S(hF)\Phi^*$ (to resemble the notation used for the orthogonal spinors). $\text{Sp}_2(r)$ acts naturally by left multiplication on this space and in an effective way, since all two-sided ideals of the algebra are trivial, and $\text{Sp}_2(r)$ generates $\check{C}^+_S(hF)$.

To the principal bundle η with group $\text{Sp}_2(r)$ we can associate a spinoplectic bundle ζ with a fiber which is isomorphic to $\check{C}'_S(hF)\Phi^*$, $\text{Sp}_2(r)$ acting by left multiplication on the fiber.

Definition 18.1.2 *Every local or global section of the vector bundle ζ of typical fiber $\check{C}'_S(hF)\Phi^*$ and structural group $\mathrm{Sp}_2(r)$, associated to the spinoplectic principal bundle will be called a* symplectic spinor *on V.*

Definition 18.1.3 *There exists a metaplectic (or, better, toroplectic) structure on V if a principal bundle η_1 with group $\mathrm{Mp}(r)$ and a principal morphism h from η_1 to the principal bundle ζ of symplectic frames can be constructed.*

Just as in Definition (18.1.1), $\mathrm{Mp}(r)$ will reduce to its subgroup $\hat{p}^{-1}(\mathrm{U}(r))$; $\mathrm{Mp}(r)$ occurring as $S^1 \times_{\mathbf{Z}_2} \mathrm{Sp}_2(r)$ will reduce to a group isomorphic to $S^1 \times_{\mathbf{Z}_2} H_e$, i.e. a torogonal group (Chapter 14, 4.1). Therefore *the notion of toroplectic structure can be identified with the notion of torogonal structure naturally associated to the almost hermitian structure (Chapter 14, 3.2).*

Finally, still relying on Chapter 14, 4.3, this metaplectic structure is the equivalent of what we called, at the beginning of this paragraph, an amorphic symplectic spinor bundle determined by the complex lagrangian bundle related to the complex structure of the tangent bundle. *Such a structure therefore always exists over the manifold V.*

Remark. As in the orthogonal case, care must be taken to distinguish between the notion of amorphic spinor and that of a spinor corresponding to a spinor structure over V. On the typical fiber of amorphic spinors we consider the action of the symplectic group, not that of a spinoplectic group. The action of the symplectic group is consistent with the conditions of (17.2.5), the non-formal viewpoint is therefore allowed for the amorphic bundles.

18.1.3   Symplectic pure spinors.

By the conventions of Chapter 17, 2.4, we can also define 'right' symplectic spinors, the space of right spinors constructed using the right ideal generated by the (e_α) will be denoted by $\Phi\check{C}_S(hF)$. In general, using the symplectic basis $\{e_{\alpha_1}, e_{\beta_1^*}\}$, we will define the real spinor spaces $\Phi_1\check{C}_S(hF)$ and $\check{C}_S(hF)\Phi_1^*$. These definitions carry over to the complex case; if $(\epsilon_\alpha, \epsilon_{\beta^*})$ is the basis derived from the (e_α, e_{β^*}) as in subsection 17.2.8, we will denote by $\check{C}'_S(hF)\Phi^*$ and $\Phi\check{C}'_S(hF)$ the associated spaces of complex symplectic spinors.

Note that the spinoplectic structures naturally induce the complex case, not the real case.

Proposition 18.1.1 *The intersection of a space of right symplectic spinors and a space of left symplectic spinors is one-dimensional, both in the real and in the complex case.*

('intersection' stands here for 'double class'.)

Proof. In the real case, for instance, we can consider $\check{C}_S(hF)\Phi^* \cap \Phi\check{C}_S(hF)$ and obviously $\Phi \cup \Phi^* = \Phi z \Phi^*$ with $z \in \mathbf{R}((h))$ if $u = z + \lambda_{HK} \cdot e^H e^{K^*}$, and every symplectic transformation will preserve this property. More generally, if $\Phi_1 \check{C}_S(hF)$ is determined by the $e_{\alpha_i^*}$ with $\gamma e_{\alpha^*} \gamma^{-1} = e_{\alpha_i^*}$, $\gamma \in \mathrm{Sp}_2(r)$, or $\gamma \Phi \gamma^{-1} = \Phi_1$ for short, $u\Phi^* = \Phi' v$ is equivalent to $u\Phi^* = \gamma \Phi \gamma^{-1}$ or $\gamma^{-1} u \Phi^* = \Phi \gamma^{-1} v$ and $u = \lambda v$, $\lambda \in \mathbf{R}((h))$. In the non-formal case, $\lambda \in \mathbf{R}$ (in the real case) or $\lambda \in \mathbf{C}$ (in the complex case). ∎

Proposition 18.1.2 *There exists a hermitian sesquilinear form $\mathcal{H}$ on the space of complex symplectic spinors $\check{C}'_S(hF)\Phi^*$ defined by*

$$\Phi\mathcal{H}(u\Phi^*, v\Phi^*)\Phi^* = \Phi\beta(\bar{\mu})v\Phi^*, \tag{1}$$

where $\beta|_{E'} = i\,\mathrm{Id}$ (Chapter 17, 1.2). $\mathcal{H}$ is invariant under the action of the group $\mathrm{Sp}_2(r)$.

Proof. We have a pure spinor and $\mathcal{H}(u\Phi^*, v\Phi^*)$ is the conjugate of $\mathcal{H}(v\Phi^*, u\Phi^*)$. Note that if u is homogeneous, $\overline{\beta(u)} = (-1)^{\deg u}\beta(\bar{u})$ and from the multiplication table

$$(\epsilon_{\alpha^*})^k(\epsilon_\alpha)^l = (\epsilon_\alpha)^l(\epsilon_{\alpha^*})^k - hilk(\epsilon_\alpha)^{l-1}(\epsilon_{\alpha^*})^{k-1} + \ldots$$
$$+ h^p(-1)^p C_l^p C_{kp}^p p! (\epsilon_\alpha)^{l-p}(\epsilon_{\alpha^*})^{k-p} + \ldots$$

it follows that

$$\mathcal{H}(\epsilon^H\Phi^*, \epsilon^H\Phi^*) = H!(h)^{|H|},$$
$$\mathcal{H}(\epsilon^H\Phi^*, \epsilon^K\Phi^*) = 0$$

where ϵ^H is analogous to the e^H introduced in Chapter 17 and h is treated as a real variable.

The basis $\epsilon^H\Phi^*/\sqrt{H!(h)^{|H|}}$ is therefore orthonormal. In the non-formal case, for the algebra $\check{C}_S(hF)_\alpha$, we get convergent series which define a pre-Hilbert norm on $\check{C}_S(hF)_\alpha\Phi^*$. The invariance of $\mathcal{H}$ under the action of $\mathrm{Sp}_2(r)$ follows at once from the condition on the spin norm. ∎

Consequence. *If there exists a spinoplectic structure on V, there also exists a field of hermitian sesquilinear forms.*

Then the local construction of $\mathcal{H}$ is global.

18.2   The three-sided inertial cocycle and the Maslov index.

18.2.1   The three-sided inertial index.

We start with purely algebraic considerations. E is an $(n = 2r)$-dimensional vector space with the symplectic form F; a lagrangian space is maximal totally isotropic for F.

Lemma 18.2.1 *Let L, L', L'' be three lagrangian spaces such that $L \cap L' = L \cap L''$, then there exist (in general, non-unique) $s \in \mathrm{Sp}(2r, \mathbf{R})$ fixing all elements of L and mapping L' onto L''.*

Proof. Let $(e_\alpha, e_{\beta*})$ be a symplectic basis for E whose vectors (e_α) lie in L. Then clearly s is expressed by :

$$\begin{cases} s(e_\alpha) = e_\alpha, \\ s(e_{\alpha*}) = a_{\alpha*}^\beta e_\beta + e_{\alpha*}, & a_{\alpha*}^\beta = a_{\beta*}^\alpha. \end{cases} \tag{1}$$

We may assume that the e_α were chosen such that $(e_1, e_2, \ldots, e_\lambda)$ generate $L \cap L'$; then $e_{\lambda+1}, \ldots, e_r, e_{1*}, \ldots, e_{\lambda*}$ generate a lagrangian L_1 for which $L_1 \cap L' = L_1 \cap L'' = 0$. Let L_2 be the lagrangian with basis $e_1, e_2, \ldots, e_\lambda, e_{(\lambda+1)*}, \ldots, e_{r*}$, then clearly

$$E = L_1 \oplus L' = L_1 \oplus L'' = L_1 \oplus L_2$$

and $L \cap L' = L \cap L'' = L \cap L_2$. But there exists a symplectic transformation θ which fixes all points of L_1, the vectors $e_1, e_2, \ldots, e_\lambda$ and sends L' to L_2; if θ' sends L'' to L_2 in a similar way, $s = \theta'^{-1}\theta$. ∎

If $L \cap L'$ is not zero, we see that s is not unique. But if $L \cap L' = L \cap L'' = 0$, (1) *proves its uniqueness and establishes a bijection between the set of lagrangians which are transversal to L and the set of symmetric $r \times r$ matrices.*

Under the assumptions of Lemma (18.2.1), it is always possible to choose the $(e_\alpha, e_{\beta*})$ such that

$$\begin{cases} s(e_\alpha) = e_\alpha, \\ s(e_{\alpha*}) = t_\alpha e_\alpha + e_{\alpha*}, & t = 0, 1 \text{ or } -1. \end{cases} \tag{2}$$

Now set $\mathcal{Z} = F(\cdot, s(\cdot))$, $\mathcal{Z}$ is a quadratic form on L' and if $L \cap L' = L \cap L'' = 0$, s is unique.

Definition 18.2.1 *If the signature of the quadratic form $\mathcal{Z}$, $\mathcal{Z}(x) = F(x, s(x))$, $x \in L'$, is (p, q) (p plus signs, q minus signs), define*

$$\mathrm{Inert}(L', L, L'') = -\mathrm{Inert}(L'', L, L') = p - q = -\sum_\alpha t_\alpha. \tag{3}$$

Remark. If the triple (L', L, L'') can be continuously deformed, keeping L' and L'' transversal to L, $\mathrm{Inert}(L', L, L'')$ remains constant if $\dim(L' \cap L'')$ remains constant. Note that if $L' \cap L'' = 0$, $e_{\alpha*} \to e_{\alpha*}$, $e_\alpha \to -e_\alpha - t_\alpha e_{\alpha*}$ gives rise to a symplectic transformation that fixes all points of L' and sends L to L'', so, when

$$L' \cap L'' = L \cap L' = L \cap L'' = 0,$$

$\mathrm{Inert}(L, L', L'') = -\mathrm{Inert}(L', L, L'')$ and Inert is antisymmetric and invariant under cyclic permutations.

If $L \cap L' = L \cap L'' = 0$ where $L' \cap L'' \neq 0$, we agree on the notations

$$\begin{aligned} \mathrm{Inert}(L, L', L'') &= -\mathrm{Inert}(L', L, L''), \\ \mathrm{Inert}(L', L'', L) &= -\mathrm{Inert}(L', L, L''). \end{aligned} \tag{4}$$

Proposition 18.2.1 *Every $\gamma \in G_S$ which fixes all points of the lagrangian is of the form $\prod_\alpha(\exp(t_\alpha(e_\alpha)^2/2))$, $t_\alpha = \pm 1$, the e_α being a basis for L; if $L \oplus L' = E$ and $L'' = \gamma(L')$, the 'signature' of the sequence of t_α, $-(\sum t_\alpha)$, is given by $\mathrm{Inert}(L', L, L'')$.*

Proof. $\hat{p}(\gamma) = s$ can be factored in terms of symplectic transvections, if $a \in L$,

$$\exp(\frac{ta^2}{2})x \exp(-\frac{ta^2}{2}) = x + hF(a, x)ta, \quad t \in \mathbf{R}^*,$$

and we may assume that $\hat{h} = 1$. a can be chosen such that $t = \pm 1$. Taking

$$\gamma = \prod_i \exp\frac{t_i(a_i)^2}{2} = \exp(\sum_i \frac{t_i(a_i)^2}{2}), \quad a_i \in L, \quad t_i = \pm 1,$$

a suitable choice of the a_i, and t_i allows us to reach every s defined by (1). In particular, we can choose the basis $\{e_\alpha, e_{\beta^*}\}$ so that s is defined by (2) and that

$$\gamma = \prod_\alpha \exp(\frac{t_\alpha(e_\alpha)^2}{2}).$$

The 'signature' of the (t_α) therefore has an intrinsic meaning. ∎

Lemma 18.2.2 *Given three lagrangians L, L' and L'', there exists a lagrangian M which is simultaneously transversal to them.*

Proof. It is easy to prove that a lagrangian L can be identified with $l = a\bar{a}^{-1}$, $a \in U(r)$, and that L is transversal to M if and only if $l - m$ is invertible, where m corresponds to M. There are infinitely many suitable M. ∎

Lemma 18.2.3 *If L_1, L_2, L_3 and L_4 are lagrangians such that L_2 is transversal to L_1, L_3, L_4 and that L_3 is transversal to L_1, L_2, L_4, then*

$$\mathrm{Inert}(L_1, L_2, L_3) - \mathrm{Inert}(L_1, L_2, L_4) + \mathrm{Inert}(L_1, L_3, L_4) - \mathrm{Inert}(L_2, L_3, L_4) = 0. \tag{5}$$

Proof. This follows from the previous remark. ∎

Now, if L_1, L_2, L_3 are arbitrary lagrangians, we choose an M which is transversal to all three of them and, in accordance with (5), we define

$$\mathrm{Inert}(L_1, L_2, L_3) = \mathrm{Inert}(L_1, L_2, M) - \mathrm{Inert}(L_1, L_3, M) + \mathrm{Inert}(L_2, L_3, M). \tag{6}$$

The indices on the right hand side remain constant if M varies while remaining transversal to L_1, L_2, L_3; the left hand side will not depend on the choice of M.

This index is antisymmetric and, by its definition, invariant under symplectic transformations. Note that (6) immediately leads to

$$\text{Inert}(L_1, L_2, L_3) = \text{Inert}(L_1, M, L_3) + \text{Inert}(L_2, M, L_1) + \text{Inert}(L_3, M, L_2) \qquad (7)$$

which displays the invariance under cyclic permutations.

Invoking Proposition (18.2.1) again, we obtain :

Proposition 18.2.2 *Given three arbitrary lagrangians* L_1, L_2, L_3, *if* $\gamma_{13} \in G_S$ *fixes all points of the lagrangian* M *which is transversal to* L_1, L_3 *and if it sends* L_1 *to* L_3; *if* γ_{21} *and* γ_{32} *have similar properties for* L_2, L_1 *and* L_3, L_2 *respectively, and if*

$$\gamma_{13} = \prod_\alpha \exp \frac{t_\alpha(e_\alpha)^2}{2}, \gamma_{21} = \prod_\beta \exp \frac{s_\beta(e_\beta)^2}{2}, \gamma_{32} = \prod_\sigma \exp \frac{\omega_\sigma(e_\sigma)^2}{2}, \quad t_\alpha, s_\beta, \omega_\sigma = \pm 1,$$

$e_1, e_2, \ldots, e_r$ *being a basis for* M, *then* $\text{Inert}(L_1, L_2, L_3) = -\sum_\alpha t_\alpha - \sum_\beta s_\beta - \sum_\sigma \omega_\sigma$.

Remark. Other definitions could be adopted; the signature could be defined as the number of minus signs (cf. Leray), but this loses some antisymmetry properties. Some authors use a signature whose sign differs from ours.

Choosing M transversal to L_1, L_2, L_3, L_4 (cf. Lemma (18.2.4)) shows immediately that (5) is valid for any four lagrangians, since it is the translation of a cohomological property. Inert is a cocycle.

We will consider the decomposition of this cocycle and define the **Z**-valued Maslov index.

18.2.2 The Maslov index.

Lemma 18.2.4 *The unitary group* $U(r)$ *acts transitively on the set of lagrangian subspaces with stabilizer* $O(r)$.

Proof. If $E = E_1 \oplus E_2$ is a symplectic Witt decomposition where $E_2 = J(E_1)$ is orthogonal to E_1, since $(Jx, y) = F(x, y)$ (Chapter 17, 2.8), let L_1 be another lagrangian; if $\{\xi_\alpha\}$ and $\{\eta_\alpha\}$ are orthogonal bases of E_1 and L_1 we send ξ_α to η_α, $J(\xi_\alpha)$ to $J(\eta_\alpha)$ and obtain a transformation which is both orthogonal and symplectic, hence unitary. $\blacksquare$

We will consider the conditions under which the image of a lagrangian G under a symplectic transformation s is a lagrangian L' transversal to L.

There exists a $\nu \in \text{Sp}_q(r)$, $q = 2, 3, \ldots, \infty$ which sends the symplectic and orthogonal basis $\{e_\alpha, e_{\beta*}\}$ of $E_1 \oplus E_2$ to the corresponding elements of a symplectic basis $\{E_\alpha, E_{\beta*}\}$ corresponding to the decomposition $E = L \oplus L'$.

Take $\gamma_0 = \nu^{-1}\gamma\nu$, with $s = \hat{p}(\gamma)$; γ_0 sends E_1 to E_2. We can choose $\hat{p}(\gamma_0) \in U(r)$, since $\hat{p}(\gamma_0)$ commutes with J, J and $\hat{p}(\gamma_0)$ can be simultaneously diagonalized; the ϵ_α, ϵ_{β^*} (Chapter 17, 2.8) are eigenvectors of J and of $\hat{p}(\gamma_0)$; using the formulas linking the (e_α, e_{β^*}) to the $(\epsilon_\alpha, \epsilon_{\beta^*})$, we see that γ_0 sends E_1 to E_2 if and only if the eigenvalues of $\hat{p}(\gamma_0)$ are pure imaginary; then $\hat{p}(\gamma_0)(e_\alpha) = t_\alpha e_{\alpha^*}$, $t_\alpha = \pm 1$. Next, $\hat{p}(\gamma)(E_\alpha) = t_\alpha E_{\alpha^*}$, and because $\hat{p}(\gamma_0)$ commutes with J, $\hat{p}(\gamma_0)(e_{\alpha^*}) = -t_\alpha e_\alpha$ and $p(\gamma)(E_{\alpha^*}) = -t_\alpha E_\alpha$.

By the proof of Proposition (17.2.9), $\exp(\sum_\alpha(-(\pi/2h)t_\alpha \epsilon_\alpha \epsilon_{\alpha^*}))$ is projected to $\hat{p}(\gamma_0)$; but if $\gamma_0' \in Sp_q(r)$ sends E_1 to E_2, $\gamma_0' = v_2\gamma_0$, where v_2 globally preserves E_2 and can be reduced to the product of a change of symplectic bases adapted to the decomposition $E_1 \oplus E_2$ and a transformation associated to an element u_2 which fixes all points of E_2, so $u_2 \in V^2(E_2)$ (Chapter 17, 2.8) and we obtain :

$$\gamma_0'' = u_2\gamma_0.$$

In the factorization (19) from Chapter 17, section 2.8, the same scalar factor $\lambda = \exp(-ih\sum a^{\alpha\alpha^*}/q)$ will appear for γ_0'', γ_0' and γ_0; where

$$\sum a^{\alpha\alpha^*} = -\frac{\pi}{2h}(\sum t_\alpha) \quad \text{and} \quad \lambda = \exp(\frac{i\pi}{q}(\frac{\sum t_\alpha}{2})),$$

λ does not change if $\sum t_\alpha/2$ increases by a multiple of $2q$. We will therefore put :

$$m_q(\gamma) = \frac{\sum t_\alpha}{2} (\mathrm{mod}\, 2q), \quad m_q(\gamma) \in \mathbf{Z}_{2q},$$

$$m_\infty(\gamma) = \frac{\sum t_\alpha}{2} \in \mathbf{Z} \quad \text{or} \quad \mathbf{Z} + \tfrac{1}{2}$$

depending on the parity of r. In the notations of Chapter 17, 2.8, we have proved :

Proposition 18.2.3

1. If $\gamma \in Sp_q(r)$, $q = 2, 3, \ldots, \infty$, sends the lagrangian L to the transversal lagrangian L', it can be factorized, up to a similarity, as $\sigma\tau_1\tau_2$ where $\hat{p}(\sigma) \in U(r)$, $\tau_i \in V^2(E_i)$,

$$\hat{p}(\sigma)(e_\alpha) = t_\alpha e_{\alpha^*},$$
$$\hat{p}(\sigma)(e_{\alpha^*}) = -t_\alpha e_\alpha, \quad t_\alpha = \pm 1, \quad \alpha = 1, 2, \ldots, r,$$
$$\sigma = \lambda \exp(a^{\alpha\beta^*}\epsilon_\alpha \epsilon_{\beta^*}) = \lambda \exp(-\sum_\alpha \frac{\pi}{2h} t_\alpha \epsilon_\alpha \epsilon_{\alpha^*}).$$

2. If $q \in \mathbf{N}^*$, $\lambda = \exp((i\pi/2q)\sum_\alpha t_\alpha)$; if $q = \infty$, $\lambda = \exp((i\pi a/2b)\sum_\alpha t_\alpha)$, a/b being an arbitrary rational number.

Definition 18.2.2 Defining $m_q(\gamma) = \sum_\alpha t_\alpha/2 (\mathrm{mod}\, 2q)$, $m_q(\gamma) \in \mathbf{Z}_{2q}$ is called the Maslov index of order q, $q \in \mathbf{N}^*$, of $\gamma \in Sp_q(r)$. Defining $m_\infty(\gamma) = \sum_\alpha t_\alpha/2 \in \mathbf{Z}$ or $\mathbf{Z} + \frac{1}{2}$, m_∞ is called the Maslov index of $\gamma \in Sp_\infty(r)$.

Remarks.

1. Note that $m_q(\gamma) = m_q(\sigma)$, $q = 2, 3, \ldots, \infty$. Since $\theta = \hat{p}(\sigma) \in U(r)$, there exists a $\delta \in \text{Spin}(2r)$ such that if p is the projection of $\text{Spin}(2r)$ on $O(2r)$, $p(\delta) = \theta$.

   $f = \epsilon_{1} \cdot \epsilon_{2^*} \ldots \epsilon_{r^*}$ is an isotropic r-vector in the orthogonal geometry and, the products being taken in the orthogonal Clifford algebra,

   $$\delta f \delta^{-1} = \exp(-\frac{i\pi}{2} \sum t_\alpha) f = \exp(-i\pi m_\infty(\gamma)) f, \qquad (8)$$

   from which, by Chapter 11,

   $$\delta f = \exp(-\frac{i\pi m_\infty(\gamma)}{2}) f = \exp(-\frac{i\pi}{2} m_2(\gamma)) f. \qquad (9)$$

2. If the lagrangians L and L' are oriented and if the mappings from L on L' preserve the orientations, the product of the t_α is 1 and $\sum t_\alpha - r$ is a multiple of 4; λ has values in the group of $\exp(2ki\pi/q)$ and a Maslov index $m'_q(\gamma) = \sum t_\alpha/2 \,(\text{mod } q)$ can be defined. $m_2(\gamma)$ determines δf up to a sign change; in the oriented case, $\sum t_\alpha/2$ is defined modulo 2 and δf is well-determined.

18.2.3   The three-sided inertial index and the Maslov index.

Definition 18.2.3 *if L_1 and L_2 are two lagrangians transversal to a lagrangian M, we define $m(L_1, L_2) = \frac{1}{2}\text{Inert}(L_1, M, L_2)$; $m(L_1, L_2)$ is called the Maslov index of the pair (L_1, L_2).*

By a previous remark, this integer or half-integer does not depend on the choice of M (as long as it is transversal to L_1 and L_2) and (7) becomes, for three arbitrary lagrangians :

$$\tfrac{1}{2}\text{Inert}(L_1, L_2, L_3) = m(L_1, L_3) + m(L_3, L_2) + m(L_2, L_1).$$

Proposition 18.2.4 *If $\gamma \in \text{Sp}_q(r)$, $q = 2, 3, \ldots, \infty$ sends the lagrangian L on the transversal lagrangian L', preserving the lagrangian M which, itself, is transversal to L and L', $m_\infty(\gamma)$ can be identified with $m(L, L')$ and $m_q(\gamma)$ is the reduction, modulo $2q$, of $m(L, L')$.*

Proof. We can choose a symplectic basis (E_α, E_{α^*}), adapted to the decomposition $E = M \oplus L$, such that $\hat{p}(\gamma)(E_\alpha)$, also denoted by $\gamma(E_\alpha)$, is equal to E_α and

$$\gamma(E_{\alpha^*}) = \theta_\alpha E_\alpha + E_{\alpha^*}, \quad \theta_\alpha = \pm 1, \quad \alpha = 1, 2, \ldots, r,$$
$$\text{Inert}(L, M, L') = -(\sum \theta_\alpha).$$

For

$$f_\alpha = E_\alpha + \theta_\alpha E_{\alpha^*} \in L',$$
$$f_{\alpha^*} = E_{\alpha^*},$$

(f_α, f_{α^*}) is a new symplectic basis; if

$$\gamma_1(f_\alpha) = f_\alpha,$$
$$\gamma_1(f_{\alpha^*}) = 2\theta_\alpha f_\alpha + f_{\alpha^*},$$

then

$$\gamma_1 \circ \gamma(f_\alpha) = -\theta_\alpha f_{\alpha^*},$$
$$\gamma_1 \circ \gamma(f_{\alpha^*}) = \theta_\alpha f_\alpha.$$

Sending (f_α, f_{α^*}) to (e_α, e_{α^*}), we see that $m_\infty(\gamma_1\gamma) = -\sum_\alpha \theta_\alpha/2$ and since γ_1 fixes the lagrangian of the f pointwise,

$$m_\infty(\gamma) = -\sum \frac{\theta_\alpha}{2}.$$

18.2.4 Cohomological problems.

On the manifold V, a field J of real endomorphisms, such that $J^2 = -\operatorname{Id}$, can be defined, thereby determining a complex and a riemannian structure for it. If a global field L of oriented lagrangians exists over V a classical result is that $T(V) = L \oplus J(L)$ and the tangent space at $x \in V$ is modeled on $E_1 \oplus E_2 = E_1 \oplus J(E_1)$. The structural group of the tangent bundle can be reduced to the orthogonal isometry group of E_1 (or E_2), i.e. $SO(r) \subset U(r)$, where $SO(r)$ is identified with $SO(r) \times SO(r)$.

Lemma 18.2.5 *The field L of oriented lagrangians defines a field of isotropic complex r-vectors for the riemannian structure.*

Proof. The subbundle defined by L has (as special cases) transition functions $u_{\alpha\beta}(x) \in U(r)$ sending E_1 to E_2, so that by formula (8) the field of isotropic r-vectors f_α over the open set $U_\alpha \cap U_\beta$ is multiplied by $\exp(-i\pi m_\infty(u_{\alpha\beta}(x)))$, $m_\infty(\hat{p}^{-1}(u_{\alpha\beta}(x)))$ can be written as $\sum_\alpha t_\alpha/2$ where $\sum_\alpha t_\alpha$ can change by multiples of 4. If $\mu_{\alpha\beta}(x)$ preserves E_1 the result is obvious.   ∎

A field L of oriented lagrangians therefore determines a spinoplectic bundle naturally associated to the complex structure. Then we see that formula (9) determines, by $\gamma = \hat{p}^{-1}(u_{\alpha\beta}(x))$, a $\mathbf{Z}_2$-valued cocycle, hence a bundle with group $\mathbf{Z}_2$ which we will call the Maslov $\mathbf{Z}_2$ bundle of the pair (L, JL) of lagrangians.

We obtain a second spinoplectic bundle, which is isomorphic to the first one if and only if the Maslov $\mathbf{Z}_2$ bundle is trivial.

Maslov bundles. The case which is usually considered corresponds to a trivial manifold $\mathbf{R}^{2r}$ with a symplectic structure and a lagrangian submanifold $\mathcal{L}$, i.e. a submanifold whose tangent space in every point is a lagrangian. If $x \in \mathcal{L}$, $(L_1)_x$ is a lagrangian over x parallel to a fixed lagrangian in $\mathbf{R}^{2r}$, L_x is the tangent lagrangian to $\mathcal{L}$ in x. Over $\mathcal{L}$ a bundle will be defined whose fiber in x is generated by the set of Maslov indices of the elements in $\operatorname{Sp}_q(r)$ mapping $(L_1)_x$ to L_x, these two lagrangians being transversal in general.

Lemma 18.2.6 *The complement in $\mathcal{L}$ of the set Σ of points x such that $(L_1)_x \cap L_x \neq$*
0 is everywhere dense in $\mathcal{L}$,

Proof. Write $V = \mathbf{R}^{2r} = E_1 \oplus E_2$, E_1 and E_2 as before, and $T_x(V) = L_x + E_2^x$, E_2^x
being parallel to E_2; there exists a symplectic transformation $\rho(x)$ such that the

$$\rho(x)(e_\alpha) = a_\alpha^{\beta^*}(x)e_{\beta^*} + e_\alpha$$

form, in general, a basis of L_x, since, in general, the matrix $a_\alpha^{\beta^*}(x)$ is regular and L_x,
E_2^x are transversal; the 'singular' points correspond to the non-transversality. As

$$a_\alpha^{\beta^*}(x) = \frac{\partial x^{\beta^*}}{\partial x^\alpha},$$

the singular points occur where $\det(\partial x^{\beta^*}/\partial x^\alpha)_{\alpha,\beta^*} = 0$, so Σ is the image of the
critical point set of a mapping which, in every point of Σ fails to be a submersion.
By the Sard theorem, $\mathcal{L} \setminus \Sigma$ is everywhere dense.

Now let $\mathcal{L}'$ be the set of $x \in \mathcal{L}$ such that $(L_1)_x \cap L_x = 0$; by Lemma (18.2.6),
$\overline{\mathcal{L}}' = \mathcal{L}$ (where $\overline{\mathcal{L}}'$ is the closure of $\mathcal{L}'$). A sheaf F with fiber $\mathbf{Z}_q$ is constructed
over $\mathcal{L}'$; it is also a trivial bundle over each open subset of $\mathcal{L}'$; F is generated by
the sections $x \to m_q(\gamma)_x$, where γ_x is constant element over each connected open
set of $\mathcal{L}'$ and sends $(L_1)_x$ to L_x. We consider the pre-sheaf $P_r F$ over $\mathcal{L}'$, then the
pre-sheaf $i(P_r F)$ where i stands for the inclusion of $\mathcal{L}'$ in $\mathcal{L}$. If U is an open set of
$\mathcal{L}$, $i^{-1}(U) = U \setminus (\Sigma \cap U)$. Over U, $i(PrF)$ consists therefore of the sections of F
over U. Applying the 'sheaf functor' F_S, we construct $F_S(i(PrF))$. If $x \in U$ and if
$s \in i(PrF)$, $[s_x]$ being the germ of s in x, it is the general element of the sheaf. This
sheaf with fiber $\mathbf{Z}_q$ can be identified with the $\mathbf{Z}_q$ bundle, hence it is the Maslov $\mathbf{Z}_q$
bundle.   ∎

This construction can be generalized if we consider two lagrangian fields $x \to L_x$
and $x \to L'_x$ over $\mathcal{L}$ in $\mathbf{R}^{2r}$. Identifying $L(x)$ with $l(x) \in \mathrm{U}(r)$ and $L'(x)$ with
$l'(x) \in \mathrm{U}(r)$, the previous reasoning can be repeated where $(L_1)_x$, L_x are replaced
by L_x, L'_x. $L(x)$ is transversal to $L'(x)$ if and only if $l(x) - l'(x)$ is invertible and
the mapping $x \to \det(l(x) - l'(x))$ from $\mathcal{L}$ to $\mathbf{C}$ is such that the complement of its
kernel is everywhere dense in $\mathcal{L}$.

Finally, a Maslov $\mathbf{Z}_q$ bundle could be defined on a manifold V with a symplectic
bundle and two fields of lagrangian subspaces L and L'. The details are left to the
reader.

18.3   Selected references.

* V. I. Arnold, *Une classe caractéristique intervenant dans les conditions de quan-*
  tification en théorie des perturbations et méthodes asymptotiques de Maslov,
  Dunod, Paris, 1972.

- J. Leray, *Analyse lagrangienne et Mécanique Quantique*, Cours du Collège de France, 1976-77.

Chapter 19

ALGEBRA DEFORMATIONS ON SYMPLECTIC MANIFOLDS.

To a $(n = 2r)$-dimensional paracompact symplectic manifold V, we associate the associative algebras $C^\infty(V)$ and $C^\infty(V, \mathbf{C})$ of differentiable real resp. complex-valued functions. F is the symplectic 2-form, $dF = 0$ and the rank of F is everywhere maximal.

In this chapter we will use the symplectic Clifford bundle to construct deformations of these algebras. Our method will prove that this problem can always be solved, i.e. that there is no obstruction to this construction.

The fundamental idea is to send the algebra $C^\infty(V)$ in the sheaf of sections of the symmetric covariant tensor bundle, which is linearly identified with the symplectic Clifford bundle; to define on the latter bundle an associative, non-commutative product, where the Planck constant is the deformation parameter; finally to return to the initial algebra.

19.1 The generalized Taylor homomorphism.

Let $x_0 \in V$ and let $(x^1, x^2, \ldots, x^n)$ be a local coordinate system on V with origin at x_0. On the tangent symplectic bundle, there exist an infinite number of torsion-free connections such that $\nabla F = 0$, two of them differ by a completely antisymmetric tensor of order 3. Let $f \in C^\infty(V)$, and $\tilde{f}$ its germ at x_0, $\tilde{C}^\infty(x_0)$ is the associative algebra of such germs, $\overset{**}{\mathsf{V}}(T_{x_0}^*)$ is the usual algebra of formal series with real coefficients constructed over the cotangent space $T_{x_0}^*$ at V in x. To every $\tilde{f} \in \tilde{C}^\infty(x_0)$ we associate $\varphi(\tilde{f}) = \hat{f} \in \overset{**}{\mathsf{V}}(T_{x_0}^*)$, defined by

$$\hat{f} = f(x_0) + (\nabla_i f)_{x_0} dx^i + \ldots + S[(\nabla_{i_1 i_2 \ldots i_k} f)_{x_0} dx^{i_1} \otimes dx^{i_2} \otimes \ldots \otimes dx^{i_k}] + \ldots \quad (1)$$

where S is a symmetrization still to be defined and ∇ is torsion-free and satisfies $\nabla F = 0$ (such a ∇ is called a *symplectic connection*).

In the flat case, we want to reobtain the ordinary Taylor expansion, its general term of order k given by

$$\sum_{q_1,q_2,\ldots,q_n} \frac{1}{q_1!q_2!\ldots q_n!}(\partial_{q_1q_2\ldots q_n}f)(dx^1)^{q_1} \vee (dx^2)^{q_2} \vee \ldots \vee (dx^n)^{q_n}$$

where

$$\partial_{q_i}f = \frac{\partial^{q_1}f}{(\partial x^1)^{q_1}}$$

etc. and $q_1 + q_2 + \ldots + q_n = k$. The general term of total degree k in (1) will be

$$\sum \frac{1}{q_1!q_2!\ldots q_n!}(\nabla_{(q_1,q_2,\ldots,q_n)}f)_{x_0}(dx^1)^{q_1} \vee (dx^2)^{q_2} \vee \ldots \vee (dx^n)^{q_n},$$

where the summation is carried out over all non-negative integers q_i such that $q_1 + q_2 + \ldots + q_n = k$ and

$$\nabla_{(q_1,q_2,\ldots,q_n)}f = \frac{q_1!q_2!\ldots q_n!}{k!}\sum(\nabla_{q_1,q_2,\ldots,q_n}f)$$

where $q_1, q_2, \ldots, q_n$ are fixed and all $k!/(q_1!q_2!\ldots q_n!)$ different derivatives occur, so that in the flat case, $\nabla_{(q_1,q_2,\ldots,q_n)}f = \partial_{q_1,q_2,\ldots,q_n}f$.

The right-hand side of (2) is in fact well-defined, independently of any local basis; if, for instance, we choose a symplectic basis in x_0, using the notations of Chapter 17, 2.5, the general term of (2) is

$$\sum \frac{1}{H!K^*!}(\nabla_{(H,K^*)}f)_{x_0}dx^H \vee dx^{K^*},$$

where there are $(n + k - 1)!/k!(n - 1)!$ terms, and it only depends on the germ of f in x_0, $\nabla_{(q_1,q_2,\ldots,q_n)}f$ is a symmetric covariant derivative of F.

Writing $\nabla_k = \nabla_{\partial/\partial x^k}$, the symmetrized derivative, denoted by $\overset{s}{\nabla}$, will satisfy

$$\overset{s}{\nabla}_k(\nabla_{(q_1,q_2,\ldots,q_n)}) = \nabla_{(q_1,q_2,\ldots,q_k+1,\ldots,q_n)}f.$$

In particular, $\overset{s}{\nabla}_i(\nabla_j f) = \nabla_{(ij)}f$ and $\nabla_{(ij)}f = \nabla_{ij}f$ if there is no torsion.

The Leibnitz formula, expressed as follows using ordinary partial derivatives,

$$\frac{1}{q_1!q_2!\ldots q_n!}\partial_{q_1q_2\ldots q_n}(fg) = \sum_{p_i\leq q_i}\frac{1}{p_1!p_2!\ldots p_n!}\frac{1}{(q_1-p_1)!(q_2-p_2)!\ldots(q_n-p_n)!}$$
$$(\partial_{p_1p_2\ldots p_n}f)(\partial_{q_1-p_1,q_2-p_2,\ldots,q_n-p_n}g)$$

remains valid by the simple formal analogy with the covariant derivatives

$$\nabla_{(q_1,q_2,\ldots,q_n)}(fg);$$

it leads to

$$\widehat{fg} = \hat{f}\hat{g}$$

so there exists a generalized Taylor homomorphism $\tilde{f} \to \hat{f}$. The kernel of this homomorphism is, in general, non-zero. To eliminate this difficulty, we will make some assumptions that will be weakened later on.

Note that if the Taylor homomorphism, computed for the usual partial derivatives, is injective, the generalized homomorphism will also be injective, since reasoning by induction proves that the sequence of $\nabla_{(q_1, q_2, \ldots, q_n)}$ is equivalent to the sequence of $\partial_{q_1 q_2 \ldots q_n} f$.

Now introduce assumption (H) :

Suppose that, for some choice of ∇, there exists a subspace $\tilde{C}^H(x_0) \neq 0$ of germs in x_0 and positive constants a and A (depending on $f \in \tilde{C}^H(x_0)$ and on x_0) such that

$$(H_1) \qquad |(\nabla_{(i_1, i_2, \ldots, i_k)} f)_{x_0}| \leq A a^k, \quad \forall k \in \mathbf{N}^*,$$
$$(H_2) \qquad \varphi|_{\tilde{C}^H(x_0)} : \tilde{f} \to \hat{f} \quad \text{is injective.}$$

It can be verified that if $\tilde{f}$ and $\tilde{g}$ satisfy H_1, so will their product (by the Leibnitz formula) and their sum. Furthermore, condition (H_1) is independent of the local basis.

Some cases where (H) can be satisfied arise if V is a flat symplectic space with the trivial connection and f an analytic function whose partial derivatives satisfy H_1, and if V is a symplectic manifold with an atlas which is compatible with (H_1) : f being chosen such that $\nabla_{(i_1, i_2, \ldots, i_q)} f = 0$ and q fixed on an open neighborhood of x_0.

19.2 Deformations of the algebra $\tilde{C}^H(x_0, \mathbf{C})$.

We consider complex-valued C^∞ functions $\tilde{f} \in \tilde{C}^H(x_0, \mathbf{C})$.

The linear isomorphism i from the symmetric algebra to the symplectic Clifford algebra over $T_{x_0}^*$ gives rise to the isomorphism $\varphi_1 = i \circ \varphi$ from $\tilde{C}^H(x_0, \mathbf{C})$ to the complexified symplectic Clifford algebra. We will write $\widehat{\tilde{f}} = \varphi_1(\tilde{f})$. Under hypothesis (H_1) we obtain an image in the non-formal complex algebra $C'_S(hF)_{1/2}$. If the product in the symmetric algebra is denoted by juxtaposition of factors and the symplectic Clifford product by $\odot$, a new multiplication is defined on $\tilde{C}^H(x_0, \mathbf{C})$ by

$$\tilde{f} * \tilde{g} = \varphi_1^{-1}(\widehat{\tilde{f}}) * \varphi_1^{-1}(\widehat{\tilde{g}}) = \varphi_1^{-1}(\widehat{\tilde{f}} \odot \widehat{\tilde{g}}) \tag{1}$$

(note that the series $\widehat{\tilde{f}} \odot \widehat{\tilde{g}}$ is convergent when the $(dx)^{ij}$ are interpreted as numbers). The product $*$ is immediately associative, but depends on the choice of the linear isomorphism i. In relation with the construction of global deformations (cf. ultra, 3 and 4), we note that if the isomorphism λ_F is used, it is 'invariant' under symplectic transformations and the product will be independent of the symplectic basis over $x_0 \in V$.

The structural group of the tangent bundle reduces to $U(r)$ and we may consider bases which are commonly called hermitian and are constructed over the vectors written as $(\epsilon_\alpha, \epsilon_{\alpha^*})$ in section 2.8 of Chapter 17. The basis changes involving matrices from the unitary group will obviously preserve the linear identifications made starting from a particular hermitian basis. The product constructed in this way is therefore 'invariant' under unitary transformations. We will make (1) explicit using such a linear isomorphism.

Denote by $(-tF)$ the symplectic form, t being a real parameter, and let $(\theta^\alpha, \theta^{\beta^*})$ be a hermitian cobasis at x_0 (a basis of $T^*_{x_0}$), $\theta^\alpha = (e_\alpha + ie_{\alpha^*})/\sqrt{2}$, $\theta^{\alpha^*} = (e_\alpha - ie_{\alpha^*})/\sqrt{2}$, take

$$\hat{f} = \hat{f}(x_0) + (\nabla_\alpha \tilde{f})_{x_0}\theta^\alpha + (\nabla_{\alpha^*}\tilde{f})_{x_0}\theta^{\alpha^*} + \ldots + \frac{1}{H!K^*!}(\nabla_{(H,K^*)}\tilde{f})_{x_0}\theta^H \vee \theta^{K^*} + \ldots \quad (2)$$

then

$$\theta^\alpha \odot \theta^{\beta^*} - \theta^{\beta^*} \odot \theta^\alpha = it\delta^{\alpha\beta^*} \quad (3)$$

and from formula (1), Chapter 17, 2.2, we find that :

$$\begin{aligned}
(\theta^{\alpha^*})^k \odot (\theta^\alpha)^l &= (\theta^\alpha)^l \odot (\theta^{\alpha^*})^k - ilkt(\theta^\alpha)^{l-1} \odot (\theta^{\alpha^*})^{k-1} + \ldots \\
&\quad + (-i)^p t^p \frac{l(l-1)\ldots(l-p+1)}{p!}k(k-1)\ldots(k-p+1)(\theta^\alpha)^{l-p} \odot (\theta^{\alpha^*})^{k-p} \\
&\quad + \ldots
\end{aligned}$$
$$(4)$$

proving that the zero degree term (for t) in $\tilde{f} * \tilde{g}$ is $\tilde{f}\tilde{g}$.

The coefficient of t in the expansion of $\widehat{f} \odot \widehat{g}$ can only come from the product of monomials

$$\frac{1}{k^*!}\frac{1}{A!B^*!}(\nabla_{(A,k^*,B^*)}\tilde{f})_{x_0}\theta^A \odot (\theta^{\alpha^*})^k \odot \theta^{B^*}$$

and

$$\frac{1}{l!}\frac{1}{C!D!}(\nabla_{(l,C,D^*)}\tilde{g})_{x_0}(\theta^\alpha)^l \odot \theta^C \odot \theta^{D^*}$$

(where the θ^{B^*} do not contain θ^{α^*} and the θ^C do not contain θ^α), reordering $(\theta^{\alpha^*})^k \odot (\theta^\alpha)^l$ and relying on (4), the other factors being considered commutative. These products have the form

$$\frac{-1}{(l-1)!(k-1)^*!A!C!B^*!D^*!}(\nabla_{(A,k^*,B^*)}\tilde{f})_{x_0}(\nabla_{(l,C,D^*)}\tilde{g})_{x_0}$$
$$(\theta^\alpha)^{l-1} \odot \theta^C \odot \theta^A \odot (\theta^{\alpha^*})^{k-1} \odot \theta^{B^*} \odot \theta^{D^*}$$

giving rise to a symmetric covariant derivative of $(\nabla_{\alpha^*}\tilde{f})(\nabla_\alpha \tilde{g})$, if we note that

$$\overset{s}{\nabla}_{p_1 p_2 \ldots p_n}(\nabla_{\alpha^*}\tilde{f}) = \overset{s}{\nabla}_{p_1 p_2 \ldots p_n}\nabla_{(0_1,\ldots,q_{\alpha^*},\ldots,0_n)}\tilde{f} = \nabla_{(p_1,p_2,\ldots,p_{\alpha^*}+q_{\alpha^*},\ldots,p_n)}\tilde{f},$$

where $q_{\alpha^*} = 1$, in a natural basis, back in the symmetric algebra.

Therefore

$$\tilde{f} * \tilde{g} = \tilde{f}\tilde{g} + tF^{\alpha^*\beta}(\nabla_{\alpha^*}\tilde{f})(\nabla_\beta \tilde{g}) + \ldots \quad (5)$$

Continuing, the coefficients of $t^2, t^3, \ldots$ will have terms

$$-\frac{1}{2!}\sum_{\alpha,\beta}(\nabla_{(\alpha^*\beta^*)}\tilde{f})(\nabla_{\alpha\beta}\tilde{g}), \quad \overset{i}{\frac{1}{3!}}\sum_{\alpha,\beta,\gamma}(\nabla_{\alpha^*\beta^*\gamma^*}\tilde{f})(\nabla_{(\alpha\beta\gamma)}\tilde{g})$$

etc. and in the flat case we have

$$\tilde{f}*\tilde{g}=\tilde{f}\tilde{g}+tF^{\alpha^*\beta}(\nabla_{\alpha^*}\tilde{f})(\nabla_{\beta}\tilde{g})+\cdots$$
$$+\frac{1}{k!}t^kF^{\alpha_1^*\beta_1}\ldots F^{\alpha_k^*\beta_k}(\nabla_{(\alpha_1^*\alpha_2^*\ldots\alpha_k^*)}\tilde{f})(\nabla_{(\beta_1\beta_2\ldots\beta_k)}\tilde{g})+\cdots \tag{6}$$

But if the symplectic manifold is arbitrary, curvature terms are added to the terms of degree $2, 3, \ldots$. For instance, using natural bases, consider

$$\nabla_{(kij)}f = \overset{s}{\nabla}_k(\nabla_{(ij)}f) = \tfrac{1}{3}(\nabla_k\nabla_{(ij)}f + \nabla_i\nabla_{(kj)}f + \nabla_j\nabla_{(ki)}f)$$

or also

$$\nabla_k\nabla_{(ij)}f - \tfrac{2}{3}R^l_{(ij)k}(\nabla_l f),$$

by the Ricci identities, R^l_{ijk} are the components of the curvature tensor. Then

$$\nabla_{(lkij)}f = \overset{s}{\nabla}_l\nabla_k\nabla_{(ij)}f = \nabla_{(lk)}\nabla_{(ij)}f$$

by the Bianchi identities; we can then continue without generating new curvature terms. Similarly, $\overset{s}{\nabla}_r\nabla_{(ijk)}f$ equals $\nabla_r\nabla_{(ijk)}f$ plus terms involving the curvature tensor and its covariant derivatives; again the Bianchi identities ensure that no new curvature terms appear by covariant derivation.

Then it is easily seen that all these correction terms vanish if the value of the product $\tilde{f}*\tilde{g}$ is taken in x_0.

19.3   Formal deformations of the algebra $C^\infty(V, C)$.

Note that both formula (6) of the previous section, valid in the flat case, and the analogous formula with curvature terms, give a computational algorithm which can be automatically extended to all germs of $C^\infty(V)$ functions, if they are considered as formal power series in t, since the verification of the associativity leads to identical computations in all cases. Hence :

Proposition 19.3.1 *On every symplectic manifold there exists in all points x_0 a formal deformation of the algebra $C^\infty_{x_0}(C)$ of germs of differentiable functions.*

The constructed deformation allows us to define the mapping $x_1 \to (\tilde{f}_{x_0}*\tilde{g}_{x_0})(x_1)$, x_1 belonging to some neighborhood of x_0. For $x_1 = x_0$, if x_0 varies, the mapping

$$x \to (\overset{?}{f}_x * \tilde{g}_x)(x),$$

denoted by $x \to (f*g)(x)$, defines an associative operation and, by a previous remark :

Proposition 19.3.2 *On every symplectic manifold, there exists a formal deformation of the associative algebra* $C^\infty(V, \mathbf{C})$ *given by the formula*

$$f * g = fg + \iota F^{\alpha^* \beta}(\nabla_{\alpha^*} f)(\nabla_\beta g) + \ldots + \frac{t^k}{k!} F^{\alpha_1^* \beta_1} \ldots F^{\alpha_k^* \beta_k} \nabla_{(\alpha_1^*, \ldots, \alpha_k^*)} f \nabla_{(\beta_1, \ldots, \beta_k)} g + \ldots$$

$$(1)$$

Proposition 19.3.3 *The product* $*$ *has the symplectic covariance property : if* θ *is a symplectic differentiable mapping,* $\theta(f * g) = (\theta f) * (\theta g)$, *where* $(\theta f)(x) = f \circ \theta^{-1}(x)$.

Proof. Let $x_1 = \theta(x_0)$:

$$(\theta(f * g))_{x_1} = (f * g)_{x_0} = \varphi^{-1}(\widehat{\widehat{f}}_{x_0} \odot \widehat{\widehat{g}}_{x_0})$$
$$(\theta f * \theta g)_{x_1} = \varphi_1^{-1}((\widehat{\widehat{\theta f}})_{x_1} \odot (\widehat{\widehat{\theta g}})_{x_1});$$

taking the composition of φ_1 with a symplectic isomorphism ψ and defining $\varphi_2 = \psi \circ \varphi_1$,

$$(\theta f * \theta g)_{x_1} = \varphi_2^{-1}(\psi(\widehat{\widehat{\theta f}})_{x_1} \odot \psi(\widehat{\widehat{\theta g}})_{x_1})$$

and we choose ψ such that $\tau(\mathcal{R}_0) = \psi(\mathcal{R}_1)$ where τ is the parallel transport for the connection ∇ and $\mathcal{R}_0, \mathcal{R}_1$ are the adapted hermitian bases in x_0 and x_1.

Since $\psi(\widehat{\widehat{\theta f}})_{x_1} = \tau \widehat{\widehat{f}}_{x_0}$,

$$(\theta f * \theta g)_{x_1} = \{\varphi_2^{-1}(\tau \widehat{\widehat{f}}_{x_0} \odot \tau \widehat{\widehat{g}}_{x_0})\}_{x_1},$$

∇ being a symplectic connection, this can be written as

$$\{\varphi_2^{-1} \circ \tau(\widehat{\widehat{f}}_{x_0} \odot \widehat{\widehat{g}}_{x_0})\}_{x_1} = \{\varphi_1^{-1}(\widehat{\widehat{f}}_{x_0} \odot \widehat{\widehat{g}}_{x_0})\}_{x_0},$$

from which the stated property follows. ∎

Summarizing, a brief description of the deformation of $C^\infty(V, \mathbf{C})$ runs as follows : the structural group of the tangent bundle is reduced to the unitary group and the isomorphism φ_1, determined by a choice of bases adapted to the almost hermitian structure, sends the sheaf of germs of differentiable functions into the sheaf of germs of sections of the complexified symplectic Clifford bundle; the zero-degree component in θ^H and θ^{K^*} of the product of germs of sections associated to $f, g \in C^\infty(V, \mathbf{C})$ then defines $f * g$.

19.4 Formal deformations of the algebra $C^\infty(V, \mathbf{R})$.

The preceding method can be applied if we replace the bases adapted to the almost hermitian structure by symplectic bases and use the 'covariant' isomorphism (for all symplectic transformations) λ_{tF}; this allows us to remain in the real field. The global existence of deformations raises no problem, so :

Proposition 19.4.1 *On every symplectic manifold, there exists a formal deforma-tion of the associative algebra $C^\infty(V, \mathbf{R})$.*

Nevertheless, the correspondence by λ_{tF} is complicated and, to have a convenient computational method, in the natural setting of theoretical mechanics, we will as-sume that the manifold possesses a global field L of real lagrangians (this holds for the usual phase space $V = T^*(M)$). We know that V has a complex structure J and that $T(V) = L \oplus JL$; a classical reasoning from differential geometry then shows that the structural group of the bundle can be reduced to $O(r) \times O(r)$. Using real cobases adapted to this decomposition of the tangent bundle, denoted by $(\varphi^\alpha, \varphi^{\alpha^*})$, formula (4) of section 19.2 is replaced by

$$
\begin{aligned}
(\varphi^{\alpha^*})^k \odot (\varphi^\alpha)^l &= (\varphi^\alpha)^l \odot (\varphi^{\alpha^*})^k - lkt(\varphi^\alpha)^{l-1} \odot (\varphi^{\alpha^*})^{k-1} + \cdots \\
&\quad + (-1)^p t^p \frac{l(l-1)\ldots(l-p+1)}{p!} k(k-1)\ldots(k-p+1)(\varphi^\alpha)^{l-p} \odot (\varphi^{\alpha^*})^{k-p} \\
&\quad + \cdots
\end{aligned}
$$

$$(1)$$

and the computations are carried out just as in the complex case, yielding formally identical results.

19.5   The Moyal product.

Let E be a real symplectic $(n = 2r)$-dimensional space with cartesian coordinates $(x^1, x^2, \ldots, x^n)$. First consider the set P of real polynomial functions in $(x^1, x^2, \ldots, x^n)$ of arbitrary finite degree.

   Let $f, g \in P$, we define

$$
\begin{aligned}
f *_{M_t} g &= \sum_0^\infty \frac{t^k}{k!} M F^k(f \otimes g), \quad \text{where} \\
F^k(f \otimes g) &= F^{i_1 j_1} F^{i_2 j_2} \ldots F^{i_k j_k} (\partial_{i_1 \ldots i_k} f) \otimes_{\mathbf{R}} (\partial_{j_1 \ldots j_k} g), \\
M(f \otimes g) &= fg,
\end{aligned}
$$

$f *_{M_t} g$ is the Moyal product of f and g for the real parameter value t. If P is replaced by the set $\hat{P}_\alpha$ of entire series in $(x^1, \ldots, x^n)$ with coefficients of the type introduced for the algebras $C_S(hF)_\alpha$, we see that the Moyal product is defined for $f, g \in \hat{P}_\alpha$.

   Using $E \otimes_{\mathbf{R}} \mathbf{R}((t)) = E'$ or by the definition of Chapter 17, the algebra $\check{C}_S(tF)$ is formal. Let $(\varphi^i, t \otimes \varphi^i = t\varphi^i)$ be a basis of $(E')^*$, the dual of E', which is a symplectic space.

   It is known that the formal algebra $\hat{P}$ is an associative algebra under the Moyal product. Let us accept this result for the moment; later we will deduce it indepen-dently in a natural way using our approach.

   Put $u(\varphi^i) = x^i$, $u(t\varphi^i) = tx^i$, u extends to a linear mapping from $(E')^*$ to P'. Since :

$$
\begin{aligned}
u(\varphi^i) *_{M_t} u(\varphi^j) - u(\varphi^j) *_{M_t} u(\varphi^i) &= tF^{ij} = tF(\varphi^i, \varphi^j) \\
u(t\varphi^i) *_{M_t} u(T\varphi^j) - u(t\varphi^j) *_{M_t} u(t\varphi^i) &= t^3 F^{ij} = tF(t\varphi^i, t\varphi^j),
\end{aligned}
$$

the universal property of symplectic Clifford algebras (Chapter 17, 1.2) implies the existence of a homomorphism $\hat{u}$ from $\check{C}_S(tF)$ to P', under the Moyal product, such that $u \circ \rho_{tF} = u$. By the remark in Chapter 17, 2.4, the two-sided ideals of $\check{C}_S(tF)$ being trivial, $\bar{u}$ must be injective. One can verify that $\bar{u}(\varphi^i)^n = (x^i)^n$, $\bar{u}(t^m(\varphi^i)^n) = t^m(x^i)^n$ and that $\bar{u}$ reaches all generators of $\hat{P}$.

The algebra $\check{C}_S(tF)$ is therefore isomorphic to the Moyal algebra.

Let us have a closer look at this isomorphism :

$$\bar{u}(\varphi^i) = x^i,$$
$$\bar{u}(\varphi^i \odot \varphi^j) = \bar{u}(\varphi^i) *_{M_t} \bar{u}(\varphi_j) = x^i *_{M_t} x^j = x^i x^j + tF^{ij},$$
$$\begin{aligned}
\bar{u}(\varphi^i \circ \varphi^j \circ \varphi^k) = x^i *_{M_t} (x^j x^k + tF^{jk}) &= x^i x^j x^k + tF^{ij}x^k + tF^{ik}x^j + tF^{jk}x^i \\
&= x^i(x^j x^k + tF^{jk}) + tF^{ij}x^k + F^{ik}x^j \\
&= x^i \bar{u}(\varphi^j \odot \varphi^k) + j^{2tF}_{x_i} \bar{u}(\varphi^j \odot \varphi^k)
\end{aligned}$$

and this relation holds for any number of factors, so that if we consider the right-hand sides, which have values in the symmetric algebra, $\bar{u}$ can be identified with the linear isomorphism λ_{2tF}.

So we have, in this setting :

Proposition 19.5.1 *The Moyal algebra can be identified with the 'Kähler-Atiyah algebra' determined by the isomorphism λ_{2tF} from the symplectic Clifford algebra to the symmetric algebra.*

19.6   Selected references.

- F. Bayen, M. Flato, C. Fronsdal, A. Lichnerowicz, D. Sternheimer, *Deformation theory and quantization I*, Annals of Physics, Vol. 111, 1978, pp. 61–110.

- A. Crumeyrolle, *Déformations d'algèbres associées a une variété symplectique : une construction effective*, Ann. Inst. H. Poincaré, Vol. 35, n. 3, 1981.

- M. De Wilde, P. B. A. Lecomte, *Existence of star-products and of formal deformations of the Poisson Lie algebra of arbitrary symplectic manifolds*, Letters in math. Physics 7, 1983, pp. 487–496.

- J. Vey, *Déformations du crochet de Poisson sur une variété symplectique*, Comment. Math. Helvetici, Vol. 50, 1975.

Chapter 20

THE PRIMITIVE IDEMPOTENTS OF THE CLIFFORD ALGEBRAS AND THE AMORPHIC SPINOR FIBER BUNDLES.

We intend to clarify the 'primitive idempotent' method of defining spinor bundles as fields of minimal ideals in some Clifford algebra. In this, the fundamental notion of 'geometric Clifford spinoriality group', similar but not at all equivalent to that of 'groupe de spinorialité' defined by the author in [1, d], will appear. Existence conditions depend on the choice of the standard left ideal and give a 'chaotic character' to the physics. The existence of primitive idempotent fields requires that very strict conditions be satisfied.

We give several examples. The primitive idempotent method offers algebraic tools, but raises a lot of geometric obstacles and cannot replace, improve or generalize the r-isotropic approach [1].

20.1 Preliminaries [1, d].

Let E be a real or complex n-dimensional vector space endowed with a non-degenerate quadratic form Q.

We denote by $C(E, Q)$ or, shorter, $C(Q)$, the Clifford algebra. If E is a real space and if Q has signature (p, q) (p positive and q negative squares), we also write $E_{p,q}$ for the space and $C_{p,q}$ for its Clifford algebra.

If $n = 2r$, $C(Q)$ is a simple central algebra. If $n = 2r + 1$, the even part $C^+(Q)$ of $C(Q)$ is a central simple algebra. We shall call denote by A either $C(Q)$, if $n = 2r$, or $C^+(Q)$, if $n = 2r + 1$.

According to the Skolem-Noether theorem, any automorphism of A is an inner automorphism. The inner automorphisms which preserve E globally are induced by the action of the non-twisted Clifford group G and will be called *geometric inner automorphisms* (or G.I.A. for short).

If J is any left or right ideal, there exists an idempotent e, which is non-unique in general, such that $J = Ae$. If the ideal is minimal, e is said to be a *primitive*

From Reports on Math. Physics n. 3, Vol. 25 (1987).

idempotent. Then there cannot exist a decomposition $e = e' + e''$ where e' and e'' are mutually annihilating non-zero idempotents ($e'e'' = e''e' = 0$).

If e and e_1 are primitive idempotents, there exists a $u \in A^*$ (the set of invertible elements in A) such that $e_1 = ueu^{-1}$, and if e and e_1 determine the ideals J and J_1 respectively, either $J \cap J_1 = J = J_1$ or $J \cap J_1 = \{0\}$.

It is well known that eAe is isomorphic to a division ring $\mathbf{K}$ and that this property characterizes the primitive idempotents.

When E is real, according to the classical periodicity properties, if e is a primitive idempotent :

$$eAe = \mathbf{R}, \quad p - q = 0, 1, 2 \,(\mathrm{mod}\, 8),$$
$$eAe = \mathbf{H}, \quad p - q = 4, 5, 6 \,(\mathrm{mod}\, 8),$$
$$eAe = \mathbf{C}, \quad p - q = 3, 7 \,(\mathrm{mod}\, 8).$$

When E is complex, $eAe = \mathbf{C}$ (cf. Chapter 5).

Notice that if $eue = \lambda$, $\lambda \in \mathbf{K} \setminus \{0\} = \mathbf{K}^*$,

$$eue = \lambda e = e\lambda, \quad \lambda = e\lambda = \lambda e$$

and $\lambda\lambda^{-1} = \lambda^{-1}\lambda = e$, where λ^{-1} is the inverse of λ in $\mathbf{K}$.

It is of importance to note that this inverse is not the inverse in A; e is its own inverse in $\mathbf{K}$, $ee = e$, but not in A, because $ee = e \neq 1$.

20.2   Amorphic spinoriality groups.

Let e be a primitive idempotent. We are looking for elements $u \in A^*$ such that $e' = ueu^{-1}$ and that both e and e' determine the same left ideal J in A.

It is evident that these elements u form a subgroup of A^*, called Σ_e, and that we can identify Σ_e and Σ_{e_1} if e and e_1 determine the same ideal J. We will call Σ_e the *stability ideal* of the left ideal $J = Ae$ for the action of inner automorphisms.

Σ_e is a 'Clifford spinoriality group (C.S.G.)'. Clearly, different C.S.G. are conjugate in A^*.

Proposition 20.2.1 *If the primitive idempotent e defines the left ideal J and if $u \in \Sigma_e$, there exists a $\lambda \in \mathbf{K}^*$ such that $eu^{-1} = \lambda$, and if $e' = ueu^{-1}$, $e' = u\lambda e$. Conversely, if one chooses $u \in A^*$ such that $eu = \lambda^{-1}e$, $\lambda \in \mathbf{K}^*$, taking $ueu^{-1} = e'$ leads to $e' = u\lambda e$ and e' determines the left ideal J if $u\lambda e = ve$, $v \in A^*$.*

Proof. If $e' = ueu^{-1}$ also determines J, $e' = \theta e$, $\theta \in A^*$ and reciprocally, then $eu^{-1} = u^{-1}\theta e = eu^{-1}e = \lambda$, $v \in A$, $\lambda \in \mathbf{K}^*$, $\lambda = \lambda e = eu^{-1}$ or $eu = e\lambda^{-1} = \lambda^{-1}e$; we obtain $e' = u\lambda e$.

The converse is easy. ∎

20.3   Idempotents and anti-involutions.

First we choose the usual anti-involution β, $\beta(x) = x$, if $x \in E$.

If e as an idempotent, so will $\beta(e)$ and if e is primitive, so too will $\beta(e)$: thus $\beta(e)$ also determines a minimal left ideal J'.

Interesting properties arise from the comparison of J and J'.

1. $J \cap J' = \{0\}$: then $e \neq \beta(e)$ and e, $\beta(e)$ determine a left ideal $\mathcal{K} = J \oplus J'$. If f denotes an idempotent determining $\mathcal{K}$, f is the sum of two primitive idempotents :

$$f = e_1 + e_2, \quad e_1 e_2 = e_2 e_1 = 0, \quad J = A e_1, \quad J' = A e_2.$$

   (a) If $e + \beta(e)$ is a non-primitive idempotent, $e + \beta(e)$ determines $\mathcal{K}$ and $e\beta(e) = \beta(e)e = 0$ according to a classical result.

   If e' also determines J, $e'\beta(e') = u\lambda(e)\beta(e)\beta(\lambda)\beta(u) = 0$. This case occurs frequently.

   (b) If $e + \beta(e)$ is not an idempotent, the previous result does not hold. This case occurs, for example, as follows : consider the algebra $C_{1,1}$ and a Witt frame $\{x_1, y_1\}$, $x_1 y_1 + y_1 x_1 = 1$; the idempotents $e = (\alpha_1 + x_1)y_1$, $\alpha_1 \neq 0$, and $y_1(\alpha_1 + x_1) = \beta(e)$ are primitive; if they verify $(e + \beta(e))^2 = e + \beta(e)$, as we have $e\beta(e) = 0$, also $\beta(e)e = 0$, but this result is wrong; the ideals corresponding to e and $\beta(e)$ have zero intersection and we must have $e + \beta(e) = 1$ if $e + \beta(e)$ is a (trivial) idempotent.

   If $e + \beta(e)$ is not an idempotent, there exists a u preserving J and a v preserving J' such that

$$ueu^{-1}v\beta(e)v^{-1} = v\beta(e)v^{-1}ueu^{-1} = 0,$$

   taking into account that $eu^{-1} = \lambda e$:

$$ev\beta(e) = 0,$$
$$\beta(e)ue = 0.$$

Reciprocally, if $v\beta(e)v^{-1}$ determines the same ideal J' as $\beta(e)$ and if $ev\beta(e) = 0$, $J = J'$ is impossible because, taking $\hat{e} = v\beta(e)v^{-1}$, $e\hat{e} = 0$, e, $\hat{e}$ determining J.

We may write $e = w\sigma\hat{e}$, $\sigma \in \mathbf{K}$ (Proposition (20.2.1)),

$$e\hat{e} = e \neq 0$$

is a contradiction.

If e' also determines J, $e' = u\lambda e$, $\beta(e) = \beta(e)\beta(\lambda)\beta(u)$ so that $ev\beta(e) = 0$ implies $e'v\beta(e') = 0$. There exists a $v \in A^*$ with $e'v\beta(e') = 0$ for any e' associated to J and v preserves J'.

2. $J \cap J' = J = J'$. $\beta(e)$ being a primitive idempotent :

$$\beta(e) = vev^{-1}, \quad v \in A^*.$$

According to Proposition (20.2.1), there exists a $\lambda \in K^*$ such that

$$\beta(e) = v\lambda e,$$

$$\beta(e)e = \beta(e) \Rightarrow \beta(e)e = e, \; e = \beta(e) \text{ and } e\beta(e) = \beta(e)e \neq 0.$$

We can state :

Proposition 20.3.1 *If e determines J and $\beta(e)$ determines J', both being minimal left ideals of $C(Q)$,*

1. *If $J \cap J' = \{0\}$, then $e \neq \beta(e)$.*

   (a) *$e + \beta(e)$ is idempotent : $e\beta(e) = \beta(e)e = 0$.*

   (b) *$e+\beta(e)$ is not idempotent : there exists a u preserving J and a v preserving J' such that $ev\beta(e) = \beta(e)ue = 0$ and $e'v\beta(e') = 0$, for any e' determining J.*

2. *If $J \cap J' = J = J'$, $\beta(e) = e$.*

3. *If $e\beta(e) = 0$, necessarily $J \cap J' = \{0\}$. More generally, if there exists a v preserving J' such that $ev\beta(e) = 0$, then $J \cap J' = \{0\}$.*

We intend to obtain simpler results, so we restrict the group A^* to its subgroup B of elements u such that $\beta(u)u = u\beta(u) \in R^*$ (in the real case) or $\beta(u)u = u\beta(u) \in C^*$ (in the complex case). Note that if $u \in B$ the set of all primitive idempotents e such that $e = \beta(e)$ or $e + \beta(e)$ is idempotent if $e \neq \beta(e)$, is preserved by the inner automorphisms defined by elements of B.

We must observe that the group Σ_e is replaced by a group S_e, and that S_e, S_{e_1} are conjugate in Σ_e (using the notations of section 2).

Note also that β commutes with the inner action of u if $u \in B$.

Proposition 20.3.2 *If $J \cap J' = J = J'$, if e, e' are primitive idempotents determining the same left ideal J and if $e' = ueu^{-1}$ with $u \in S_e$, then $e = e'$.*

($\beta(e) = e$ implies $\beta(e') = e'$ because β commutes with u.)
Proof. We have $\beta(e)\beta(\lambda)\beta(u) = u\lambda e$ (Proposition (20.2.1)),

$$e\beta(\lambda)\beta(u) = u\lambda e = e\beta(\lambda)\beta(u)e = e\beta(\lambda)\beta(u)\beta(e),$$

but $eu = \lambda^{-1}e = e\lambda^{-1}$, so

$$\begin{aligned} \beta(\lambda)\beta(u)\beta(e) &= \beta(e), \\ e\beta(e) &= u\lambda e, \\ e &= e'. \end{aligned}$$

Remark 1. *If e and e' both determine J, if $\beta(e)$ determines J' and if $\beta(e')$ determines J'', $J'' \neq J'$ except if $e = e'$.*
Since $\beta(e') = \beta(u^{-1})\beta(e)\beta(u)$, if $\beta(e')$ determines the same left ideal as $\beta(e)$, there exists a $\sigma \in \mathbf{K}$ such that

$$\beta(e)\beta(u^{-1}) = \sigma\beta(e)$$

(Proposition (20.2.1)), but

$$\beta(u)\beta(e) = \beta(\lambda^{-1})\beta(e);$$

multiplying the first equality by the second and vice versa, we obtain :

$$\beta(e) = \sigma\beta(\lambda^{-1})\beta(e), \quad \sigma = \beta(\lambda),$$
$$\beta(u)\beta(e)\beta(u^{-1}) = \beta(e) = \beta(e'),$$

then $e = e'$. We can now summarize our results :

Proposition 20.3.3 *If the primitive idempotent e determines the left ideal J and if $\beta(e)$ determines the left ideal J',*

1. *If $e \neq \beta(e)$ and if $e + \beta(e)$ is an idempotent, then*

$$J \cap J' = \{0\}, \quad e\beta(e) = \beta(e)e = 0.$$

If $e' \neq e$, $e' = ueu^{-1}$, $u \in S_e$ also determines J, and if $\beta(e')$ determines J'', then $J \cap J'' = \{0\}$; if $e' \neq e$, $J' \cap J'' = \{0\}$.

2. *If $e = \beta(e)$, $J \cap J' = J = J'$, $\forall e'$ such that $e' = ueu^{-1}$, $u \in S_e$, if e' also determines J, $e' = e$.*

Remarks 2. In case 2, the group S_e appears as one of the $u \in B$ such that $eu = ue$.
    In case 1, $e' = \theta e$ with $\theta \in \mathbf{K}^*$, $\theta \neq 1$ is impossible : in fact, $e' = \theta e$ yields

$$\beta(e') = \beta(e)\beta(\theta) = \beta(\theta)\beta(e),$$

$\beta(e')$ and $\beta(e)$ correspond to the same left ideal, and that contradicts the remark if $\theta \neq 1$.
    We shall give below an example with $e' = e$ (case 1).
    We can choose $\tilde{\beta}$ instead of β, where $\tilde{\beta} = \beta \circ \alpha$, α is the main involution of $C(Q)$: $\alpha(x) = -x$, $x \in E$.
    When $u \in B$, put $\beta(u)u = u\beta(u) = N(u) \in \mathbf{R}^*$ or $\mathbf{C}^*$; if $e = \beta(e)$, in the notations of section 2,

$$eu^{-1} = \lambda e, \quad ue = N(u)\beta(\lambda)e,$$
$$ueu^{-1} = N(u)\beta(\lambda)\lambda e,$$
$$e' = N(u)\beta(\lambda)\lambda e,$$
$$N(u)\beta(\lambda)\lambda = 1.$$

If $u \in G$, [2] u defines a G.I.A. and $u \in B$: with the elements of G, we obtain a subgroup σ_e of Σ_e and S_e : σ_e is called a *geometric Clifford spinoriality group* (*G.C.S.G.*). σ_e is defined, a priori, modulo a conjugation in Σ_e, when $J = Ae$ is given.

But $\sigma_e = \sigma_{e'}$ if $e' = ve$ is any idempotent in Ae.

In fact, if $u \in \sigma_e$, $eu = \lambda^{-1}e = e\lambda^{-1}$, $\lambda \in K^*$,

$$veu = v\lambda^{-1}e,$$
$$e'u = (v\lambda^{-1}v^{-1})e' = e'ue' = \sigma e', \quad \sigma \in K^*,$$
$$e'u = \sigma e' = e'\sigma.$$

Conversely, if $e'u = \sigma e' = e'\sigma$,

$$veu = \sigma ve,$$
$$eu = (v^{-1}\sigma v)e = eue = \lambda^{-1}e = e\lambda^{-1}, \quad \lambda \in K^*.$$

The proof works also with S_e, $S_{e'}$ in the same situation. We can speak of σ_e (or S_e) as the subgroup of G (or B) whose elements preserve the left ideal $J = Ae$.

20.4   Examples and comments.

20.4.1   The real space $E_{1,3}$.

We have $p - q = -2 (\bmod 8)$ and the example belongs to the quaternionic type. The spinor space dimension is 2 over $\mathbf{H}$ and 8 over $\mathbf{R}$.

Let (e_0, e_1, e_2, e_3) be a frame of $E_{1,3}$ with $(e_0)^2 = 1$, $(e_i)^2 = -1$, $i = 1, 2, 3$; it is easy to prove that

$$h = \frac{1 + e_0}{2}, \quad h_1 = \frac{1 + e_0 e_3}{2}, \quad \frac{1 + e_1 e_2 e_3}{2}$$

are primitive idempotents and that they are not equivalent in G. We notice that $\beta(h) = h$ and $\tilde{\beta}(h_2) = h_2$, but that $\tilde{\beta}(h_1) \neq h_1$ and $\beta(h_1) \neq h_1$, $\beta(h_1) + h_1 = 1$, so h_1 corresponds to the case 1,(a) and h_1, h_2 to the case 2 of Proposition (20.3.1).

Actually we can check :

$$\{huh, u \in C(Q)\} = \{(ae_1e_2 + be_2e_3 + ce_3e_1 + d)h, \ a, b, c, d \in \mathbf{R}\},$$
$$\{h_2uh_2, u \in C(Q)\} = \{(ae_1e_2 + be_2e_3 + ce_3e_1 + d)h_2, \ a, b, c, d \in \mathbf{R}\},$$
$$\{h_1uh_1, u \in C(Q)\} = \{(a + be_1 + ce_2 + de_1e_2)h_1, \ a, b, c, d \in \mathbf{R}\}$$

where the division ring $\mathbf{H}$ appears.

According to the classical properties of the simple algebras, $C_{1,3}$ is the direct sum of two left ideals (and of two right ideals as well); for example $(1 - e_0)/2 = \hat{h}$ and h give such a decomposition.

---

[2] Note that $u \in G^+$ if $n = 2r + 1$, because $A^* = C^+(Q)$.

First we are looking for Σ_h. A real frame for spinor space is

$$h, e_1e_2h, e_2e_3h, e_3e_1h, e_Nh, e_N(e_1e_2h), e_N(e_2e_3h), e_N(e_3e_1h),$$

with $e_N = e_0e_1e_2e_3$.

We look for $u \in A^*$ such that

$$hu = \lambda^{-1}h, \quad \lambda \in \mathbf{H}^*,$$
$$u = h\alpha + \hat{h}\alpha_1, \quad \alpha, \alpha_1 \in A,$$
$$hu = h\alpha = \lambda^{-1}h = h\lambda^{-1}.$$

We can then write

$$u = h\alpha + \hat{h}(\gamma + e_N\delta),$$

α, γ, δ are quaternions (of the form $ae_1e_2 + be_2e_3 + ce_3e_1 + d$), commuting with h and $\hat{h}$, and $\hat{h}e_N = e_N h$.

If u' is the inverse of u,

$$u' = h\alpha' + \hat{h}(\gamma' + e_N\delta'), \quad uu' = 1,$$

and then

$$h\alpha\alpha' + \hat{h}(\gamma + e_N\delta)h\alpha' + \hat{h}(\gamma + e_N\delta)\hat{h}(\gamma' + e_N\delta') = 1.$$

A right multiplication by h gives

$$h\alpha\alpha' = h, \quad \alpha' = \alpha^{-1}$$

(necessarily $\alpha \neq 0$); a right multiplication by $\hat{h}$ gives likewise :

$$\hat{h}(\gamma + e_N\delta)\hat{h}(\gamma' + e_N\delta')\hat{h} = \hat{h}, \quad \gamma' = \gamma^{-1}$$

(because $\gamma \neq 0$).

Taking these results into account :

$$h + \hat{h}(\gamma + e_N\delta)h\alpha^{-1} + \hat{h} + \hat{h}\gamma e_N\delta' = 1,$$
$$\hat{h}e_N h\delta\alpha^{-1} + \hat{h}e_N\gamma\delta' = 0,$$
$$\delta' = -\gamma^{-1}\delta\alpha^{-1}.$$

We obtain if $h' = uhu^{-1}$,

$$h' = (1 + e_N\delta\alpha^{-1})h,$$

but the coefficient of h must be invertible if h and h' define the same ideal, and it is easy to check that $\delta = 0$ and $h' = h$.

Σ_h is the set of elements u belonging to $h\mathbf{H}^* \oplus \hat{h}\mathbf{H}^*$, and because $h\hat{h} = \hat{h}h = 0$ we can state : Σ_h, the *C.S.G.* of $h = (1 + e_0)/2$, is isomorphic to $\mathbf{H}^* \times \mathbf{H}^*$. It's also the stability group of h.

We obtain the same results with h_2.

Remark 3. If we are interested in σ_h, using only the G.I.A. since $u \in \Sigma_h$ is written :

$$u = h(ae_1e_2 + be_2e_3 + ce_3e_1 + d) + \hat{h}(a'e_1e_2 + b'e_2e_3 + c'e_3e_1 + d'), \quad a, b, \ldots, d' \in \mathbf{R},$$

and the Lie algebra of G contains only terms in e_1e_2, e_2e_3, e_3e_1 and 1. σ_h is isomorphic to $\mathbf{H}^*$.

Now consider Σ_{h_1}. Note that $(\dfrac{1 - e_3}{\sqrt{2}})h(\dfrac{1 + e_3}{\sqrt{2}}) = h_1$ and we have the spinor frame

$$h_1, e_1h_1, e_2h_1, e_1e_2h_1, e_0h_1, e_0e_1h_1, e_0e_2h_1, e_0e_1e_2h_1,$$

we take : $u = h_1\alpha + \hat{h}\alpha_1$, $\hat{h}_1 = \dfrac{1 - e_0e_3}{2}$, $\alpha \in \mathbf{H}$ again, and $\alpha_1 = \gamma + e_0\delta$, $\alpha, \gamma, \delta \in \mathbf{H}$,

$$e_0\delta = \delta_1e_0, \quad \delta_1 \in \mathbf{H},$$
$$e_0h_1 = \hat{h}_1e_0.$$

The same computation gives $\alpha^{-1} = \alpha'$, $\gamma^{-1} = \gamma'$, we come to

$$h_1(\delta\alpha^{-1} + \gamma\delta') = 0,$$
$$\delta' = -\gamma^{-1}\delta\alpha^{-1},$$
$$h_1' = uh_1u^{-1} = (1 + e_0\delta\alpha^{-1})h_1.$$

The invertibility condition gives $\delta = 0$, Σ_{h_1} is isomorphic to $\mathbf{H}^* \times \mathbf{H}^*$ and Σ_{h_1} is also the stability group of h_1.

These results were expected. But what about σ_{h_1} ? Taking into account the Lie algebra of G and

$$h_1 = \frac{1 + e_0e_3}{2}, \quad \hat{h}_1 = \frac{1 - e_0e_3}{2},$$

$v = Ae_0e_3 + Be_1e_2 + C$, $A, B, C \in \mathbf{R}$, is the general element of the abelian Lie algebra of σ_{h_1}; if the upper index 0 stands for the connected component, we can state :

With the real space $E_{1,3}$, *the A.C.S.G.* σ_h^0 *and* $\sigma_{h_1}^0$ *have real dimensions 4 and 3 respectively.*

After the projection p onto the orthogonal group :

$$p(\sigma_h^0) = O(3)$$

and $p(\sigma_{h_1}^0)$ is a two-dimensional abelian group.

20.4.2   The real space $E_{3,1}$.

$p - q = 2$ and the Clifford algebra $C_{3,1}$ belongs to a real type. The real spinors could be called 'Majorana spinors' and belong to a four-dimensional real space; $C_{3,1}$ is the direct sum of four spinor spaces.

We choose $(e_0)^2 = (e_1)^2 = (e_2)^2 = 1$, $(e_3)^2 = -1$ and the primitive idempotent

$$h = \tfrac{1}{4}(1 + e_0)(1 + e_0 e_1 e_3) = \tfrac{1}{4}(1 + e_0)(1 + e_1 e_3),$$

$\beta(h) \neq h \neq \tilde{\beta}(h)$, $\mathbf{K} = \mathbf{R}$.

We have to determine u such that

$$hu = \lambda^{-1}h, \quad \lambda \in \mathbf{R}^*.$$

We put $u = ha_0 + \sum_1^3 h_i a_i$, h, h_1, h_2, h_3 are a complete system of primitive idempotents :

$$h_1 = \tfrac{1}{4}(1 - e_0)(1 - e_0 e_1 e_3), \quad h_2 = \tfrac{1}{4}(1 - e_0)(1 + e_0 e_1 e_3),$$
$$h_3 = \tfrac{1}{4}(1 + e_0)(1 - e_0 e_1 e_3),$$

$hu = ha_0$, because $hh_i = 0$ and $hu = \lambda^{-1}h$. We can choose $a_0 \in \mathbf{R}_+^*$, $a_0 = \lambda^{-1}$.

If $u'u = uu' = 1$, necessarily $u' = ha_0^{-1} + \sum_1^3 h_i a_i'$,

$$uu' = h + (\sum_1^3 h_i a_i)ha_0^{-1} + (\sum_1^3 h_i a_i)(\sum_1^3 h_i a_i') = 1$$

and we have an analogous relation for $u'u$.

According to the direct sum $h + h_1 + h_2 + h_3 = 1$, we must satisfy one of the following conditions :

$$h_i a_i (ha_0^{-1} + \sum_1^3 h_j a_j') = h_i,$$
$$h_i a_i' (ha_0 + \sum_1^3 h_j a_j) = h_i.$$

Note :

$$h(e_0 e_2) = (e_0 e_2)h_1, \quad h_2(e_0 e_2) = (e_0 e_2)h_3, \quad (e_0 e_2)h = h_1(e_0 e_2),$$
$$h(e_1 e_2) = (e_1 e_2)h_3, \quad h_2(e_1 e_2) = (e_1 e_2)h_1, \quad (e_1 e_2)h = h_3(e_1 e_2),$$
$$h(e_0 e_1) = (e_0 e_2)h_2, \quad h_2(e_0 e_1) = (e_0 e_1)h, \quad (e_0 e_1)h = h_2(e_0 e_1).$$

Consider the h_2 line, where the a_k^j, $k = 1, 2, 3, 4$, are real :

$$h_2(\alpha + \beta e_0 e_2 + \gamma e_1 e_2 + \delta e_0 e_1)(ha_0^{-1} + \sum_1^3 h_j(a_1^j + a_2^j e_0 e_2 + a_3^j e_1 e_2) + a_4^j e_0 e_1) = h_2$$

and we calculate $\alpha, \beta, \gamma, \delta$.

$$
\begin{aligned}
h_2 = \; & h_2 \delta a_0^{-1} e_0 e_1 && + \; \gamma h_2 e_1 e_2 && (a_1^1 + a_2^1 e_0 e_2 + a_3^1 e_1 e_2 + a_4^1 e_0 e_1) \\
& && + \; \alpha h_2 && (a_1^2 + a_2^2 e_0 e_2 + a_3^2 e_1 e_2 + a_4^2 e_0 e_1) \\
& && + \; \beta h_2 e_0 e_2 && (a_1^3 + a_2^3 e_0 e_2 + a_3^3 e_1 e_2 + a_4^3 e_0 e_1)
\end{aligned}
$$

giving :

$$\alpha a_2^1 - \beta a_2^3 - \gamma a_3^1 = 1,$$
$$\alpha a_2^2 + \beta a_1^3 - \gamma a_4^1 = 0,$$
$$\alpha a_3^2 + \beta a_4^3 + \gamma a_1^1 = 0,$$
$$\alpha a_4^2 - \beta a_3^3 + \gamma a_2^1 + \delta a_0^{-1} = 0.$$

The determinant

$$\Delta_2 = \begin{vmatrix} a_1^2 & -a_2^3 & -a_3^1 \\ a_2^2 & a_1^3 & -a_4^1 \\ a_3^2 & a_4^3 & a_1^1 \end{vmatrix}$$

cannot vanish; α, β, γ, δ define a_2.

We obtain similarly the coefficients of a_1 and a_3.

Finally, as $h' = uhu^{-1} = u\lambda h = ua_0^{-1}h$,

$$h' = [1 + (\sum_1^3 h_i a_i)a_0^{-1}]h.$$

Calling β' the coefficient of $e_0 e_2$ in a_1 and γ'' the coefficient of $e_1 e_2$ in a_3, we obtain :

$$h' = [1 + a_0^{-1}(\beta' e_0 e_2 + \delta e_0 e_1 + \gamma'' e_1 e_2)]h$$

and the invertibility condition gives $\delta = \beta' = \gamma'' = 0$.

The real dimension of Σ_h is 10.

Now we consider only $u \in G$ (the Clifford group) and we determine σ_h (at least its connected component of the identity). $\mathcal{L}(G)$ is the Lie algebra of G.

If $v \in \mathcal{L}(\sigma_h)$, according to section 2, $hv = \lambda h$, $\lambda \in \mathbf{R}$

$$(1 + e_0)(1 + e_0 e_1 e_3)v = \lambda(1 + e_0)(1 + e_0 e_1 e_3),$$

we obtain easily :

$$v = ae_0(e_1 + e_3) + be_2(e_1 + e_3) + c(e_3 e_1) + d + \lambda$$

and σ_h is four-dimensional.

Instead of checking $uhu^{-1} = h'$, we calculate

$$v \in \mathcal{L}(\sigma_h), \quad vh - hv = (v - \lambda)e$$
$$vh - hv = (ae_0(e_1 + e_3) + be_2(e_1 + e_3) + c(e_3 e_1 + 1))h.$$

More particularly, if $v \in \mathcal{L}(\mathrm{Pin}(Q))$, $d + \lambda = 0$ and

$$v = (ae_0 + be_2)(e_1 + e_3) + ce_3 e_1, \quad (vh - hv \neq 0),$$

is a general element of the Lie algebra of a group σ_h'.

If we consider the stability group of h, if v belongs to its Lie algebra, $vh - hv = 0$, and the connected component of the stability group of h reduces to the identity.

Remark 4. In the paper [1, f] we defined a group called 'Groupe de spinorialité élargi réduit' (reduced enlarged spinoriality group). It is easy to verify that (at least for its connected component), σ_h' is isomorphic to this group, $c = 0$ corresponds to our reduced spinoriality group.

20.4.3   An important remark about the stability group of the primitive idempotents.

Suppose we change of primitive idempotent to define the same minimal left ideal.
For example in $C_{3,1}$, we choose :

$$h' = (1 + e_1 + e_3)h(1 - e_1 - e_3) = (1 + e_1 + e_3)h,$$

$1 + e_1 + e_3$ and $1 - e_1 - e_3$ belong to Σ_h, h' determines the same ideal as h.
We intend to compare the connected stability groups of h and h' in the connected component of G. For h this set is the identity.
We search for $v \in \mathcal{L}(G)$ such that

$$vh' - h'v = 0, \quad v \in \mathcal{L}(\sigma_{h'}).$$

It is easy to verify that $\mathcal{L}(\sigma_{h'}) = \mathcal{L}(\sigma_h)$.
Take $v = (ae_0 + be_2)(e_1 + e_3) + ce_3e_1 + k$ where $a, b, c, k \in \mathbf{R}$.

$$vh' - h'v = v(1 + e_1 + e_3)h - (1 + e_1 + e_3)hv = 0,$$
$$vh - hv = (e_1 + e_3)hv - v(e_1 + e_3)h = -c(1 - e_0)(e_1 + e_3)$$
$$\tfrac{1}{2}(ae_0 + be_2)(1 - e_0)(e_1 + e_3) = -c(1 - e_0)(e_1 + e_3),$$
$$c = a/2, \quad b = 0$$

$v = ae_0(e_1 + e_3) + (a/2)e_3e_1 + k$ is the general element of the stability group's Lie algebra for h'. Thus the stability group of h' differs from the identity.
We state the crucial result :

Two primitive idempotents can define the same minimal left ideal but have different stability groups in G.

This explains why the primitive idempotent fields are, in general, inappropriate tools to define amorphic spinor bundles (cf. below).

Nevertheless, we have obtained interesting results about idempotents that are preserved by some anti-involution β or $\tilde{\beta}$: if $\beta(e) = e$ or if $\tilde{\beta}(e) = e$, the group σ_e is the stability group of e and if e' also determines the minimal ideal Ae, $\sigma_{e'} = \sigma_e$ is the stability group of e'.

We insist that two distinct minimal left ideals Ae, Ae' give rise to groups Σ_e and $\Sigma_{e'}$ which are always conjugate in A^* but not necessarily conjugate in G (cf. the example above with $h, h_1, h_2 \in C_{1,3}$).

20.4.4

It is particularly important to define the anti-involutions γ, commuting with a G_1-action, G_1 being a subgroup of G (or G^+ if n is odd) such that $\gamma(e) = e$ for some idempotent. We shall examine this problem in the section 7 of this chapter. In bundle problems we shall be confronted with particular G.I.A. satisfying some particular conditions. The most important case is given by G.I.A. preserving a direct ortho-gonal decomposition $E_{p,q} = E_p \oplus E_q$ in positive and negative parts. There exists an automorphism α' in $C_{p,q}$ such that $\alpha'(e_j) = e_j$ if $Q(e_j) = 1$ and $\alpha'(e_j) = -e_j$ if $Q(e_j) = -1$. The anti-involution β composed with α' gives a new anti-involution β', but β' commutes only with the G_1-action where G_1 is the subgroup of G or G^+ preserving the above orthogonal decomposition. We shall prove in section 7 that any anti-involution γ, commuting with a G_1-action (G_1 preserving E, subgroup of G or G^+ is n is odd) is β, $\tilde{\beta}$, or like β'.

For example, in $C_{1,3}$, with the orthogonal decomposition associated to the frame e_0, e_1, e_2, e_3, the idempotent $h = \frac{1}{4}(1 + e_0)(1 + e_1 e_3)$ satisfies $\beta(h) = h$. Naturally, Proposition (20.3.2) is applicable with β'.

20.4.5   Real and complex minkowskian spinors.

Again consider the space $E_{1,3}$: a spinor frame (after choice of the particular idem-potent $(1 + e_0)/2$ is given by :

$$1 + e_0, \qquad e_2 e_3(1 + e_0), \qquad e_3 e_1(1 + e_0), \qquad e_1 e_2(1 + e_0),$$
$$e_N(1 + e_0), \quad e_N e_2 e_3(1 + e_0), \quad e_N e_3 e_1(1 + e_0), \quad e_N e_1 e_2(1 + e_0),$$

where the quaternionic frame $1, e_2 e_3, e_3 e_1, e_1 e_2$ appears modulo $(1 + e_0)$. (Some au-thors call these spinors 'biquaternions'.)

Here $(e_N)^2 = -1$; if one substitutes i for e_N, a frame for 4-dimensional com-plex space is obtained. This complex space is nothing but a spinor space for the complexified Clifford algebra $C_{1,3}^{\mathbb{C}}$.

Note that spinor spaces are defined modulo a representation equivalence. We can choose instead of $h = (1 + e_0)/2$ the primitive idempotent $h_1 = (1 + e_0 e_3)/2 = ((1 - e_3)/\sqrt{2})h((1 + e_3)/\sqrt{2})$ and h is not geometrically equivalent to h_1.

Substituting the complex number i for e_N, we obtain the idempotent e :

$$e = (\frac{1 + e_0 e_3}{2})(\frac{1 - i e_1 e_2}{2})$$

and if we define the Witt basis

$$y_1 = \frac{e_0 + e_3}{2}, \quad y_2 = \frac{e_1 + i e_2}{2}, \quad x_1 = \frac{e_0 - e_3}{2}, \quad x_2 = \frac{i e_2 - e_1}{2},$$

$$e = -x_1 x_2 y_1 y_2,$$

and according to the previous construction real spinors determined by the idem-potent $(1 + e_0 e_3)/2$ correspond to the complex spinors defined by the idempotent $(-x_1 x_2 y_1 y_2)$ where the two-dimensional isotropic vector $f = y_1 y_2$ appears (cf. the author's methods [1] or Chapter 5 above).

Observe that these two kinds of spinors are not geometrically equivalent (non covariant in the Lorentz sense).

Now we consider the spaces $E_{3,1}$ and $E_{3,1}^{\mathbf{C}}$,

$$x_1 = \frac{e_1 - e_3}{2}, \quad y_1 = \frac{e_1 + e_3}{2}, \quad x_2 = \frac{e_0 - ie_2}{2}, \quad y_2 = \frac{e_0 + ie_2}{2}$$

is a Witt basis and

$$x_1 x_2 y_1 y_2 = \left(\frac{1 + e_1 e_3}{2}\right)\left(\frac{1 + ie_0 e_2}{2}\right),$$

if we replace i by e_N :

$$\left(\frac{1 + e_1 e_3}{2}\right)\left(\frac{1 + e_N e_0 e_2}{2}\right) = \left(\frac{1 + e_1 e_3}{2}\right),$$

but $\dfrac{1 + e_1 e_3}{2}$ is a non-primitive idempotent

$$\frac{1 + e_1 e_3}{2} = \left(\frac{1 - e_0}{2}\right)\left(\frac{1 + e_1 e_3}{2}\right) + \left(\frac{1 + e_0}{2}\right)\left(\frac{1 + e_1 e_3}{2}\right) = h + h_2$$

using the notations of 4.2.

The **C**-spinor space is four-dimensional over **C**, eight-dimensional over **R**. From two examples of **R**-spinor spaces (determined by h and h_2) we obtain the realification of the **C**-spinor space determined by $\frac{1}{4}(1 + e_1 e_3)(1 + e_N e_0 e_2)$, and the **C**-spinor space itself is determined by $\frac{1}{4}(1 + e_1 e_3)(1 + ie_0 e_2)$ or by the isotropic 2-vector $y_1 y_2$.

Grosso modo we obtain the complex case changing suitably e_N into i. Note nevertheless that this method gives a particular idempotent for $C_{3,1}^{\mathbf{C}}$ and, as there exists an infinite set of idempotents giving the same ideal as the isotropic 2-vector $y_1 y_2$, the idempotent method and the author's 2-isotropic method are not at all equivalent : only the latter is a 'geometric' method.

20.5   Amorphic spinor bundles.

Before entering this subject it is important to note that Clifford algebras possess two kinds of properties : *they are semi-simple (or simple) associative algebras (and idempotents are connected with this property), but they also are 'geometric algebras' arising from an orthogonal space (the spinor group comes from that).* The bundle problems are tied to G or $O(p, q)$ actions and are therefore geometric, so we work with the geometric inner automorphism (G.I.A.) preserving globally the standard space E.

20.5.1   Minimal left ideal bundles.

Let V or $V_{p,q}$ be a connected pseudo-riemannian manifold with a regular pseudo-metric of signature (p, q).

There exists a vector bundle $\mathrm{Clif}(V)$ whose fibers are isomorphic to $C_{p,q}$, with structural group $\mathrm{Clif}(Q)$ (or $\mathrm{Clif}(p,q)$), natural extension of $O(Q)$ (or $O(p,q)$) or natural extension of the improper inner action if G or $\mathrm{Pin}(Q)$.

We consider a vector subbundle of which the fibers are minimal left ideals of the $\mathrm{Clif}(V)$ fibers.

The typical fiber for that subbundle is a minimal left ideal of the standard algebra $C_{p,q}$, but note that there does not exist any left action of G or $\mathrm{Pin}(Q)$, or $\mathrm{Spin}(Q)$, on the fibers, there only exists an improper inner action for these groups. This subbundle is called an *amorphic spinor bundle* (A.S.B).

If we call σ_e a *geometric Cliffordian spinoriality group* (G.C.S.G.), this projection (non skew) onto $O(p,q)$, $p(\sigma_e)$ will be called an *orthogonal spinoriality group* (O.S.G.) (the primitive idempotent e defines a standard minimal ideal in $C_{p,q}$).

A standard result about the fibrations gives the statement :

Proposition 20.5.1 *There exists an A.S.B. with standard fiber Ae if and only if the structural group $O(p,q)$ of the tangent bundle $T(V)$ is reducible to the orthogonal spinoriality group $p(\sigma_e)$.*

Remark 5. Note that this definition is, a priori, completely different from the one about 'spinor structures' associated to a covering of the bundle of pseudo-orthogonal frames. No proper action of a group such as $\mathrm{Pin}(Q)$ or $\mathrm{Spin}(Q)$ is involved.

On the contrary, the 'maximal isotropic subspace' method, giving also bundles of minimal left ideals, and developed by the author in [1] (or Chapter 14 above), is strongly related to the covering principal bundle approach.

If $\beta(e) = e$ or $\tilde{\beta}(e) = e$ or $\beta'(e) = e$, i.e. if e is invariant under an anti-involution which commutes with the structural group, the group σ_e is the stability group of e (Proposition (20.3.2)) and we have :

Proposition 20.5.2 *If the primitive idempotent e determining the standard fiber Ae is invariant under an anti-involution commuting with the structural group of $T(V)$, the existence of an A.S.B. modeled by Ae is equivalent to the existence of a global field of primitive idempotents in the Clifford bundle.*

Here we can ask whether, in the Clifford algebra A, there exists a primitive idempotent e' and an anti-involution γ commuting with the $p(G)$-action (if n is even) or the $p(G^+)$-action (if n is odd), such that $\gamma(e') = e'$. We can also consider some subgroup of G or G^+.

Remark 6. A priori, the existence of an A.S.B. does not imply the existence of a field of primitive idempotents patterned by some standard primitive idempotent e.

20.5.2 Examples of A.S.B.

a) First consider the space $E_{1,3}$.

If we choose the standard idempotent $h = (1 + e_0)/2$, $\beta(h) = h$, $J \cap J' = J = J'$ and σ_h is the stability group of h, σ_h preserves e_0 and preserves globally the space (e_1, e_2, e_3). If there exists on $V_{1,3}$ an A.S.B. modeled by h, the structural group of $T(V)$ reduces to $O(3)$ and there exists a 'time-like' vector field and a primitive idempotent field.

If we choose now the idempotent $h_2 = (1 + e_1 e_2 e_3)/2$, $\tilde{\beta}(h_2) = h_2$, the group σ_{h_2} preserves h_2 and $e_1 e_2 e_3$, the existence of an A.S.B. implies a space orientation for $V_{1,3}$ and the reduced structural group is $O(1) \times SO(3)$.

Finally, we take $h_1 = (1 + e_0 e_3)/2$, $\beta'(h_1) = h_1$, β' is an anti-involution of which the type was defined in subsection 4.4, compatible with the group $O(1) \times O(3)$. If we have an A.S.B. there exists a field of hyperbolic planes (corresponding to $e_0 e_3$) and if we suppose that $V_{1,3}$ is orientable, $p(\sigma_{h_1})$ admits the Lie algebra determined by $e_1 e_2$, the structural group of $T(V)$ is reducible to S_1.

b) Consider the space $E_{3,1}$ and the primitive idempotent

$$h = \tfrac{1}{4}(1 + e_0)(1 + e_2 e_3) = \tfrac{1}{4}(1 + e_0)(1 + e_0 e_1 e_3).$$

Taking into account the reduction to $O(1) \times O(3)$, there exists an anti-involution β such that $\beta(h) = h$ and $p(\sigma_h)$ considered here in $O(1) \times O(3)$ is the stability group of h.

If we suppose that $V_{3,1}$ is orientable (we want to use our previous results) we obtain $p(\sigma_h) = \mathrm{Id}$.

The existence of an A.S.B. modeled on h implies the triviality of the tangent bundle. (We find again this result using $h' = (1 + e_1 + e_3)h$, $v = ae_0(e_1 + e_3) + (a/2)e_3 e_1 + k$ is a general element of the Lie algebra of the stability group; $k = 0$ is possible, $a = 0$ is necessary.)

These examples prove that if V carries a pseudo-riemannian structure :

Proposition 20.5.3 *The existence of an A.S.B. over V is not connected with the intrinsic topological properties of the manifold V.*

20.5.3 Connections over A.S.B. : amorphic spin-euclidean connections (A.M.S.E. connections).

First suppose that there exists a standard idempotent with $\beta(e) = e$, and choose Ae as a standard fiber. We have, according to classical terminology, a $p(\sigma_e)$-structure in the bundle of $T(V)$-frames and, according to our results above, a bundle of minimal left ideals determined by a global field of idempotents : abusively we note this field $x \to e_x$ and $\beta(e_x) = e_x$, $\forall x \in V$.

If $u \in \sigma_e$, $ueu^{-1} = e$ and v belongs to the Lie algebra $\mathcal{L}(\sigma_e)$ if and only if

$$ve - ev = 0.$$

Still according to classical results there exists a principal connection ∇ with a $\mathcal{L}(\sigma_e)$-valued form w such that $e_x w(X_x) - w(X_x)e_x = 0$ (A.M.S.E. connection) (abusively $ew(X) - w(X)we = 0$) for any vector field X over V.

Proposition 20.5.4 *If* $e = \beta(e)$, $X \in D^1(V)$, $\nabla_X e = 0$ *if* $e : x \to e_x$ *is the fundamental field of idempotents and* ∇ *the principal* $p(\sigma_e)$*-connection.*

Proof. Indeed, let γ be a curve with $(d\gamma/dt)_{t=0} = X(\gamma(0))$ and $(e_1, e_2, \ldots, e_n)_{\gamma(t)}$ an orthonormal frame obtained by ∇-parallel translation of $(e_1, e_2, \ldots, e_n)_{\gamma(0)}$ along γ. If T is any vector field and if $T(\gamma(t)) = T^\alpha(\gamma(t))e_\alpha(\gamma(t))$, it is classical that

$$(\nabla_X T)_{\gamma(0)} = (\frac{dT^\alpha}{dt})_{t=0}(e_\alpha)_{\gamma(0)}$$

and our connection ∇ extends naturally to the Clifford algebra. If ∇ preserves the amorphic spinor space field, the translated of $(e)_{\gamma(0)}$ is $(e)_{\gamma(t)}$ and $\nabla_X e = 0$. ∎

Remark 7. We have the same result with $\tilde{\beta}$ and β'; the last case supposing a previous reduction of the structural group.

Now consider the case where we do not know if there exists a primitive idempotent field over V.

If $u \in \sigma_e$, $ueu^{-1} = e'$, $eu = e\lambda^{-1} = \lambda^{-1}e$, $\lambda \in K^*$, if e is the selected standard idempotent, in the notations of Proposition (20.2.1).

In the Lie algebra $\mathcal{L}(\sigma_e)$:

$$ev = eb = be,$$
$$ve - ev = (v - b)e.$$

Nevertheless, we can locally define a field $x \to e_x$, the local fiber is $A(x)e_x$. Suppose that $x = \gamma(t)$; the translated of $(e)_{\gamma(t)} = E_{\gamma(t)}$ are such that

$$(e)_{\gamma(t)} = \theta_{\gamma(t)} E_{\gamma(t)},$$

and we have (abusively)

$$\nabla_X e = W(X)e$$

where $W(X)$ is a local Clifford field.
$\theta_{\gamma(t)} = u(\gamma(t))\lambda(\gamma(t))$ and the derivative for $t = 0$ gives $W = u'(0) + \lambda'(0)$.

Remark 8. $\nabla_X e = W(X)e$ implies $e\nabla_X e = 0$. From $e^2 = e$ we obtain :

$$\begin{aligned}\nabla_X e^2 &= (\nabla_X e)e + e(\nabla_X e) = \nabla_X e \\ &= W(X)e + e\nabla_X e = W(X)e\end{aligned}$$

hence $e\nabla_X e = 0$.

Conversely, if $e \nabla_X e = 0$, as

$$\nabla_X(\psi e) = \nabla_X(\psi e^2) = \psi(\nabla_X e)e + \psi(e\nabla_X e) + (\nabla_X \psi)e,$$

$$\nabla_X(\psi e) = W(X)e, \quad \forall \psi.$$

One finds this result in [2].

Note that this property only has a local meaning, $x \to e_x$ being local.

It appears that the best tool for the definition of amorphic spinor bundles is not the idempotent field but the geometric Clifford spinoriality groups σ_e. Also, the global existence of fields of primitive idempotents greatly constraints the topology of the manifold and may depend on the standard idempotent e, according to subsection 4.3.

20.5.4 Spinor fields and exterior algebra fields.

A Clifford algebra is linearly isomorphic to an exterior algebra, but there is, unfortunately, no canonical isomorphism[3]; so we must stress the conventional character of any particular isomorphism. The most 'natural' identification consists in the choice of orthonormal frames, so that it becomes possible to speak of the homogeneous components of a Clifford number; the transition functions of the tangent or cotangent bundles have values in the orthogonal group and allow a linear identification of the Clifford bundle and the exterior bundle.

But what about the amorphic spinor bundles and the exterior bundle ? Can one somehow identify a spinor field with an exterior algebra field ?

If we consider an amorphic spinor field carrying the orthogonal (improper) representation of the reduced group σ_e, there is no problem, but we are confined to the reduced group.

Physicists usually consider ordinary spinor fields carrying the faithful $\text{Pin}(Q)$ or $\text{Spin}(Q)$ representation; after convenient reduction of the $\text{Pin}(Q)$ group to a subgroup H, the identification could be possible and coherent, but we will be restricted to the group H (cf. Chapters 11 and 14). With the $\text{Pin}(Q)$ group, the identification between ordinary spinor fields and elements of minimal left ideals, endowed with the faithful representation of $\text{Pin}(Q)$ by left multiplication, is not coherently possible.

The following example clarifies the situation : consider the space $E_{3,1}$ with primitive idempotent $e = \frac{1}{4}(1 + e_0)(1 + e_1 e_2)$ (cf. above), and the spinor :

$$\psi = (\alpha + \beta e_0 e_2 + \gamma e_1 e_2 + \delta e_0 e_1)e,$$

$e_0 \in \text{Pin}(Q)$, the improper orthogonal action of $\text{Pin}(Q)$ gives :

$$e_0 \psi e_0^{-1} = \psi, \quad (e_0^{-1} = e_0),$$
$$e_0 \psi = (\alpha + \gamma e_0 e_1 e_2)e$$

---

[3]There are infinitely many Kähler-Atiyah isomorphisms, cf. Chapter 3, 1.5.

and there is no equivariant action respecting the identification.

We summarize the situation as follows :

- The identification between Clifford algebra fields and exterior algebra fields is convenient for orthonormal frames (different conventions lead to inextricable checking).

- The identification between amorphic and ordinary spinor fields restricts us to some spinoriality group and $\text{Pin}(Q)$ or $\text{Spin}(Q)$ spinors in the ordinary meaning are not exterior form fields.

- A lot of spinoriality groups appear (or may appear) with amorphic spinor bundles of which the existence conditions have no intrinsic topological meaning for the manifold under consideration (a lot of covariance groups are possible in the same algebraic context).

Finally, we examine the possible identification between the de Rham laplacian and the Dirac laplacian.

We consider here the pseudo-riemannian cotangent bundle $T^*(V)$, (E^α), $\alpha = 1, \ldots, n$ is a local orthonormal frame.

To the p-form θ with components $\theta_{\alpha_1\alpha_2\ldots\alpha_p}$ we associate the Clifford algebra field

$$E^{\alpha_1} E^{\alpha_2} \ldots E^{\alpha_p} \theta_{\alpha_1\alpha_2\ldots\alpha_p}$$

and if we are working with a convenient amorphic spinoriality group the identification extends to the spinors.

The checking of the Dirac laplacian is easier if we use the E^α as the components of the 'fundamental tensor-spinor' γ $(\nabla\gamma = 0)$; if we introduce

$$E^\alpha \nabla_\alpha \psi = P(\psi)$$

we easily obtain

$$P^2(\psi) = E^\alpha E^\beta \nabla_\alpha \nabla_\beta \psi = \Delta\psi,$$

Δ is the Dirac laplacian.

But

$$E^\alpha \nabla_\alpha \psi = E^\alpha \wedge \nabla_\alpha \psi + i_{E^\alpha}(\nabla_\alpha \psi),$$

(i_{E^α} is the inner product) and according to an elementary result :

$$E^\alpha \nabla_\alpha \psi = d\psi - \delta\psi,$$

(∇ is a torsion-free connection) where d is the usual exterior derivative and δ the coderivative

$$P^2(\psi) = (d - \delta)^2 \psi = -(d\delta + \delta d)\psi,$$

because $d^2 = \delta^2 = 0$.

If we put $\Delta_{DR} = -(d\delta + \delta d)$, Δ_{DR} is the de Rham laplacian and

$$P^2\psi = \Delta\psi = \Delta_{DR}\psi,$$

where we can identify the spinor field ψ with an exterior algebra field.

Remark 9. There exists a close relation between this situation and some developments in electromagnetism theory where both laplacians are used. In [1, e] however, we are using the isotropic approach instead of the idempotent approach and we have a special case of the last situation by a translation of some results.

20.6   Epilogue.

It appears that amorphic spinor structures lead to many geometric difficulties, particularly if one uses primitive idempotent fields on manifolds. Mathematicians are free to introduce and to study these intricate situations, but what about the meaning of physical theories constructed from them ? Amorphic spinor fields possess a 'chaotic' and 'exotic' character. What is the advantage ? Some people would like to work with differential forms, and these are a really nice tool, but we have proved in this paper that some identifications give rise to big obstacles; differential forms and Pin(Q)-spinors are very different objects, both locally and globally, so we should not confuse them.

It is a very good idea to consider spinors as elements of minimal left ideals in Clifford algebras, but as complex vector spaces are essential in physical theories, the best point of view is to consider complexified vector spaces : then it is possible to preserve particular properties coming from the signature (the converse is a popular idea, but a wrong one, cf. the author's theory of conjugation, Majorana spinors and sesquilinear forms over spinor space [1, d], [1, g] and Chapter 10). In this complexified context the 'r-isotropic' method developed by the author is not only an algebraic method but chiefly a geometric one : it gives a field of left ideals as in the idempotent approach, but these spinor spaces are closely connected with coverings of the principal tangent (or cotangent) frame bundles (any pseudo-riemannian bundle will do). Then the best definition of spinor structures in the wider sense can be given.

If we want to obtain the real case, we introduce Majorana spinors covariant under the Clifford group action (G-action) if $p - q = 0, 2\,(\text{mod}\,8)$, or under the special Clifford group (G^+) action if $p - q = 0, 1, 2, 6, 7\,(\text{mod } 8)$. (cf. [1, g] and Chapter 10). We then have the covariance according to the Lorentz action and the theory satisfies the 'relativistic credo'.

Except if one admits the philosophical idea of a 'chaotic physics', it is reasonable to drop the idempotent approach, but are physicists and mathematicians reasonable people ?

20.7   Complements.

20.7.1

a) *We are looking for the anti-involutions γ of the real algebra $C(Q)$ which commute with the orthogonal inner action of the Clifford group G, or some subgroup G_1 of G, if $n = \dim E$ is even.*

$\forall g \in G, \forall x \in C(Q)$ we have

$$\gamma(gxg^{-1}) = g\gamma(x)g^{-1}$$
$$\gamma(x) = \gamma(g)g\gamma(x)g^{-1}\gamma(g^{-1}).$$

As $\gamma(x)$ is any element y of the central algebra $C(Q)$, necessarily $\gamma(g)g \in \mathbf{R}^*$.

If we choose $g = x \in E$, $x^2 \neq 0$, then $x \in G$,

$$\gamma(x)x = \hat{\gamma}, \quad \hat{\gamma} \in \mathbf{R}^*,$$
$$\gamma(x) = \mu x, \quad \mu \in \mathbf{R}^*,$$

$\gamma^2 = \mathrm{Id}$ implies $\gamma(x) = \pm 1$, so the eigenvalues of γ are ± 1. If $\gamma(x) = x, \forall x \in E$, we obtain the involution β. If $\gamma(x) = -x, \forall x \in E$, we obtain the involution $\tilde{\beta}$.

If there exist two proper subspaces E_1, E_2, eigenspaces for 1 and -1 :

$$x = x_1 + x_2, \quad x_i \in E_i, \quad i = 1, 2.$$

If γ and g commute :

$$\gamma g(x_1) = g(x_1), \quad g(x_1) \in E_1,$$
$$\gamma g(x_2) = -g(x_2), \quad g(x_2) \in E_2,$$

g preserves E_1 and E_2, but if g can be any element of G, this case is impossible.

In even dimension, there exists in $C(Q)$ only the anti-involutions β and $\tilde{\beta}$ commuting with the orthogonal action of G.

Nevertheless, if we consider a subgroup G_1 whose elements preserve E_1 and E_2, if γ preserves E, γ commutes with the orthogonal action of G_1 only if G_1 is a subgroup preserving the direct decomposition $E = E_1 \oplus E_2$ and γ reduces to β or $\tilde{\beta}$ over each E_i (cf. subsection 4.4 above).

Note that if we consider $G^+ \subset G$, the preceding method fails (if $x \in G$, $x \notin G^+$).

b) *We are looking for γ which commutes with the G^+ action in $C(Q)$, $n = 2r$.*

$C(Q)$ is simple, so any anti-involution is of the form :

$$\gamma(x) = \beta(bxb^{-1}), \quad b \in C^*(Q),$$

and $\gamma^2 = \mathrm{Id}$ implies $\beta(b) = \pm b$. The commutation with any $g \in G^+$ is equivalent to $bg = \pm gb$, [4] (we choose $g \in \mathrm{Spin}(Q)$, one sees that the condition is admissible).

---

[4]Here we use $\beta(g) = \pm g$; notice that $\beta'(g) = \pm g$ can be wrong if β' is defined as in subsection 4.4 ! But $\beta'(g) = \pm g$ does hold if $g \in G_1$.

Notice that if g is any element of G^+,

$$bg = gb;$$

in fact, if we consider an orthonormal frame $(e_1, e_2, \ldots, e_n)$ in E, if $k \neq 0, 1, -1$, $e_2 + ke_3$ is non-isotropic; also if

$$be_1e_2 = \epsilon e_1 e_2 b, \quad \epsilon = \pm 1$$

and $kbe_1e_3 = -\epsilon ke_1e_3b$, we obtain the contradictory result :

$$be_1(e_2 + ke_3) = \epsilon e_1(e_2 - ke_3)b.$$

Hence $bg = \epsilon gb$, $\epsilon = 1$ or (-1), ϵ is the same for any g; but $bg = -gb$ is impossible (consider $be_1e_3 = -e_1e_3b$ and $be_2e_3 = -e_2e_3b$, implying $be_1e_3 = e_1e_3b$).

If $b \in C^+(Q)$, b belongs to the center of $C^+(Q)$, $b = \lambda + \mu e_N$, $\lambda, \mu \in \mathbf{R}$, $e_N = e_1e_2 \ldots e_n$. If $b \in C(Q)$, $b = b_0 + b_1$, $b_0 \in C^+(Q)$, $b_1 \in C^-(Q)$. $(b_0 + b_1)g = g(b_1 + b_0)$ implies $g_0b = bg_0$ and $b_1g = gb_1$. $b_0 = \mathbf{R} \oplus \mathbf{R}e_N$, we prove that b_1 vanishes :

$$b_1 = \sum \lambda^{i_1 \ldots i_m} e_{i_1} e_{i_2} \ldots e_{i_m},$$

m even. If $e_{k_1} e_{k_2} \ldots e_{k_m} \neq 1$, there exists a e_i which anticommutes with this term, say $e_{K_m} = e_i$. $e_ib_1 - b_1e_i = 0$ implies :

$$2\lambda^{K_1 K_2 \ldots K_m} e_i e_{K_1} \ldots e_{K_m} + \sum \mu^{j_1 \ldots j_l} e_{j_1} \ldots e_{j_l} = 0$$

where the $\sum$-block does not contain $e_i e_{K_1} \ldots e_{K_m}$, then $\lambda^{K_1 \ldots K_M} = 0$ for any non-empty sequence $(K_1, K_2, \ldots, K_m)$. Finally $b = \lambda + \mu e_N$ and since we have

$$\beta(e_N) = (-1)^r e_N,$$

if r is even, $\gamma = \beta$ or $\tilde{\beta}$; if r is odd, $\gamma = \beta$ or $\tilde{\beta}$ or another case is possible, using invertible elements $b = \lambda + \mu e_N$ with $\lambda\mu \neq 0$. This last class of anti-involutions do not preserve E.

Remark. Method b) also works in case a) and gives $bx = -xb$, $x \in E$, $x^2 \neq 0$, $b = e_N$ if $b \notin \mathbf{R}^*$.

If we choose $b \in C^+(Q)$ with $\beta(b) = \pm b$, the subgroup G_1 of G given by the set of $g \in G$ such that $bg = \pm gb$, is the maximal subgroup of G commuting with $\gamma : \gamma(x) = \beta(bxb^{-1})$. For example, the $\tilde{\beta}'$ anti-involution subsection 4.4 corresponds to $b = e_1e_2 \ldots e_p$, G_1 preserves $(e_1, e_2, \ldots, e_p)$, hence also the orthogonal subspace $(e_{p+1}, \ldots, e_n)$.

Hence the choice of $b \in C^*(Q)$ determines the γ-commuting group G_1; on the contrary, the choice of G_1 can determine several γ (cf. G^+ above).

20.7.2

Now consider the odd-dimensional case, $n = 2r + 1$, and the simple central algebra $C^+(Q)$, the anti-involutions γ commuting with the action of G^+.

β and $\tilde{\beta}$ are identical, and we define another anti-involution γ according to $\gamma(x) = \beta(bxb^{-1})$. The case b) applies, but $C^+(Q)$ is central and $b \in \mathbf{R}^*$.

In odd dimension, there exists in $C^+(Q)$ only the anti-involution $\beta = \tilde{\beta}$ commuting with the orthogonal action of G^+.

The remark at the end of the first subsection extends naturally to the odd-dimensional case.

20.7.3

We are looking for a pair (γ, e) of a primitive idempotent e and an anti-involution γ, such that $\gamma(e) = e$ for a given $C(Q)$, $n = 2r$ and E is a real space.

We put : $\beta(e) = beb^{-1}$, $\gamma(x) = \beta(bxb^{-1})$; $\gamma^2(x) = x$ implies

$$\beta(b^{-1})bxb^{-1}\beta(b) = x, \quad \forall x \in C(Q)$$

and the necessary condition

$$\beta(b) = \epsilon(b), \quad \epsilon = \pm 1,$$
$$\gamma(e) = \beta(beb^{-1})$$

but $\beta(e) = beb^{-1}$ implies $e = \beta(b^{-1})\beta(e)\beta(b) = \beta(beb^{-1})$ and $\gamma(e) = e$.

Conversely, if $\gamma'(e) = e$, γ' an anti-involution, there exists a $b' \in C^*(Q)$ with $\gamma'(x) = \beta(b'xb'^{-1})$, $\gamma'(e) = e$ implies $\beta(b'eb'^{-1}) = e$ and $\beta(e) = b'eb'^{-1}$, $\gamma'^2(x) = x$ implies $\beta(b') = \epsilon b'$ and we can state :

Any anti-involution γ, such that $\gamma(e) = e$, is defined according to $\gamma(x) = \beta(bxb^{-1})$, if b satisfies $\beta(e) = beb^{-1}$ with $\beta(b) = \epsilon b$, $\epsilon = \pm 1$.

Moreover we have the remarkable property :

If there exists such a pair (γ, e), (γ', e') has the same property, if e' is any primitive idempotent in $C(Q)$.

$$e' = ueu^{-1},$$
$$\beta(ueu^{-1}) = \beta(u^{-1})\beta(e)\beta(u) = \beta(u^{-1})bu^{-1}(ueu^{-1})ub^{-1}\beta(u),$$
$$\beta(e') = b'e'b'^{-1}$$

with $b' = \beta(u^{-1}bu^{-1}u)$,

$$\beta(b') = \epsilon\beta(u^{-1})bu^{-1} = \epsilon b',$$
$$\gamma'(x) = \beta(b'xb'^{-1}), \quad \gamma'(e') = e'.$$

It is sufficient to examine a particular primitive idempotent e_1 in $C(Q)$. If $\beta(e_1) = be_1b^{-1}$ with $\beta(b) = \epsilon b$, $\gamma(e_1) = e_1$ and $\gamma'(e') = e'$.

Note that b such that $\beta(e) = beb^{-1}$ is not unique. The subgroup (proper or improper) G_1 of $\text{Pin}(Q)$ is defined according to the supplementary condition

$$bg = \pm gb, \quad \forall g \in G_1.$$

It depends on $b, b', \ldots$ or on $e, e', \ldots$ associated to $\gamma, \gamma', \ldots$.

If e belongs to Lounesto's list it is easy to see that $\beta'(e) = e$. Then there exist γ' such that $\gamma'(e') = e'$ for any idempotent e'.

20.8   Selected references.

1. A. Crumeyrolle

    (a) *Structures spinorielles*, Ann. Inst. H, Poincaré, Vol. XI, n. 1, 1969, pp. 19–55.

    (b) *Spin fibrations over manifolds and generalized twistors*, Proc. Symp. in pure mathematics, A.M.S. Stanford, 1973, Vol. 27, part I.

    (c) *Fibrations spinorielles et twisteurs généralisés*, Period. Math. Hungarica, Vol. 6(2) 1975, pp. 143–171.

    (d) *Algèbres de Clifford et spineurs*, Cours Fac. Sc. de Toulouse, 1974.

    (e) *Théorie d'E.D. en spin maximum 1*, Ann. Inst. H. Poincaré, Vol. 22, n. 1, 1975, pp. 43–61.

    (f) *Schémas d'E.D. en spin 1/2*, ibid., Vol. 23, n. 3, 1975, pp. 259–274.

    (g) *Conjugation in spinor spaces—Majorana and Weyl spinors*, Proc. Winter School, 1986, Srni (CSSR).

2. W. Graf, *Differential forms as spinors*, Ann. Inst. H. Poincaré, Vol. 24, n. 1, 1978.

3. P. Lounesto, *On primitive idempotents of Clifford algebras*, Helsinki 1977, Report HTKK, A 113.

Chapter 21

SELF-DUAL YANG-MILLS FIELDS AND THE
PENROSE TRANSFORM IN THE SPINOR CONTEXT.

We use particular properties of spinor geometry in dimension 4, in order to give a new approach for the introduction of self-dual connections in Yang-Mills theory (instantons) and the Penrose transform. Our approach essentially uses pure spinors and associated totally isotropic spaces; twistors are not employed. Our method shows that the Penrose transform is nothing but a far and rather hidden resurgence of results of E. Cartan. We finally indicate a possible generalization of the instantons, called by us 'spintantons', which is connected to our definition of the enlarged spinor structures. *Here we consider C^∞ structures only.*

21.1 Introduction and basic notions.

21.1.1 Yang-Mills fields ([8, 6, 5]).

The simplest Maxwell theory makes use of a skew-symmetric covariant tensor field F in the Minkowski space $E_{1,3}$:

$$F = F_{\alpha\beta}(\theta^\alpha \wedge \theta^\beta)$$

which is assumed to satisfy the equations for a non-inductive medium :

$$\nabla_\beta F^{\alpha\beta} = J^\alpha \tag{1}$$

$$dF = 0 \tag{2}$$

where J is the electric current and ∇ is the pseudo-riemannian connection.

Locally, $F = dA$, $A = A_\alpha\theta^\alpha$ and we can use $A - \partial_\alpha\psi$ instead of A, where ψ is a scalar function.

From Geometrodynamics Proceedings, 1985, World Scientific Publish. Singapore—edited by A. Prastaro.

If we consider a principal bundle with U(1) as a structural group over $E_{1,3}$ and a section $x \to u(x)$, then $x \to \exp(i\varphi(x)) \cdot u(x)$ is also a section. If A defines a connection ω, the components A'_α, A_α are related by

$$A'_\alpha = A_\alpha + i\partial_\alpha\varphi = A_\alpha - \partial_\alpha\psi$$

with $\psi = -i\varphi$. The curvature form is $\Omega = d\omega + \frac{1}{2}[\omega, \omega] = dA$ and (1), if $J = 0$, can be expressed as $d(*\Omega) = 0$ (* is the Hodge operator for the volume form $\eta_{\alpha\beta\gamma\delta}$). We can obtain (1) from a variational principle using the Lagrangian $\mathcal{L} = \frac{1}{4}F_{\alpha\beta}F^{\alpha\beta}$, varying the A_α.

The Yang-Mills fundamental idea was to change U(1) into SU(2).

More generally, one considers at present a principal bundle over the manifold V ($E_{1,3}$ or the Einstein space-time $V_{1,3}$ as well) with structural group G, ω is a principal connection, with curvature form Ω.

If σ, σ' are two sections such that

$$\sigma'(x) = \sigma(x)g(x), \quad x \in V, \quad g(x) \in G,$$

then

$$\omega' = \mathrm{Ad}\,(g^{-1})\omega + g^{-1}dg,$$
$$\omega = (A^I_\alpha\theta^\alpha)G_I,$$

G_I is a frame in the Lie algebra $\mathcal{L}(G)$,

$$\Omega = \Omega^I_{\lambda\mu}G_I(\theta^\lambda \wedge \theta^\mu).$$

Usually G is a linear group (real or complex). In the real case one takes the lagrangian

$$\mathcal{L} = \frac{1}{4}\Omega^a_{b\lambda\mu}\Omega^{b\lambda\mu}_a,$$

varying the $A^a_{b\lambda}$ components of ω, the Y.-M. equations appear to be

$$\nabla_\mu\Omega^{a\lambda\mu}_b + [A_\lambda, \Omega^{\lambda\mu}]^a_b = 0 \tag{3}$$

(where the covariant derivation only involves λ and μ). Often one takes into account the vector-valuedness of the curvature and (3) is translated to

$$D(*\Omega) = 0. \tag{4}$$

With a non-linear group, we choose

$$\mathcal{L} = \frac{1}{4}g^{\alpha\lambda}g^{\beta\mu}B(\Omega_{\alpha\beta}, \Omega_{\lambda\mu})$$

where g is the fundamental metric and B is the Killing form. If the group G is semi-simple, B is non-degenerated and if G is compact, B is a negative definite form.

Mathematicians often prefer to write

$$\mathcal{L} = -\tfrac{1}{4}||\Omega||^2$$

or $-\tfrac{1}{4}||\Omega||^2 + L(\varphi, D_\mu\varphi)$ if another field φ interacts with Ω.

In the physicist's jargon : the connection becomes the 'gauge potential', the curvature becomes the 'gauge field', G is the 'gauge group' and an isomorphism of the principal bundle a 'gauge transformation'. If we choose different cross-sections, a change of cross-sections is a 'gauge change'.

21.1.2   Self-dual and anti self-dual fields.

V is a C^∞ riemannian or pseudo-riemannian oriented manifold with $n = 2k$ dimensions and $\wedge^p$ (resp. $\Gamma(\wedge^p)$) the bundle of p-forms (resp. the space of p-form cross-sections). The star (or Hodge) operator $*$ is defined by

$$\alpha \wedge (*\beta) = (\alpha/\beta)\eta$$

where (α/β) is the scalar product of the differential forms, η the n-volume form or the orientation :

$$\eta = \epsilon\sqrt{|g|}dx^1 \wedge dx^2 \wedge \ldots \wedge dx^n$$

locally, $\epsilon = \pm 1$;

$$* : \wedge^p \to \wedge^{n-p}.$$

In practice, with the local components :

$$(*\alpha)^{i_{p+1},\ldots,i_n} = \eta^{i_1,\ldots,i_n}\alpha_{i_1,\ldots,i_p}$$

(modulo a normalization coefficient). We recall the classical results :

$$* * (-1)^{p(n-p)} = \begin{cases} \text{Id}, & \text{in the riemannian case,} \\ \epsilon_1 \text{ Id} & \text{in the pseudo-riemannian case,} \end{cases}$$

where $\epsilon_1 = 1$ if $\det g > 0$ or (-1) if $\det g < 0$.

If $k = 2$, $*^2 = \text{Id}$ in the riemannian case and $-\text{Id}$ in the minkowskian case, and $p = 2$, giving the direct sum

$$\wedge^2 = \wedge^2(+) \oplus \wedge^2(-)$$

corresponding to the spectral decomposition for the star operator.

The bundle $\wedge^2$ is the direct sum of self-dual and anti self-dual forms.

Note that $*$ is conformally invariant.

It is then natural to consider connections with curvature form Ω such that $*\Omega = \pm\Omega$ in the riemannian case, or $*\Omega = \pm i\Omega$ in the minkowskian case. The Y.-M. condition $D(*\Omega)$ is then automatically satisfied, $D\Omega = 0$ being the Bianchi identity.

21.1.3   Instantons ([1, 5]).

Some physicists are interested in connections giving a lagrangian with compact support so that the action integral is always defined over any domain with boundary in $\mathbf{R}^4$. Sending $\mathbf{R}^4$ to S^4 by stereographic projection, one considers a C^∞ vector bundle (with complex rank r, base S^4) associated to a principal bundle with group SU(r) and provided with a connection whose curvature is self-dual : $*\Omega = \Omega$ (self-dual connection). This bundle is called an instanton. If $*\Omega = -\Omega$, the curvature is anti self-dual. A change of orientation exchanges both cases.

It is difficult to find a physical meaning for such a scheme (a pseudo-particle ?) except for changing t (the time), to it in the quadratic minkowskian form, but unfortunately many topological difficulties arise from this algebraic modification. Probably it would be better to consider the compactified Minkowski space isomorphic to $S^3 \times_{\mathbf{Z}_2} S^1$ and thus define the 'minkowskian instanton' over this last base ($*\Omega = i\Omega$).

A bundle with self-dual connection satisfies topological conditions, for example, for a hermitian bundle the first Pontryagin number is positive ([1]).

With elliptic signature self-dual connections give an absolute minimum for the Yang-Mills lagrangian : this property explains why, in spite of their artificial nature, many people are interested in instantons.

21.2   The Penrose transform as a resurgence of Cartan's work.

Grosso modo the Penrose transform consists in the one to one association between an instanton and a certain vector bundle over $P_3(\mathbf{C})$; usually the authors interested in this subject rely their approach on the rather fancy notion called 'twistor' ([7]) and use the Pauli-Dirac matrix formalism. This method appears rather as an intricate and heavy artifice. As a matter of fact, in the modern frame, and according to the ideas of E. Cartan ([2]), the Penrose transform is really like a far resurgence of the important notions described in ([2, 3, 4a] and Chapter 8 above) and called 'pure spinors'. In 4 and 6 dimensions any even or odd spinor is also a pure spinor ([4b]) and this remark is crucial for the following developments.

We choose the minkowskian signature, but our results are also valid for the elliptic signature.

21.2.1   Pure minkowskian and conformo-minkowskian spinors and associated spaces.

Our notations are like in the preceding paper [4b] and the chapters above.

$E_{1,3}$ is the Minkowski space with orthonormal frame (e_1, e_2, e_3, e_4), $(e_1)^2 = 1$, we define

$$x_1 = \frac{e_1 + e_4}{2}, \quad x_2 = \frac{ie_2 + e_3}{2}, \quad y_1 = \frac{e_1 - e_4}{2}, \quad y_2 = \frac{ie_2 - e_3}{2}$$

(a Witt frame) and $f = y_1 y_2$.

$uf = u^+ f + u^- f$ is a minkowskian spinor.

The maximal totally isotropic subspaces (m.t.i.s.) attached to $u^+ f$ and $u^- f$ (pure spinors) respectively, are constituted by the set of points $x \in E_{1,3}^{\mathbf{C}}$ such that $x(u^+ f) = 0$ or $x(u^- f) = 0$. Thus two families of m.t.i.s. appear (even and odd ones). If

$$u^- f = (ax_1 + bx_2)f \tag{1}$$
$$u^+ f = (a' + b'x_1 x_2)f \tag{2}$$

the m.t.i.s. are defined, according to [4b], by

$$x = (ax_1 + bx_2)s + (ay_2 - by_1)t, \tag{3}$$

for the odd m.t.i.s.,

$$x = (b'x_1 + a'y_2)s + (a'y_1 - b'x_2)t, \tag{4}$$

for the even m.t.i.s.

Two m.t.i.s. with different parity intersect and have an isotropic straight line in common ([3]).

The m.t.i.s. are bijectively attached to the pair of projective spinors represented by $(u^- f, u^+ f)$; this pair bijectively defines a homogeneous straight line in $E_{1,3}^{\mathbf{C}}$, ([3, 4a] and Chapter 12 above).

Now we consider $E_{2,4}$ and $E_{2,4}^{\mathbf{C}}$ with orthonormal frame

$$(e_0, e_1, e_2, e_3, e_4), \quad (e_0)^2 = 1, \quad (e_5)^2 = -1.$$

$$x_0 = \frac{e_0 + e_5}{2}, \quad y_0 = \frac{e_0 - e_5}{2}, \quad \hat{f} = y_0 y_1 y_2,$$

and projective spinors, with determined parity, are again in bijective correspondence with 3-complex m.t.i.s.

Let $\hat{u}^+ \hat{f} = (a_0 + a_1 x_0 x_1 + a_2 x_0 x_2 + a_3 x_1 x_2)\hat{f}$ be an even conformo-minkowskian spinor, the points ξ lying in the associated m.t.i.s. (π) are defined by :

$$\xi \hat{u}^+ \hat{f} = 0, \quad \xi \in E_{2,4}^{\mathbf{C}}$$

according to a classical result. If $\xi = \alpha x_0 + \beta x_1 + \gamma x_2 + \alpha' y_0 + \beta' y_1 + \gamma' y_2$, we obtain, if $a_0 \neq 0$,

$$\xi = (\frac{\beta' a_1 + \gamma' a_2}{a_0})x_0 + (\frac{\gamma' a_3 - \alpha' a_1}{a_0})x_1 - (\frac{\alpha' a_2 + \beta' a_3}{a_0})x_3 + \alpha' y_0 + \beta' y_1 + \gamma' y_2. \tag{5}$$

Now we introduce the injective image $\mathcal{E}$ of $E_{1,3}$ in $E_{2,4}$ by means of the 'isotropic transform' v (cf. Chapter 12, 2.1) :

$$v(x) = \frac{x^2 - 1}{2}e_0 + x + \frac{x^2 + 1}{2}e_5 = x^2 x_0 + x - y_0. \tag{6}$$

It is easy to verify that $\mathcal{E}$ is contained in the isotropic cone of $E_{2,4}$; modulo a negligible set this image represents the compactified Minkowski space isomorphic to $S^1 \times_{\mathbf{Z}_2} S^3$. v extends naturally to the complexified space $E_{1,3}^{\mathbf{C}}$, $\mathrm{Im}(v)$ becomes $\mathcal{E}'$.

(π) defined by (5) is contained in $E_{2,4}^{\mathbf{C}}$ and its intersection with it corresponds to $\alpha' = -1$. We call $(\pi_1) = \pi \cap \epsilon'$, (π_1) is the set of points such that :

$$\xi = \frac{\beta' a_1 + \gamma' a_2}{a_0} x_0 + \frac{\gamma' a_3 + a_1}{a_0} x_1 - \frac{\beta' a_3 - a_2}{a_0} x_2 + \beta' y_1 + \gamma' y_2 - y_0. \tag{7}$$

According to the isotropic transform the points of (π_1) arise from the points of a totally isotropic affine plane in $E_{1,3}^{\mathbf{C}}$, called (π_2) and defined by

$$x = \frac{\gamma' a_3 + a_1}{a_0} x_1 + \frac{a_2 - \beta' a_3}{a_0} x_2 + \beta' y_1 + \gamma' y_2 \tag{8}$$

and the parallel homogeneous plane (π_3) by

$$x = \beta'(-a_3 x_2 + a_0 y_1) + \gamma'(a_3 x_1 + a_0 y_2). \tag{9}$$

Remark. In these computations, we have assumed $a_0 \neq 0$. If $a_0 = 0$, we can assume that $a_3 \neq 0$ and obtain analogous results; (9) becomes

$$x = a + 3(\beta x_1 + \gamma x_2).$$

Finally, after the choice of a frame in $E_{2,4}$ and $E_{1,3}$, the set $\tilde{\Sigma}^+$ constructed by the projective even spinors for $E_{2,4}^{\mathbf{C}}$, that we can identify with the set of projective spinors for $E_{1,3}^{\mathbf{C}}$, defines the space $P_3(\mathbf{C})$, and the points corresponding to $a_0 = a_3 = 0$ constitute the space $P_1(\mathbf{C})$.

We write $\mathcal{O}(P_3(\mathbf{C})) = P_3(\mathbf{C}) \setminus P_1(\mathbf{C})$..

If we compare (9) and (4), we see that (π_3) is the more general even m.t.i.s. in $E_{1,3}^{\mathbf{C}}$, and we obtain :

Proposition 21.2.1 *After the choice of a frame in $E_{1,3}$ and $E_{2,4}$, we can associate to any element of $\mathcal{O}(P_3(\mathbf{C}))$ an even totally isotropic affine plane (π_1) in $E_{2,4}^{\mathbf{C}}$ and an even totally isotropic affine plane (π_2) in $E_{1,3}^{\mathbf{C}}$. Thus $\mathcal{O}(P_3(\mathbf{C}))$ is in one to one correspondence with the set of the even totally isotropic affine planes of $E_{1,3}^{\mathbf{C}}$.*

The same proposition holds for the odd planes.

Remark 2. We shall use below the elementary remark that an even affine plane in $E_{1,3}^{\mathbf{C}}$ is an equivalence class of pairs (x, π_3), $x \in E_{1,3}^{\mathbf{C}}$, π_3 an even m.t.i.s. in $E_{1,3}^{\mathbf{C}}$, where (x, π_3) and (x', π_3') are equivalent if and only if $\pi_3 = \pi_3'$ and $x - x' \in \pi_3$.

21.2.2   Self-dual connections and vector bundles.

We consider fibrations with a structural group acting effectively in the typical fiber. G_2^+ denotes the grassmannian of even m.t.i.s. in $E_{1,3}^C$. The fundamental 4-form of $E_{1,3}$ is defined by the dual frame of (e_1, e_2, e_3, e_4), written

$$\theta^1 \wedge \theta^2 \wedge \theta^3 \wedge \theta^4.$$

According to subsection 1.2 we say that a plane in $E_{1,3}^C$, defined by a decomposable 2-form Φ is self-dual (resp. anti self-dual) if

$$*\Phi = i\Phi \quad \text{resp.} \quad *\Phi = -i\Phi.$$

We write down that :

$$\theta^1 \wedge \theta^3 \xrightarrow{*} \theta^4 \wedge \theta^1 \xrightarrow{*} \theta^3 \wedge \theta^1,$$
$$\theta^3 \wedge \theta^4 \xrightarrow{*} \theta^2 \wedge \theta^1 \xrightarrow{*} \theta^4 \wedge \theta^3,$$
$$\theta^4 \wedge \theta^2 \xrightarrow{*} \theta^3 \wedge \theta^1 \xrightarrow{*} \theta^2 \wedge \theta^4.$$

Proposition 21.2.2 *Every even m.t.i.s. in $E_{1,3}^C$ is anti self-dual. Every odd m.t.i.s. in $E_{1,3}^C$ is self dual.*

Proof. Indeed, from (9), x is determined by the pair $-a_3x_2 + a_0y_1, a_3x_1 + a_0y_2$, hence, after a dual identification by the bivector :

$$\begin{aligned}\Phi = {} & i(a_0^2 + a_3^2)(\theta^1 \wedge \theta^2) + (a_3^2 - a_0^2)(\theta^1 \wedge \theta^3) + 2a_0a_3(\theta^1 \wedge \theta^4) \\ & + i(a_0^2 - a_3^2)(\theta^2 \wedge \theta^4) + 2ia_0a_3(\theta^2 \wedge \theta^3) - (a_0^2 + a_3^2)(\theta^3 \wedge \theta^4)\end{aligned}$$

defining the plane (π_3). $*\Phi = -i\Phi$ is immediate.   ∎

The same proof works with odd m.t.i.s.

Lemma 21.2.1 *$*\Omega = i\Omega$ is equivalent to Ω null over any even m.t.i.s. (π_3).*

Proof. We associate to (π_3) a conjugate plane $\overline{\overline{\pi_3}}$, using a conjugation taking into account the signature. It is easy to prove that $\overline{\overline{\pi_3}}$ and π_3 have the same parity. The planes (π_3) and $(\overline{\overline{\pi_3}})$ carry a 'real Witt frame' and the computation of Φ above shows that this form is a linear combination of

$$\theta^1 \wedge \theta^3 + i\theta^2 \wedge \theta^4, \quad \theta^1 \wedge \theta^4 + i\theta^2 \wedge \theta^3, \quad \theta^4 \wedge \theta^3 + i\theta^1 \wedge \theta^2$$

constituting an anti self-dual frame; supposing that Ω is null over (π_3) is equivalent to say that the three components of Ω over the anti self-dual frame are zero, then Ω is a self-dual form.   ∎

Coming back to the fibration problems, we put in correspondence $(x, \pi_3) \in E_{1,3}^C \times G_2^+$ and $\theta \in \mathcal{O}(P_3(\mathbf{C}))$, $\theta = \varphi(x, \pi_3)$, φ is only a surjective map; (ξ) being a vector bundle with $\mathcal{O}(P_3(\mathbf{C}))$ as a base, $\varphi^*(\xi)$ is a bundle over $E_{1,3}^C \times G_2^+$, with $\varphi^*(\xi)$ we can construct a bundle over $E_{1,3}^{\mathbf{C}}$ if and only if, for any points $z, z' \in \mathcal{O}(P_3(\mathbf{C}))$, $z = \varphi(x, \pi_3)$, $z' = \varphi(x, \pi_3')$, the fibers at x deduced from ξ_z and $\xi_{z'}$ are the same. It is then easy to see that such a z' describes any right line from z if z is fixed. (ξ) is trivial over any such line.

$\varphi^*(\xi)$ restricted to $E_{1,3}^{\mathbf{C}}$ has a natural parallelism between x and x', if $x - x' \in \pi_3$, for some even plane (π_3). According to standard results it is as much as to say that there exists a connection with null curvature form along the even m.t.i.s. (π_3). According to Lemma (21.2.1), the curvature is self-dual.

We can state :

Proposition 21.2.3 *A vector bundle over $\mathcal{O}(P_3(\mathbf{C}))$ inducing a trivial bundle over any right line defines also a vector bundle over $E_{1,3}^{\mathbf{C}}$ with the same rank, having a self-dual connection and satisfying the Y.-M. condition.*

Conversely, let (η) be a rank r complex vector bundle with structural group, over $E_{1,3}^{\mathbf{C}}$, with a self-dual connection.

We call trivializing distinguished open set $(\mathcal{U})$ of $E_{1,3}^{\mathbf{C}}$ an open set with a field of even affine m.t.i.s. such that

$$x \rightarrow \pi_2(x), \quad \eta/\mathcal{U} \simeq \mathcal{U} \times \mathbf{C}^r.$$

When x describes $(\mathcal{U})$ we can associate to it a point $z \in \mathcal{O}(P_3(\mathbf{C}))$ and z describes an open set $(\mathcal{V})$ in $\mathcal{O}(P_3(\mathbf{C}))$. At z we construct ξ_z isomorphic to the η_x fiber. Evidently we need an identification between ξ_z and $\xi_{z'}$ if

$$z = \varphi(x, \pi_2(x)), \quad z' = \varphi(x, \pi_2'(x)),$$

the constructed bundle is trivial over any straight line.

If $(x, \pi_2(x))$ and $(x', \pi_2(x'))$ give the same z, $\pi_3(x) = \pi_3(x')$ and $x' - x \in \pi_3(x)$, but the self-dual connection has a curvature form Ω which vanishes over $\pi_3(x)$, and we can identify η_x and $\eta_{x'}$, the construction of the fiber at z is coherent. The differentiability comes if we consider that we have a bundle over the set of equivalence classes (x, π_3), because η is null over any even (π_3) and the classes are identified with the points of $\mathcal{O}(P_3(\mathbf{C}))$. Then :

Proposition 21.2.4 *For any vector bundle, with a structural group, and $E_{1,3}^{\mathbf{C}}$ base, carrying a Y.-M. self-dual connection, we can associate a vector bundle over $\mathcal{O}(P_3(\mathbf{C}))$ with the same rank; this bundle is trivial over any straight line.*

Now we intend to prove that $\mathcal{O}(P_3(\mathbf{C}))$ can be replaced by $P_3(\mathbf{C})$ in Propositions (21.2.3) and (21.2.4).

Indeed, let $\hat{u}_0^+ \hat{f} \neq 0$ be a pure spinor, it is always possible to find an open set $(\mathcal{O})$ in the space of projective spinors, which contains $\hat{u}_0^+ \hat{f}$ and such that $(\mathcal{O})$ is in bijective correspondence with an open set $(\mathcal{O}_1)$ in $E_{1,3}^{\mathbf{C}} \times G_2^+$. To show this, we determine a new Witt frame, taking

$$\begin{cases} \sigma(y_i) = y_i, \\ \sigma(x_i) = \alpha_i^j y_j + x_i, \end{cases} \quad i = 0, 1, 2, \quad \alpha_i^j = -\alpha_j^i \tag{10}$$

a_0 then becomes

$$a_0' = a_0 + \alpha_0^1 a_1 + \alpha_0^2 a_2 + \alpha_0^3 a_3, \quad a_1' = a_1, \quad a_2' = a_2, \quad a_3' = a_3 \tag{11}$$

and because the a_i, $i = 1, 2, 3$ are not all zero we can obtain $a_0' \neq 0$.

This change of frames can be interpreted as a transformation by an element γ belonging to the reduced special Clifford group $G_0'^+ = \exp(\wedge^2 \hat{F})$, $\hat{F} = (x_0, x_1, x_2)$ (cf. [4a] and Chapter 8, 1.2),

$$\gamma \hat{u}^+ \hat{f} \gamma^{-1} = \gamma \hat{u}^+ \gamma^{-1} \hat{f}$$

because $G_0'^+$ is in the connected component of a spinoriality group ([4a]). γ transforms a neighborhood $(\mathcal{O})$ of $\hat{u}_0^+ \hat{f}$ with $a_0 = 0$ into an open set $\mathcal{O}_1$ where the new coefficient a_0 is different from 0.

If (ξ) is a vector bundle (with a structural group) over $P_3(\mathbf{C})$, naturally associated to a principal bundle, the construction of $\varphi^*(\xi)$ above is independent of the choice of a frame, modulo the change (10) : because this change replaces $\varphi^*(\xi)$ by $\varphi^*(\gamma^*(\xi))$, where γ^* is a principal bundle isomorphism. However, constructing a bundle over all of $P_3(\mathbf{C})$, coming from the bundle (η) above (cf. the proof of Proposition (21.2.4)) requires the gluing of bundles obtained over the four open sets corresponding to $a_0 \neq 0$ or, if $a_0 = 0$, $a_i \neq 0$, $i = 1, 2, 3$; if, for example, $a_2 \neq 0$ in a point of $P_3(\mathbf{C})$ reached within two frames R_0 $(a_0 \neq 0)$ and R_1 $(a_1 \neq 0)$ we have to change simply the coordinates of $\mathbf{C}^4 \times (G_2^+)$, X^1 for $(a_0 + \alpha_0^1 a_1 X^1)/a_0$, X^2 for $(a_0 + \alpha_0^1 a_1 X^2)/a_0$, Y^1 for Y^1 and Y^2 for Y^2 (in (11) $\alpha_0^1 \neq 0$, $\alpha_0^2 = \alpha_1^3 = 0$). This yields the gluing isomorphism. Finally, the bundles being considered through some covariant, permitted isomorphisms[2] we can state :

Theorem 21.2.1 *There is a natural one to one correspondence between :*

1. *complex vector bundles with structural group, base $P_3(\mathbf{C})$, trivial over every line of $P_3(\mathbf{C})$, and*

2. *bundles with the same rank over $E_{1,3}^{\mathbf{C}}$ having a self-dual connection.*

Remarks.
a) With the euclidean space E_4 and $E_4^{\mathbf{C}}$, nothing changes in these developments, except for the definition of the self-dualty ($*\Omega = \Omega$ instead of $*\Omega = i\Omega$). Theorem

---

[2]Some authors, forgetting these peculiarities, have caused confusion.

(21.2.1) holds in the euclidean case. It is the same over the $E_{2,2}$ space with neutral signature, but in that case we can define real spinors directly and consider $P_3(\mathbf{R})$ instead of $P_3(\mathbf{C})$.

b) A bundle over $E_{1,3}^{\mathbf{C}}$ is also a bundle over $E_{1,3} \times E_{1,3}$, we can then consider bundles over $E_{1,3}$ (or E_4) in Theorem (21.2.1).

Now we have :

Theorem 21.2.2 *There is a natural one to one correspondence between :*

1. *complex vector bundles with structural group, base $P_3(\mathbf{C})$, trivial over the real lines of $P_3(\mathbf{C})$ and*

2. *bundles with the same rank over $S^3 \times S^1$ (or S^4) having a self-dual connection.*

Proof. The proof uses stereographic projection, it is almost the same for both cases : choose for example S^4. By means of two stereographic projections φ_1 and φ_2 with antipodal centers, one puts over S^4 an atlas with two charts U_1 and U_2, each U_i diffeomorphic to a topological disk of $\mathbf{R}^4$. If there exists over $\mathbf{R}^4$ a Y.-M. bundle with self-dual curvature, there corresponds to it two bundles ξ_1 and ξ_2 with the same property over U_1 and U_2, because φ_1 and φ_2 are conformal transformations. ξ_1 and ξ_2 have isomorphic restrictions to $U_1 \cap U_2$, a gluing method allows to obtain a unique bundle ξ over S^4 with the stated property.   ∎

Remark. According to a) in the previous remark, we have the same result with $S_2 \times S_2$ and $P_3(\mathbf{R})$.

Proposition (21.2.4), Theorems (21.2.1) and (21.2.2) express the Penrose transform [5] *with smooth hypotheses only.*

Often the rank of the bundle (ξ) is two and the structural group is $\mathrm{SU}(2,\mathbf{C})$.

Evidently, any reduction of the structural group gives more particular structure for the corresponding bundle over $P_3(\mathbf{C})$: starting from $\mathrm{SU}(2,\mathbf{C})$ one obtains a quaternionic structure over the fibers ([1, 5]).

21.2.3   Spintantons.

'Spintanton' is coined for spin-instanton. Suppose that the curved space-time oriented manifold $V_{1,3}$ has an enlarged spinor structure. If the complexified tangent space $T_{\mathbf{C}}(V_{1,3})$ has a field of even m.t.i.s., $V_{1,3}$ satisfies our previous hypothesis and conversely ([4a, 4c] and Chapter 14, 4.1).

Let $(\mathcal{U})$ be an open set in $V_{1,3}$ trivializing the tangent bundle and let σ be the associated spinoriality projective bundle (fibers of the enlarged spinor bundle are minimal ideals in the tangent Clifford algebra and define projective complex space in three dimensions). It is always possible to increase the tangent bundle by means of a trivial bundle with complex rank 1, giving again in every fiber $E_{1,3}^{\mathbf{C}}$ an immersion in $E_{2,4}^{\mathbf{C}}$.

Suppose we also have over $V_{1,3}$ a vector bundle η with structural group G acting effectively on it. We locally reobtain the situation described in Propositions (21.2.3) and (21.2.4). On account of the equivariance of the spinor group action and the special Lorentz group action which transforms even m.t.i.s. into one another, we can associate to the bundle (η) with self-dual connection a complex bundle (ξ) over the projective enlarged spinor bundle (σ). Bundle (ξ) will be trivial over any subset of σ identifiable to some $\mathcal{U} \times D$, where D is a straight line in $P_3(\mathbf{C})$—an intrinsic property because the spinor group acts linearly in the spinor space—and where $\mathcal{U}$ is an open trivializing set, homeomorphic with a pseudo-euclidean disk.

It is clear that the results hold for any riemannian or pseudo-riemannian manifold in four dimensions, admitting an enlarged spinor structure.

Definition 21.2.1 *We shall call 'spintanton' any G-complex vector bundle (η) over a manifold in four dimensions $(V_4, V_{1,3}, V_{2,2}, V_{3,1})$ having an enlarged spinor structure and carrying a G-self-dual connection.*

Hence its properties are described by the geometry of certain vector bundles over an enlarged spinor bundle of typical fiber $P_3(\mathbf{C})$.

It is very well known that spheres have strict spinor structures, but there exists an important peculiarity : indeed S^4 has a unique class of enlarged spinor structures, because it is possible to construct over S^4 a global field of m.t.i.s. in $T_{\mathbf{C}}(V_4)$. The enlarged spinor bundle is identifiable with $P_3(\mathbf{C}) \times S^4$. The same property holds for $S_3 \times S_1$, because $T(S_3 \times S_1)$ is trivializable and also for $S_2 \times S_2$ because S_2 is carrying a global field of isotropic directions ([4, d] and Chapter 14, 6.1).

Finally, a 'spintanton' over S_4, $S_3 \times S_1$, $S_2 \times S_2$ reduces to an instanton, proving the soundness of the generalization.

The situation is particularly simple for a space-time $V_{1,3}$ admitting a strict spinor structure because $T(V_{1,3})$ is parallelizable and again a bundle over $P_3(\mathbf{C})$ appears, corresponding to a self-dual connection.

Spintantons over $V_{1,3}$ and the associated bundle over σ clearly have a physical meaning.

21.3   References.

1. M. F. Atiyah, R. S. Ward, Communications in math. physics, n. 55, 1977, pp. 117–124.

2. E. Cartan, *Leçons sur la théorie des spineurs*, Hermann, Paris, 1938.

3. C. Chevalley, *The algebraic theory of spinors*, Columbia University Press, New-York, 1954.

4. A. Crumeyrolle

    (a) *Algèbres de Clifford et spineurs*, Toulouse, 1974.

    (b) Annales I.H.P. Section A. Vol. 34, 1981, pp. 351–372.

    (c) Periodice Math. Hungarica, Vol. 6(2), 1975, pp. 143–171.

    (d) Kodai. Math. Journal 7, 1984.

5. A. Douady, J. Verdier, *Les équations de Yang et Mills*, Astérisque 71–72, Séminaire E.N.S. 1977-78 (Ed. Soc. Math. de France).

6. H. Kerbrat, Ann. I.H.P., Vol. 21 A, n. 4, p. 333.

7. R. Penrose, R. S. Ward, *Twistors for flat and curved space-time*, General relativity and Gravitation 2, p. 283–328, Plemum Press, 1980.

8. G. N. Yang, R. C. Mills, Phys. Rev. t. 96, 1954, p. 191.

Chapter 22

SYMPLECTIC STRUCTURES, COMPLEX STRUCTURES, SYMPLECTIC SPINORS AND THE FOURIER TRANSFORM.

22.1 Introduction.

We have directed our attention to the problems raised by the Fourier transform on curved symplectic manifolds. The generalization of the remarkable tool offered by this transform in $\mathbf{R}^n$ to such manifolds is awkward and leads to the difficulty known as the 'Maslov class' or 'index'. It appears that a geometrization of this transform is required before it can be turned into a global tool. In the process, the complex structure J adapted to the symplectic form and to an associated pseudo-riemannian form will prove to be closely linked with the Fourier transform, which can be identified, up to some constant factor, with the lifting $\tilde{J}$ of J to a symplectic spin group. Then the geometrization can be carried out at once, if a few conditions, which hold in all usual cases, are satified; *the geometric Fourier transform is the natural action of $\tilde{J}$ on the symplectic spinors.*

The 'nice' geometric properties of the Fourier transform can then easily be explained, along with their relation to the Heisenberg group and algebra, as was pointed out already by many authors, and especially by R. Howe ([3]). We give some examples of computations carried out using this approach (the laplacian, the Hermite functions, Hecke's formula, covariance problems). But this method seems most promising in the setting of geometric quantization, when applied to the deformation of Poisson algebras.

We have concentrated our effort on the algebraic and geometric aspect rather than on those problems that seemed closer to analysis, leaving them to the specialists; we will only indicate the relation with functional spaces, distribution theory and differential operators—several of these could lead to important extensions which the reader may carry out.

Extract from 'Journal of Geometry and Physics', Vol. 2, n. 3, 1985, Pitagora Ed., Bologna.

22.2   Elementary notions.

Let E be a real $(2r)$-dimensional vector space with a symplectic form F. The endomorphism J of E satisfies $J^2 = -1$ and endows it with a complex structure.

Definition 22.2.1 *J is said to be adapted to F in signature (p,q), $p + q = r$, if there exists a non-degenerate Hermitian form $\eta : E \times E \rightarrow \mathbf{C}$ of signature (p,q), such that $F = -\operatorname{im}(\eta)$.*

Writing $\eta(x,y) = \operatorname{re}(\eta(x,y)) + i \operatorname{im}(\eta(x,y))$, we see that $(x,y) \rightarrow -\operatorname{im}(\eta(x,y))$ is a symplectic form on E and that $(x,y) \rightarrow \operatorname{re}(\eta(x,y))$ is a symmetric quadratic form of signature $(2p, 2q)$.

Note that $\operatorname{re}(\eta(ix,y)) = -\operatorname{im}(\eta(x,y)) = F(x,y)$, $\operatorname{re}(\eta(x,y)) = -F(Jx,y)$ and $\eta(x,y) = -F(Jx,y) - iF(x,y)$ or, more simply,

$$\operatorname{re}(\eta(x,y)) = (x,y)$$

where

$$(Jx,y) = F(x,y), \quad (Jx, Jy) = (x,y), \quad F(Jx, Jy) = F(x,y), \tag{1}$$

and J is orthogonal and symplectic.

A maximal totally isotropic subspace of E (for the symplectic form F) is said to be *lagrangian*; such a space must have real dimension r.

Proposition 22.2.1 *Let L be a lagrangian which is non-isotropic for the pseudo-metric $(\cdot,\cdot)$; if J is adapted to F, $J(L)$ is a supplementary lagrangian of L and is orthogonal to it.*

Proof. Indeed, if $x \in L$ and if $y = J(x')$, $x' \in L$, $(Jx', x) = F(x', x) = 0$ since L is a lagrangian. L being non-isotropic for $(\cdot, \cdot)$ and $J(L)$ orthogonal to L, these spaces are supplementary. ∎

Proposition 22.2.2 *Let $E = L \oplus L'$, where L and L' are lagrangians and g is a scalar product of signature (p,q) on L; then there exists a unique complex structure J on L, adapted to F for the signature (p,q) and such that $J(L) = L'$ and that the restriction to L of the hermitian form with imaginary part given by $(-F)$ is g.*

Proof. If J exists, g is transported to L' by the isometry J, so that L' is orthogonal to L; then necessarily $F(x,y) = g(Jx,y)$, $\forall x, y \in E$; hence $J(x)$ follows from $F(x,y) = g(Jx,y)$, $J^2 = -1$ will hold and we will take $\eta(x,y) = g(x,y) - iF(x,y)$ for J to be adapted to F in the signature (p,q). ∎

In practice, we consider a symplectic basis (e_α, e_{β^*}), $\alpha, \beta = 1, 2, \ldots, r$, adapted to the decomposition $E = L \oplus L'$ and such that

$$F(e_\alpha, e_{\beta^*}) = \delta_{\alpha\beta}, \quad F(e_\alpha, e_\beta) = F(e_{\alpha^*, \beta^*}) = 0,$$
$$g(e_\alpha, e_\alpha) = 1, \quad \alpha = 1, \ldots, p, \quad g(e_\alpha, e_\alpha) = -1, \quad \alpha = p+1, \ldots, p+q$$

and $g(e_\alpha, e_{\beta^*}) = 0$.

$F(Je_\alpha, e_{\beta^*}) = -g(e_\alpha, e_{\beta^*}) = 0$, hence $Je_\alpha \in L'$.

$F(Je_\alpha, e_\alpha) = -g(e_\alpha, e_\alpha) = \pm 1$ hence

$$\begin{array}{lll} Je_\alpha = e_{\alpha^*}, & Je_{\alpha^*} = -e_\alpha, & \alpha = 1, \ldots, p, \\ Je_\alpha = -e_{\alpha^*}, & Je_{\alpha^*} = e_\alpha, & \alpha = p+1, \ldots, p+q. \end{array}$$

By Proposition (22.2.2), a symplectic space E can be constructed with J adapted to F in the signature (p, q); given L, the choice of L' determines L uniquely. Proposition (22.2.1) shows that if L is an arbitrary lagrangian which is non-isotropic for $(\cdot, \cdot)$, an orthogonal basis $(e_1, e_2, \ldots, e_r)$ can be chosen in L and completed with elements $(e_{1^*}, e_{2^*}, \ldots, e_{r^*})$ to a symplectic basis which is adapted to the decomposition $E = J \oplus J(L)$.

Note that $U(p, q)$ is the set of elements of $Sp(2r, \mathbf{R})$ which commute with J; it is also the set of elements of $O(2p, 2q)$ with determinant 1 which commute with J, since

$$F(u(x), u(Jy)) = F(x, Jy), \quad \forall x, y \in E$$

means that $u \in Sp(2r, \mathbf{R})$ and

$$F(u(x), J(u(y))) = F(x, Jy), \forall x, y \in E$$

means that $u \in O(2p, 2q)$.

If χ_α stands for a real number, 1 if $\alpha = 1, 2, \ldots, p$ and (-1) if $\alpha = (p+1), \ldots, (p+q)$, it will be convenient to define

$$\epsilon_\alpha = \frac{e_\alpha - i\chi_\alpha e_{\alpha^*}}{\sqrt{2}}, \quad \epsilon_{\alpha^*} = \frac{e_\alpha + i\chi_\alpha e_{\alpha^*}}{\sqrt{2}}.$$

Also note that

$$(\epsilon_\alpha, \epsilon_\beta) = (\epsilon_{\alpha^*}, \epsilon_{\beta^*}) = 0, \quad (\epsilon_\alpha, \epsilon_{\beta^*}) = \delta_{\alpha\beta},$$

$$F(\epsilon_\alpha, \epsilon_\beta) = F(\epsilon_{\alpha^*}, \epsilon_{\beta^*}) = 0, \quad F(\epsilon_\alpha, \epsilon_{\beta^*}) = 0 \ (\alpha \neq \beta), \quad F(\epsilon_\alpha, \epsilon_{\alpha^*}) = i\chi_\alpha,$$

$$\begin{array}{lll} J(\epsilon_\alpha) = i\epsilon_\alpha, & J(\epsilon_{\alpha^*}) = -i\epsilon_{\alpha^*}, & \eta(\epsilon_{\alpha^*}, \epsilon_{\alpha^*}) = \chi_\alpha, \\ J(e_\alpha) = \chi_\alpha e_{\alpha^*}, & J(e_{\alpha^*}) = -\chi_\alpha e_\alpha. \end{array}$$

22.3 Symplectic recalls.

As in our previous articles and works, we introduce several symplectic Clifford algebras. h being an indeterminate (but we may interpret it up to a constant factor, as the physicist's Planck constant), we take the quotient of the tensor algebra over the module

$$E' = E \otimes_{\mathbf{K}} \mathbf{K}[[h]], \quad \mathbf{K} = \mathbf{R} \text{ or } \mathbf{C},$$

by the two-sided ideal generated by the elements

$$x \otimes y - y \otimes x - hF(x,y), \quad x,y \in E.$$

This gives us the associative algebra $C_S(hF)$ (also called a *Weyl algebra*), which will be 'enlarged' in several ways :

- $\check{C}_S(hF)$, whose general element is a formal series in the powers of e_α, e_{β^*}, $\alpha, \beta = 1, 2, \ldots, r$, with coefficients in $\mathbf{K}[[h]]$, called the *formal symplectic Clifford algebra*. We will also use the field $\mathbf{K}((h))$ of rational fractions.

- $C_S(hF)^\alpha$, for which h is fixed and the general element is

$$\hat{u} = M(H, K^*) \frac{e^H e^{K^*}}{(H! K^*!)^{1/2+\alpha}},$$

where α is a real number, $\alpha > 0$, H and K^* are multi-indices,

$$e^H = (e_1)^{h_1} (e_2)^{h_2} \ldots (e_r)^{h_r},$$

$|H| = \sum h_i$, $H! = h_1! h_2! \ldots h_r!$, the h_i being non-negative integers, and from a certain rank onwards,

$$|M(H, K^*)| \le \sigma(\hat{u}) \rho(\hat{u})^{|H|+|K^*|},$$

where $\sigma(\hat{u})$ and $\rho(\hat{u})$ are positive constants.

- $\check{C}_S(hF)_T$, the truncated algebra, obtained as the quotient of $\check{C}_S(hF)$ by the ideal of multiples of h^N, where N is a fixed positive integer, $N > 1$; the coefficients of its elements are polynomials in h of degree not exceeding $(N-1)$.

When h is fixed, the truncated algebras give rise to algebras whose elements are series with coefficients in $\mathbf{K}$; if h is then understood to be 'very small', we obtain the viewpoint of non-standard analysis.

If $a \in E$, $t \in \mathbf{K}$, for all $x \in E$,

$$\exp(ta^2) x \exp(-ta^2) = x + 2ht F(a, x) a$$

where $\exp(ta^2)$ can be considered as an element of $\check{C}_S(hF)$, $\check{C}_S(hF)_T$ or $C_S(hF)^\alpha_T$.

We have previously introduced the symplectic spinors as a projective limit of a sequence of left ideals. This can be done for each of the algebras defined here.

The spaces of symplectic spinors are constructed starting from a symplectic basis (e_α, e_{β^*}) and can be linearly identified with a symmetric algebra over the space of e_α, $\alpha = 1, \ldots, r$, where the meaning varies according to the chosen symplectic Clifford algebra.

For the sequel, it is advantageous to consider a space of symplectic spinors as the quotient of the symplectic Clifford algebra by the left ideal generated by the (e_{α^*}), the representation being the left multiplication followed by the quotient.

This convention allows us to identify the multiplication by e_{α^*} with $(-h\partial/\partial q^\alpha)$, and e_α with q^α, replacing $(e_\alpha)^k$ by $(q^\alpha)^k$, q^α being a coordinate variable; these identifications follow from

$$(e_{\alpha^*})(e_\alpha)^k = (e_\alpha)^k(e_{\alpha^*}) - hk(e_\alpha)^{k-1}.$$

This convention leads us to interpret the Clifford algebra $C_S(hF)$ as the 'Weyl algebra' of differential operators with constant coefficients, for a space $\mathbf{R}^r \oplus (\mathbf{R}^r)^*$, the symplectic spinors becoming series in the symplectic algebra of $\mathbf{R}^r$.

The defining conditions of $C_S(hF)^\alpha$ will then ensure the analyticity and, for $\alpha > 1/2$, we obtain functions with rapidly decreasing derivatives. Note that these function spaces are also algebras.

22.4 Some algebraic results.

In this section we first take $h = 1$; if p is the projection of any symplectic group or subgroup onto the symplectic group, we can prove that

$$p \circ \exp(-\frac{\pi}{2}\sum_\alpha(\chi_\alpha \epsilon_\alpha \epsilon_{\alpha^*})) = J \tag{1}$$

which can easily be verified using the expansion of the exponential.

More generally, $U(p,q)$ lies in the set of images, under p, of the elements

$$\exp \lambda \exp(a^{\alpha\beta^*} \epsilon_\alpha \epsilon_{\beta^*})$$

in the metaplectic group, where $\lambda \in \mathbf{C}$ and $a^{\alpha\beta^*} \in \mathbf{C}$ (no summation intended).

The Lie algebra of $U(p,q)$ is defined by the elements $\sum_{\alpha,\beta} a^{\alpha\beta^*} \epsilon_\alpha \epsilon_{\beta^*}$, $U(p,q)$ acts on the space of (e_{α^*}) and since

$$\sum_{\alpha,\beta} a^{\alpha\beta^*}(\epsilon_\alpha\epsilon_{\beta^*})\epsilon_{\gamma^*} - \epsilon_{\gamma^*}\sum_{\alpha,\beta} a^{\alpha\beta}(\epsilon_\alpha\epsilon_{\beta^*}) = i\chi_\gamma(\sum_\alpha a^{\gamma\alpha^*}\epsilon_{\alpha^*}),$$

expressing that the matrix of $i\chi_\gamma\sum a^{\gamma\alpha^*}\epsilon_{\alpha^*}$ is antihermitian for η, we get

$$\bar{a}^{\alpha\beta^*} = a^{\beta\alpha^*}.$$

We are interested in the Lie algebra of $O(p,q)$, whose elements are characterized by $a^{\alpha\beta^*} = -a^{\beta\alpha^*}$. Writing

$$a^{\alpha\beta^*}\epsilon_\alpha\epsilon_{\beta^*} = \tfrac{1}{2}a^{\alpha\beta^*}(\epsilon_\alpha\epsilon_{\beta^*} - \epsilon_\beta\epsilon_{\alpha^*})$$
$$= \tfrac{1}{2}ia^{\alpha\beta^*}(\chi_\beta e_\alpha e_{\beta^*} - \chi_\alpha e_\beta e_{\alpha^*}).$$

We have proved :

Proposition 22.4.1 *The Lie algebra of* $O(p,q)$ *is isomorphic to the set of linear combinations with pure imaginary (resp. real) coefficients of the* $\epsilon_\alpha \epsilon_{\beta^*} - \epsilon_\beta \epsilon_{\alpha^*}$ *(resp. of the* $\chi_\beta e_\alpha e_{\beta^*} - \chi_\alpha e_\beta e_{\alpha^*}$*).*

A simple computation in the symplectic Clifford algebra yields :

$$e_\alpha e_{\beta^*} \left(\sum_\gamma \chi_\gamma (e_\gamma)^2 \right) - \left(\sum_\gamma \chi_\gamma (e_\gamma)^2 \right) e_\alpha e_{\beta^*} = -2\chi_\beta e_\alpha e_\beta$$

from which we deduce that $\sum_\gamma \chi_\gamma (e_\gamma)^2$ commutes with

$$\chi_\beta e_\alpha e_{\beta^*} - \chi_\alpha e_\beta e_{\alpha^*},$$

and the same holds for $\sum_\gamma \chi_\gamma (e_{\gamma^*})^2$ and $\sum_\gamma e_\gamma e_{\gamma^*}$.

Putting

$$U = \tfrac{1}{2} \sum_\gamma \chi_\gamma (e_\gamma)^2, \quad V = \tfrac{1}{2} \sum_\gamma \chi_\gamma (e_{\gamma^*})^2, \quad W = \frac{r - 2\sum_\gamma e_\gamma e_{\gamma^*}}{2}, \tag{2}$$

it is easy to verify that

$$[U, V] = -W, \quad [U, W] = -2U, \quad [V, W] = 2V. \tag{3}$$

Hence we obtain the Lie algebra of $SL(2, \mathbf{R})$ and we can state :

Proposition 22.4.2 *The* U, V, W *defined by (3) are the elements of a basis for the Lie algebra of a subgroup of* $Sp(2r, \mathbf{R})$ *which is isomorphic to* $SL(2, \mathbf{R})$ *(or* $SO(2,1)$*); the elements of the connected component of the identity in* $O(p,q)$ *commute with this subgroup.*

Hence $O(p,q)$ and $SL(2, \mathbf{R})$ form a pair of subgroups in $Sp(2r, \mathbf{R})$ such that each one is the centralizer of the other, a well-known result (cf. [3]).

In a frame which is adapted to the decomposition in two supplementary lagrangians, U, V, W are, in terms of the components of the 'metric' and symplectic tensors, given by

$$U = \tfrac{1}{2} \sum_{\alpha,\beta} g^{\alpha\beta} e_\alpha e_\beta, \quad V = \tfrac{1}{2} \sum_{\alpha,\beta} g^{\alpha^* \beta^*} e_{\alpha^*} e_{\beta^*}, \quad W = \frac{r}{2} - \sum_{\alpha,\beta} F^{\alpha\beta^*} e_\alpha e_{\beta^*},$$

where $F^{\alpha\beta^*}$ stands for $g^{\alpha\lambda} g^{\beta^* \mu^*} F_{\lambda\mu^*}$.

Proposition (22.4.2) can be related to the following well-known property : every symplectic space (E, F) of dimension $2r$, with a complex structure J adapted to F in the signature (p, q), $p + q = r$, can be (non-uniquely) identified with the tensor product of a two-dimensional symplectic space and an r-dimensional pseudo-euclidean space with a metric of signature (p, q).

The U, V, W correspond to the Lie algebra of $\mathrm{Sp}(2, \mathbf{R}) \otimes \mathrm{Id}_r = \mathrm{SL}(2, \mathbf{R}) \otimes \mathrm{Id}_r$ and the elements of $O(p, q)$ are identified with $\mathrm{Id}_2 \otimes O(p, q)$, which offers a better explanation for the commutation properties expressed in Proposition (22.4.2).

Recalling the correspondence with q^α and p_α, we associate

$$
\left.
\begin{array}{ll}
U & \text{and} \quad \frac{1}{2} g_{\alpha\beta} q^\alpha q^\beta \\[2mm]
V & \text{and} \quad \frac{1}{2} g^{\alpha\beta} \dfrac{\partial^2}{\partial q^\alpha \partial q^\beta} \\[2mm]
W & \text{and} \quad \frac{1}{2} r - q^\alpha \dfrac{\partial}{\partial q^\alpha}
\end{array}
\right]
\tag{4}
$$

for the natural action on the symplectic spinors.

We recognize terms representing :

- the pseudo-euclidean norm,

- the laplacian,

- the Liouville field.

At the level of the Lie algebra of the metaplectic group $\mathrm{Mp}(r)$, J comes from $-(\pi/2) \sum_\alpha \chi_\alpha \epsilon_\alpha \epsilon_{\alpha^*}$, and from $-(\pi/4) \sum_\alpha \chi_\alpha ((e_\alpha)^2 + (e_{\alpha^*})^2)$ at the level of the Lie algebra of the spin group $\mathrm{Sp}_2(r)$.

Hence J is obtained by p from $\exp(-(pi/2)(U + V))$ and we have the :

Proposition 22.4.3 J 'lifts' to the operator

$$
\tilde{J} = \exp(-(\pi/2)(U + V))
\tag{5}
$$

and, in terms of the infinitesimal operators on symplectic spinors, it corresponds to

$$
-\frac{\pi}{4}\left(g_{\alpha\beta} q^\alpha q^\beta + g^{\alpha\beta} \frac{\partial^2}{\partial q^\alpha \partial q^\beta}\right)
\tag{6}
$$

i.e. up to a factor, to the sum of the Beltrami operator ρ^2 and the laplacian Δ.

22.5 The Fourier transform and the complex structure J.

To ease the comparison between our results, the classical results and those of other authors such as [3], we will adopt the table

$$
e_\alpha e_{\beta^*} - e_{\beta^*} e_\alpha = \frac{1}{2\pi i} \delta_{\alpha\beta}
\tag{1}
$$

to define our symplectic Clifford algebra; this means that we take $h = 1/(2\pi i)$ in formula (1) of 22.2.

On the infinitesimal level, J comes from $-i\pi^2(U+V)$, or, in terms of differential operators on the spinors, from

$$\frac{-\pi}{4}(2\pi\rho^2 - \frac{\Delta}{2\pi}).$$

W must be changed to $(i/2\pi)(\sum_\gamma e_\gamma e_{\gamma^*} - r/2)$ with

$$[U,V] = -W, \quad [U,W] = \frac{U}{2\pi^2}, \quad [V,W] = \frac{-V}{2\pi^2}. \tag{2}$$

Then we put $\tilde{J}_1 = \exp[-(\pi i/4)(2\pi\rho^2 - \Delta/(2\pi))]$, which is just $\tilde{J}$ using the new conventions.

Remark. If we use $h/(2\pi i)$ in (1), $-i\pi^2(U+V)$ becomes $(-i\pi^2/h)(U+V)$, $(-\pi i/4)(2\pi\rho^2 - \Delta/(2\pi))$ becomes $(-\pi i/4h)(2\pi\rho^2 - h^2\Delta/(2\pi))$ and $h^2/(2\pi^2)$ replaces $1/(2\pi^2)$ in (2).

A symplectic spinor being associated to a sequence function f of the q^α, if $\mathcal{F}^{p,q}$ stands for the Fourier transform :

$$\mathcal{F}^{p,q}(f)(y) = \int_{\mathbb{R}^{p+q}} f(x) e^{-2\pi i(x \cdot y)} x$$

the result obtained explicitly in [3] yields

$$\tilde{J}_1 = \exp\frac{-\pi i(p-q)}{4}\mathcal{F}^{p,q}.$$

Assuming that the Fourier transform of f exists or, equivalently, that the product by $\tilde{J}_1$ has a meaning (for instance when $\alpha > 1/2$, in the algebra $C_S(F)^\alpha$, the transform of f is in $C_S(F)^\alpha$), we can state the

Proposition 22.5.1 *Up to the constant factor* $\exp((-\pi i/4)(p-q))$, *the Fourier transform* $\mathcal{F}^{p,q}$ *can be identified with the lifting* $\tilde{J}_1$ *of the complex structure* J *which is adapted to the symplectic structure* F *in the signature* (p,q) *in the symplectic spin group* $\text{Sp}_2(p+q)$, *acting naturally on the space of symplectic spinors, by left multiplication.*

As a verification, we consider $f(x) = e^{-\pi(x \cdot x)}$, which corresponds to

$$\exp(-\pi \sum_\alpha (e_\alpha)^2)\Phi^* = \exp(-2\pi U)\Phi^*$$

in the symplectic spinor space $\check{C}_S(F)\Phi^*$; its Fourier transform is

$$\exp(\frac{\pi i(p-q)}{4})\exp[-\frac{\pi i}{4}(2\pi\rho^2 - \frac{\Delta}{2\pi})]\exp(-2\pi U)\Phi^*,$$

The U, V, W correspond to the Lie algebra of $\mathrm{Sp}(2,\mathbf{R}) \otimes \mathrm{Id}_r = \mathrm{SL}(2,\mathbf{R}) \otimes \mathrm{Id}_r$ and the elements of $O(p,q)$ are identified with $\mathrm{Id}_2 \otimes O(p,q)$, which offers a better explanation for the commutation properties expressed in Proposition (22.4.2).

Recalling the correspondence with q^α and p_α, we associate

$$
\left.
\begin{array}{ll}
U & \text{and} \quad \frac{1}{2}g_{\alpha\beta}q^\alpha q^\beta \\[2mm]
V & \text{and} \quad \frac{1}{2}g^{\alpha\beta}\dfrac{\partial^2}{\partial q^\alpha \partial q^\beta} \\[3mm]
W & \text{and} \quad \dfrac{r}{2} - q^\alpha \dfrac{\partial}{\partial q^\alpha}
\end{array}
\right\}
\tag{4}
$$

for the natural action on the symplectic spinors.

We recognize terms representing :

- the pseudo-euclidean norm,

- the laplacian,

- the Liouville field.

At the level of the Lie algebra of the metaplectic group $\mathrm{Mp}(r)$, J comes from $-(\pi/2)\sum_\alpha \chi_\alpha \epsilon_\alpha \epsilon_{\alpha\bullet}$, and from $-(\pi/4)\sum_\alpha \chi_\alpha ((e_\alpha)^2 + (e_{\alpha\bullet})^2)$ at the level of the Lie algebra of the spin group $\mathrm{Sp}_2(r)$.

Hence J is obtained by p from $\exp(-(pi/2)(U+V))$ and we have the :

Proposition 22.4.3 J 'lifts' to the operator

$$
\tilde{J} = \exp(-(\pi/2)(U+V))
\tag{5}
$$

and, in terms of the infinitesimal operators on symplectic spinors, it corresponds to

$$
-\frac{\pi}{4}\left(g_{\alpha\beta}q^\alpha q^\beta + g^{\alpha\beta}\frac{\partial^2}{\partial q^\alpha \partial q^\beta}\right)
\tag{6}
$$

i.e. up to a factor, to the sum of the Beltrami operator ρ^2 and the laplacian Δ.

22.5 The Fourier transform and the complex structure J.

To ease the comparison between our results, the classical results and those of other authors such as [3], we will adopt the table

$$
e_\alpha e_{\beta\bullet} - e_{\beta\bullet} e_\alpha = \frac{1}{2\pi i}\delta_{\alpha\beta}
\tag{1}
$$

to define our symplectic Clifford algebra; this means that we take $h = 1/(2\pi i)$ in formula (1) of 22.2.

Convergence problems in this setting may be tackled using an algebra $C_S(hF)^\alpha$ or truncation modulo h^N; such computations arise in the Maslov method of stationary phase and are most interesting for applications to physics.

By the remark at the end of section 2, we see that if $\alpha > 1/2$, using a Taylor expansion, our definition of the Fourier transform on the symplectic spinors coincides exactly with the classical definition in the set of analytic functions represented by series in $C_S(hF)^\alpha$ (e_α is replaced by q^α).

Another generalization consists in the replacement of $\mathbf{R}^r \oplus (\mathbf{R}^r)^*$ by the cotangent bundle to a real r-dimensional manifold with a pseudo-riemannian structure of signature (p,q), $p+q = r$, $M = T^*(V)$ has a canonical symplectic structure if J is a complex structure in the sense of Definition (22.5.1), its existence can be verified as in Proposition (22.2.2), since there exist global fields of supplementary lagrangians on $T^*(V)$, the structural group of the cotangent bundle reducing to $\mathrm{U}(p,q)$, and in this case, even to $\mathrm{O}(p,q)$, $g^{\alpha\beta^*}\epsilon_\alpha\epsilon_{\beta^*}$ has a global meaning, just as

$$\tilde{J}_1 = \exp(-i\pi^2 g^{\alpha\beta^*}\epsilon_\alpha\epsilon_{\beta^*})$$

in the symplectic Clifford algebra of M.

If symplectic spinor fields related to the real symmetric algebra of V can be defined, (a sufficient condition for this is the existence of a field of real oriented lagrangians, cf. [1, b] and Chapter 18), restricting $T^*(V)$ to V we will consider a function in $C^\infty(V)$ (given locally as $f(q^1,\ldots,q^r)$) and associate a sheaf $\varphi(f)$ of enlarged symmetric algebras to it by the following natural procedure :

At every point $x \in V$ we construct a formal series in the algebra $\overset{**}{\vee}T_x^*(V)$ whose terms are the symmetrized covariant derivatives for a torsion-free symplectic connection (cf. [1] and Chapter 19). This sheaf of symmetric algebras is also a symplectic Clifford algebra sub-sheaf, hence a symplectic spinor sub-sheaf. $\tilde{J}_1$ (or $\tilde{J}_h$) is a section of the symplectic Clifford algebra sheaf acting naturally by left multiplication on $\varphi(f)$:

$$\varphi(f) \to \exp(\frac{\pi i(p-q)}{4})\tilde{J}_1 \cdot \varphi(f)$$

is a global formal Fourier transform if we restrict ourselves to the algebras $\check{C}_S(hF)$.

If non-isomorphic symplectic spinor structures exist on M, we will obtain different global Fourier transforms, and the set of these global Fourier transforms is parametrized by $H^1(M, \mathbf{Z}_2)$, whose elements are just the Maslov classes, modulo $\mathbf{Z}_2$ (cf. [1, b] and Chapter 18).

Clearly, if we replace the $\mathrm{Sp}_2(r)$ symplectic spinor structures by the $\mathrm{Sp}_q(r)$ structures or by the metaplectic structures, a $\mathbf{Z}_q$-valued or even S^1-valued Maslov index will arise ([1, b]).

The difficulties related to the Maslov index appear as soon as one wants to 'glue' the classical Fourier transforms, which are particular local expressions of the global transform just defined and whose existence depends on that of a symplectic spinor structure for $T^*(V)$—their existence is guaranteed if supplementary fields of real oriented lagrangians can be defined (the reduction of the structural group to $SO(p,q)$).

Summarizing the principal results :

Proposition 22.5.2 *If, over a real r-dimensional manifold V with pseudo-riemannian structure of signature (p,q) $(p+q=r)$, symplectic spinor fields for the cotangent bundle $T^*(V)$ with a complex structure J adapted to the metric can be defined, there exist generalized global formal Fourier transforms which are obtained by the left multiplication (modulo $\exp(\pi i(p-q)/4)$), by a lifting $\tilde{J}_h$ of J in the symplectic spinor sheaf for every function $f \in C^\infty$.*

Note that this transform is no longer formal if the sheaf associated to f is, for instance, in the algebra $C_S(hF)_x^\alpha$ at every point x. Then we completely recover, locally, the classical interpretation, if $\alpha > 1/2$.

Remarks.

1. If the polynomial $e^{2\pi x^2}(d^n/dx^n)(e^{-2\pi x^2})$, x being a real variable, is called the nth degree Hermite polynomial, it will be naturally related to

$$\exp(2\pi e_\alpha^2)(e_{\alpha^*})^n \exp(-2\pi e_\alpha^2)$$

in the space of symplectic spinors, relying on the correspondence principle.

The same holds for $\exp(4\pi U)P(e_{\alpha^*})\exp(-4\pi U)$ where $P(e_{\alpha^*})$ stands for

$$P(e_{1^*}, e_{2^*}, \ldots, e_{r^*}),$$

which is associated to

$$e^{2\pi(x\cdot x)}P(-\frac{1}{2\pi i}\frac{\partial}{\partial a^\alpha})e^{-2\pi(x\cdot x)} = P(x), \quad x \in \mathbf{R}^r;$$

the set of all $e^{-\pi(x\cdot x)}P(x)$ is total in $\mathcal{L}_{\mathbf{C}}^2(\mathbf{R}^r)$ ([2]), it is constructed from the

$$\exp(-2\pi U)\exp(4\pi U)\tilde{J}P(e_\alpha)\tilde{J}^{-1}\exp(-4\pi U) = \exp(2\pi U)P(e_{\alpha^*})\exp(-4\pi U)$$

applying the correspondence principle.

So our approach immediately explains these results; we will say more about this later. We can also understand now why, using symplectic spinors, we are able to recover all the essential algebraic properties of the Fourier transform.

2. In this setting, we could also try to reobtain some integral transforms having an inversion property and a Parseval-Plancherel theorem.

We will replace $\tilde{J}_h$ by an element of the symplectic Clifford algebra, and, more precisely, by an element γ of the group $Sp_2(r)$.

The commutation with the action of $O(p, q)$ will in general be lost, but the inversion property and the Parseval-Plancherel theorem will be preserved, for if $\gamma \in \mathrm{Sp}_2(r)$, $\beta(\bar\gamma)\gamma = 1$. Nevertheless, we will not necessarily obtain an integral transform (this, in fact, may already occur for elements in the covering of $\mathrm{SL}(2, \mathbf{R})$ ([6, p. 198])).

The refinement of these results offers an interesting research subject.

3. The action of the symplectic Clifford algebra on spinor space corresponds to that of a set of differential operators. The elements of the symmetric algebra generated by the (e_{α^*}) correspond to constant coefficient differential operators.

22.6   Bilinearity and sesquilinearity.   The 'symplectic spin' Parseval-Plancherel theorem.

Consider the symplectic spinors, whose space is linearly isomorphic to the symmetric algebra of the e_α, $\alpha = 1, \ldots, r$. They correspond to the Schrödinger representation. We will consider either $\check{C}_S(hF)\Phi^*$ or $C_S(hF)^\alpha \Phi^*$, depending on the interest in the formal case.

β being the main antiautomorphism satisfying $\beta|_{E_{\mathbf{C}}} = i\,\mathrm{Id}$, we write symbolically, in terms of double classes :

$$\Phi\beta(u)v\Phi^* = \Phi B(u\Phi, v\Phi^*)\Phi^* \tag{1}$$

where the $u\Phi$, $v\Phi^*$ have real or complex coefficients. $B(u\Phi, v\Phi^*)$ is a scalar[2] since we take the quotient by the (e_{α^*}) to the right and by the (e_α) to the left. The scalar result is in $\mathbf{K}[[h]]$ or in $\mathbf{C}$ depending on whether the algebra used is formal or not.

It is easy to see that :

$$B(e^{H^*}\Phi, e^K \Phi^*) = 0 \quad \text{if} \quad H \neq K,$$
$$B(e^{H^*}\Phi, e^H \Phi^*) = (-ih)^{|H|} H!.$$

A duality between the space of Φ spinors and the space of Φ^* spinors appears here. Note that the convergence conditions on $v\Phi^* \in C_S(hF)^\alpha \Phi^*$ can be satisfied for elements such that $u\Phi \notin C_S(hF)^\alpha \Phi^*$, e.g. $u\Phi = \exp t(e_{\gamma^*})^2 \Phi$, which act as 'distributions' in relation to them.

It will prove interesting to define a positive definite hermitian sesquilinear 'form'; we choose $h = 1/(2\pi i)$ because we want to compare with a Fourier transform, but then it is better to set

$$E_\alpha = \exp(-\frac{i\pi}{4})e_\alpha, \quad E_{\alpha^*} = \exp(-\frac{i\pi}{4})e_{\alpha^*},$$

---

[2]In fact, on a manifold, it is a density of weight 1.

the elements E_α, E_{α^*} are 'real' and have the multiplication table

$$E_\alpha E_{\beta^*} - E_{\beta^*} E_\alpha = \frac{1}{2\pi} \delta_{\alpha\beta}.$$

There exists an obvious isomorphism between the symplectic Clifford algebras constructed over the (e_k) and those constructed over the (E_k), so that we consider the symplectic Fourier transform determined by the multiplication by $\exp(\pi i(p-q)/4)\tilde{J}_1$ where $\tilde{J}_1$ is obtained from

$$J(E_\alpha) = \chi_\alpha E_{\alpha^*}, \quad J(E_{\alpha^*}) = -\chi_\alpha E_\alpha,$$

J is adapted to F in the signature (p, q), $F(E_\alpha, E_{\beta^*}) = \delta_{\alpha\beta}/(4\pi)$,

$$F(E_\alpha, J E_\alpha) = \frac{\chi_\alpha}{4\pi}, \quad \text{etc.}$$

Let β be the main antiautomorphism deduced from

$$\beta(E_\alpha) = -iE_{\alpha^*}, \quad \beta(E_{\alpha^*}) = iE_\alpha,$$

the symplectic spinors being determined by the (E_α).

We will write

$$\beta(u\Phi^*) = \Phi\beta(u),$$
$$\beta(\overline{u\Phi^*}) = \Phi\beta(\bar{u}),$$

where $^-$ stands for the complex conjugate. We form

$$\Phi\beta(\bar{u})v\Phi^* = \Phi\mathcal{H}(u\Phi^*, v\Phi^*)\Phi^*, \tag{2}$$

$\mathcal{H}(u\Phi^*, v\Phi^*)$ is a scalar and it is easy to verify that

$$\mathcal{H}(e^H\Phi^*, e^K\Phi^*) = 0, \quad \text{if} \quad K \neq H,$$
$$\mathcal{H}(e^H\Phi^*, e^H\Phi^*) = |h|^{|H|}H!.$$

$\mathcal{H}$ is positive definite hermitian sesquilinear. It is invariant under all elements of the group $\mathrm{Sp}_2(r)$ and, in particular, under $\tilde{J}_1$ (the symplectic Parseval-Plancherel theorem).

Note that the exponential elements $\exp ta^2$, $\exp tb^2$ can, for certain values of t, lead to divergent series for B and $\mathcal{H}$ if h is a complex number; in that case formal series in h must be considered; h being a pure imaginary indeterminate for the (e_k) or a real indeterminate for the (E_k).

Important remark. We could have chosen the symplectic spinors constructed with the $\widehat{\epsilon_\alpha}$ or with the $\widehat{\epsilon_{\alpha^*}}$, where

$$\widehat{\epsilon_{\alpha^*}} = \frac{e_\alpha + ie_{\alpha^*}}{\sqrt{2}}, \quad \widehat{\epsilon_\alpha} = \frac{e_\alpha - ie_{\alpha^*}}{\sqrt{2}},$$

and, if $h = 1$, we can verify that

$$(\widehat{\epsilon_\alpha}, \widehat{\epsilon_\beta}) = (\widehat{\epsilon_{\alpha^*}}, \widehat{\epsilon_{\beta^*}}) = 0, \quad J(\widehat{\epsilon_\alpha}) = i\chi_\alpha \widehat{\epsilon_\alpha}, \quad J(\widehat{\epsilon_{\alpha^*}}) = -i\chi_\alpha \widehat{\epsilon_{\alpha^*}},$$
$$(\widehat{\epsilon_\alpha}, \widehat{\epsilon_{\alpha^*}}) = \chi_\alpha, \quad (\widehat{\epsilon_\alpha}, \widehat{\epsilon_{\beta^*}}) = 0, \quad \text{if } (\alpha \neq \beta),$$
$$F(\widehat{\epsilon_\alpha}, \widehat{\epsilon_\beta}) = F(\widehat{\epsilon_{\alpha^*}}, \widehat{\epsilon_{\beta^*}}) = 0, \quad F(\widehat{\epsilon_\alpha}, \widehat{\epsilon_{\beta^*}}) = i\delta_{\alpha\beta}.$$

Next, setting

$$X = \sum_\alpha \tfrac{1}{2}(\widehat{\epsilon_\alpha})^2, \quad Y = \sum_\alpha \tfrac{1}{2}(\widehat{\epsilon_{\alpha^*}})^2, \quad Z = \frac{i}{2}(\sum_\alpha 2\widehat{\epsilon_\alpha}\widehat{\epsilon_{\alpha^*}} - ir) \tag{3}$$

we see that

$$[X, Y] = Z, \quad [X, Z] = -2X, \quad [Y, Z] = 2Y \tag{4}$$

which determine a Lie algebra isomorphic to that of $SL(2, \mathbf{C})$ or $SO(3,1)$ (or the Lorentz group). Note that X, Y, Z are linear combinations of $\sum_\alpha(e_\alpha)^2$, $\sum_\alpha(e_{\alpha^*})^2$ and $\sum_\alpha(e_\alpha e_{\alpha^*})$.

Let us construct a spinor space over the symmetric algebra of the (ϵ_α) (which correspond to the Bargmann-Fock representation). If we want to easily compare our results with the classical Fourier transform, we choose $h = -1/(2\pi)$, the correspondence principle assign the product by $\widehat{q^\alpha}$ to $\widehat{\epsilon_\alpha}$ and the derivation

$$-\frac{1}{2\pi i}\frac{\partial}{\partial q^\alpha}$$

to $(\widehat{\epsilon_{\alpha^*}})$.

The Fourier transform then corresponds, up to a constant factor, to the action of $\exp[-(i\pi/4)(2\pi\rho^2 - \Delta/(4\pi))]$ on the space of polynomials in q^α, or to the left multiplication by

$$\hat{J}_1 = \exp[-\frac{\pi^2 i}{2}(\sum_\alpha \chi_\alpha(\widehat{\epsilon_\alpha})^2 + \sum_\alpha \chi_\alpha(\widehat{\epsilon_{\alpha^*}})^2)].$$

Using the antiautomorphism β for which $\beta|_{E_\mathbf{C}} = i\,\mathrm{Id}$, we can also construct a hermitian sesquilinear 'density form' $\mathcal{K}$:

$$\Phi\beta(\bar{u})v\Phi^* = \Phi\mathcal{K}(u\Phi^*, v\Phi^*)\Phi^*. \tag{5}$$

It is easy to verify that

$$\mathcal{K}(\widehat{\epsilon}^H\Phi^*, \widehat{\epsilon}^K\Phi^*) = 0, \quad \text{if } K \neq H,$$
$$\mathcal{K}(\widehat{\epsilon}^H\Phi^*, \widehat{\epsilon}^H\Phi^*) = |h|^{|H|}H!.$$

To have a Parseval-Plancherel theorem, we must replace $\hat{J}_1$ by $\tilde{J}_1$ ($\tilde{J}_1$ is the transform of $\hat{J}_1$ under $e_\alpha \to \widehat{\epsilon_\alpha}$, $e_{\alpha^*} \to \widehat{\epsilon_{\alpha^*}}$).

$\mathcal{H}$ clearly is invariant under the action of the symplectic spin group $Sp_2(r)$.

Finally, we see that from the geometric standpoint, we could define a spin Fourier transform by the product with the lifting $\tilde{J}$ of the complex structure (modulo an isomorphism of the symplectic Clifford algebra which, restricted to the spinors, is an isometry for the hermitian sesquilinear forms associated to their spaces, which may differ). We know that $\tilde{J}$ commutes with the elements of the lifting of the unitary group $U(p, q)$.

Remark. If we consider two arbitrary symplectic spinor spaces with general elements $u\Phi_1^*$ and $u\Phi^*$ (associated to some ϵ_α or ϵ_{α^*} constructed, as before, from real elements similar to e_α, e_{α^*}), we can easily define a sesquilinear form on their product : the transitive action of the symplectic group on the set of lagrangian subspaces reduces this problem to the previous case. The ambiguity of the symplectic transformation associating two given lagrangians does not cause any serious problems because the introduced sesquilinear forms correspond to each other under the action of $Sp_2(r)$. In the literature, equivalent computations are known as 'pairing' ([5]).

22.7   The laplacian and the symplectic geometry.

We once again put $h = 1$. $\sum \chi_\alpha (e_{\alpha^*})^2$ corresponds to the laplacian Δ (generalized to the signature (p, q)) when its action on the symplectic spinors, interpreted as polynomials in q^α, $\alpha = 1, \ldots, r$, is considered.

Taking the exponential, $\gamma = \exp(\sum \chi_\alpha (e_{\alpha^*})^2 t)$ is a symplectic transvection and

$$\gamma u \Phi^* = \gamma u \gamma^{-1} \gamma \Phi^* = \gamma u \gamma^{-1} \Phi^*.$$

$\Delta u \Phi^* = 0$ is equivalent to $\gamma u \Phi^* = u \Phi^*$ for all $t \in \mathbf{R}$.

Indeed, $\gamma u \Phi^* = u \Phi^*$ implies

$$\gamma u \gamma^{-1} \Phi^* = u \Phi^*,$$

from which, taking the derivative at the identity,

$$
\begin{aligned}
(\sum \chi_\alpha (e_{\alpha^*})^2 u - u \sum \chi_\alpha (e_{\alpha^*})^2) \Phi^* &= 0, \\
\sum \chi_\alpha (e_{\alpha^*})^2 u \Phi^* &= 0.
\end{aligned}
$$

Conversely, if $\sum \chi_\alpha (e_{\alpha^*})^2 u \Phi^* = 0$, $\gamma u \Phi^*$ is constant, but for $t = 0$, $\gamma(0) = 1$ and the result follows. Hence :

Proposition 22.7.1 *The harmonic symplectic spinors $u\Phi^*$ are the invariant spinors under the natural action of the one-parameter group*

$$t \to \exp(\sum \chi_\alpha (e_{\alpha^*})^2 t).$$

Remark. If we consider any second order constant coefficient differential operator (in the $\partial/\partial q^\alpha$)

$$\sum A^{\alpha\beta}\frac{\partial^2}{\partial q^\alpha\partial q^\beta},$$

the same property obviously holds for the one parameter symplectic transformations subgroup

$$t \to \exp[(\sum_{\alpha,\beta} A^{\alpha\beta}e_{\alpha^*}\cdot e_{\beta^*})t].$$

Consider the quotient of the algebra $\check{C}_S(hF)$ by the two-sided ideal (S) generated by $(\sum(e_{\alpha^*})^2 - 1)$.

Every homogeneous polynomial in the (e_{α^*}) can be written as

$$Q(e_{\alpha^*}) = P_0(e_{\alpha^*}) + (\sum(e_{\alpha^*})^2)P(e_{\alpha^*})$$

which, iterated, gives

$$P_0(e_{\alpha^*}) + (\sum(e_{\alpha^*})^2)P_1(e_{\alpha^*}) + (\sum(e_\alpha^*)^2)^2 P_2(e_{\alpha^*}) + (\sum(e_{\alpha^*})^2)^3 P_3(e_{\alpha^*}) + \ldots$$

where $P_0, P_1, P_2, P_3, \ldots$ cannot be divided by $\sum(e_{\alpha^*})^2$.

Applying $\gamma(1) = \exp\sum(e_{\alpha^*})^2$ to $P_i(e_{\alpha^*})$ by

$$\gamma(1)P_i(e_{\alpha^*})\gamma(1)^{-1},$$

then taking the quotient by S, we see at once that $P_i(E_{\alpha^*})$ is invariant under $\gamma(1)$, because $\gamma(1) = 1$ modulo S.

We have obtained, in very little time, a classical result on the decomposition of polynomials in terms of harmonic factors $P_i(q^\alpha)$ and of $\sum(q^\alpha)^2$ (cf. [2]). In particular, if q^α is real (taking $q = 0$), the restriction to the unit sphere S^{n-1} of a polynomial in n variables is equal to the restriction of some harmonic polynomial.

The symplectic spinors obtained in the quotient algebra by S are Fourier series in r variables, when e_{α^*} is replaced by q^α.

Clearly this will also hold for the symplectic spinors constructed with the $(\epsilon_\alpha, \epsilon_{\alpha^*})$, the reasoning being the same.

The above remark allows us to give a meaning to the notion of Fourier series on a quadric.

Remarks.

1. We also see that the norm obtained from $\mathcal{K}$ is not the usual Hilbert space norm on L^2, since the $(e^H\Phi^*)$ do not form an orthonormal basis in this space.

2. Δ is self-adjoint for $\mathcal{K}$: this follows at once from the interpretation of Δ as $\sum e_{\alpha^*}$ (up to a real coefficient).

3. Δ being self-adjoint for $\mathcal{K}$, a classical reasoning shows that every symplectic spinor can be decomposed as $u\Phi^* = v\Phi^* + \Delta(w\Phi^*)$, where $v\Phi^*$ is harmonic. Recalling the interpretation of Δ, we know that every polynomial in the (e_{α^*}) is of the form

$$Q(e_{\alpha^*}) = P_0(e_{\alpha^*}) + (\sum(e_{\alpha^*})^2)P(e_{\alpha^*}),$$

where $P_0(e_\alpha)$ is harmonic, and we reobtain the property established just before using S.

Now we will apply the correspondence principle $e_\alpha \to q^\alpha$, $e_{\alpha^*} \to -\frac{\partial}{\partial q^\alpha}$ $(h = 1)$ to study some invariance and covariance problems for the laplacian.

Consider the elements (C) :

$$\begin{cases} \chi_\alpha e_\alpha e_{\beta^*} - \chi_\beta e_\beta e_{\alpha^*}, & \text{(infinitesimal rotations)} \\ \sum_\alpha e_\alpha e_{\alpha^*}, & \text{(infinitesimal homotheties)} \\ e_{\alpha^*}, & \text{(infinitesimal translations)} \\ \frac{1}{2}\sum_\alpha \chi_\alpha(e_\alpha)^2 e_{\gamma^*} - \chi_\gamma e_\gamma \sum_\gamma (e_\alpha e_{\alpha^*}) = \Phi_\gamma, & \text{(infinitesimal accelerations)} \end{cases}$$

each of which corresponds to an element from (C') :

$$\begin{cases} -q^\alpha \chi_\alpha(\partial/\partial q^\beta) + q^\beta \chi_\beta(\partial/\partial q^\alpha) = q_\beta(\partial/\partial q^\alpha) - q_\alpha(\partial/\partial q^\beta), \\ -q^\alpha(\partial/\partial q^\alpha), \\ -\partial/\partial q^\alpha, \\ \sum_\alpha q_\gamma q^\alpha(\partial/\partial q^\alpha) - \frac{1}{2}\sum_{\alpha,\beta} g^{\alpha\beta} q_\alpha q_\beta(\partial/\partial q^\gamma). \end{cases}$$

It is easy to verify directly that the elements of C form a Lie algebra such that (C') is a 'representation' of it in terms of vector fields in $\mathbf{R}^r$, with a metric of signature (p, q).

The Lie algebra of the complete conformal group $C_{p,q}$ can be recognized. Δ is 'represented' by $\sum \chi_\alpha(e_{\alpha^*})^2$.

We already noted that

$$\Delta(\chi_\alpha e_\alpha e_{\beta^*}) - (\chi_\beta e_\beta e_{\alpha^*})\Delta = 0, \quad \Delta e_{\alpha^*} - e_{\alpha^*}\Delta = 0,$$

so the laplacian is invariant for the Lie algebra of infinitesimal isometries.

Then

$$\Delta(\sum_\alpha e_\alpha e_{\alpha^*}) = [\sum_\alpha e_\alpha e_{\alpha^*} - 2]\Delta$$

is a covariance property of Δ under homotheties; finally, the 'accelerations' give

$$\Delta\Phi_\beta - \Phi_\beta\Delta = 2\chi_\beta e_\beta\Delta + (r - 2)e_{\beta^*}$$

which is a covariance property only when $r = 2$.

22.8   Some other applications.

As examples, and to illustrate particular advantages of our approach, we will prove some results which, in the literature, are established using analytical techniques.

First take $h = 1/(2\pi i)$ (and the spinors constructed on the symmetric space of the (e_α), all χ_α being 1).

Lemma 22.8.1 *The natural action of* $\exp(-\pi \sum(e_\alpha)^2 t) = \exp(-2\pi U)t$ *sends* e_{α^*} *to* $e_{\alpha^*} - 2\pi h t e_\alpha = e_{\alpha^*} + i t e_\alpha$.

Proof. Just compute

$$\exp(-\pi \sum (e_\alpha)^2 t) e_{\alpha^*} \exp(\pi \sum (e_\alpha)^2 t).$$

∎

Taking $t = 1$, e_{α^*} becomes $e_{\alpha^*} + i e_\alpha = \sqrt{2}\epsilon_{\alpha^*}$. P being a real homogeneous harmonic polynomial of degree k, we will prove *Hecke's formula* :

$$\mathcal{F}(P(x)e^{-\pi(x\cdot x)}) = (-i)^k P(x) e^{-\pi(x\cdot x)}.$$

Note that if we order by e_{α^*}, e_α,

$$\exp(2\pi U t) P(e_{\alpha^*}) \exp(-2\pi U t) = P(e_{\alpha^*})$$

(recall that $P(e_{\alpha^*})$ stands for $P(e_{1^*}, e_{2^*}, \ldots, e_{r^*})$), modulo terms in $(e_{\alpha^*})^\lambda (e_\alpha)^\mu$ (no term involving the (e_{α^*}) only, of degree less than k, occurs on the right-hand side because P is harmonic (Proposition (22.7.1))).

On the other hand, the action of $\exp(2\pi U t)$ extends to an algebra automorphism and hence

$$\exp(2\pi U t) P(e_{\alpha^*}) \exp(-2\pi U t) = P(e_{\alpha^*} - i t e_\alpha),$$
$$P(e_{\alpha^*}) \exp(-2\pi U t) = \exp(-2\pi U t)(P(e_{\alpha^*}) + (-it)^k P(e_\alpha) + R(e_\alpha, e_{\beta^*}))$$

where $R(e_\alpha, e_{\beta^*}) = Q(e_\alpha)$, $\deg Q < k$, modulo a left multiple of the e_α^*,

$$P(e_{\alpha^*}) \exp(-2\pi U)\Phi^* = (-i)^k P(e_\alpha) \exp(-2\pi U)\Phi^* + Q(e_\alpha)\exp(-2\pi U)\Phi^*.$$

But $Q(e_\alpha)$ vanishes identically, for if we consider the algebra antiautomorphism extending $e_\alpha \to e_{\alpha^*}$, $e_{\alpha^*} \to e_\alpha$,

$$P(e_\alpha - i e_{\alpha^*}) = (-i)^k P(i e_\alpha + e_{\alpha^*}) = (-i)^k P(e_{\alpha^*} - i e_\alpha),$$

P being real.

If a $Q(e_\alpha)$ occured in $P(e_{\alpha^*} - ie_\alpha)$, a term involving only the (e_{α^*}) and of the same degree as $Q(e_\alpha)$ would appear in $P(e_\alpha - ie_{\alpha^*})$, relying on by the exchange of e_α and e_{α^*}, and this was proved impossible before. Hence

$$
\begin{aligned}
P(e_{\alpha^*})\exp(-2\pi U)\Phi^* &= (-i)^k P(e_\alpha)\exp(-2\pi U)\Phi^* \\
\tilde{J}_1 P(e_{\alpha^*})\tilde{J}_1^{-1}\tilde{J}_1 \exp(-2\pi U)\Phi^* &= (-i)^k \tilde{J}_1 P(e_\alpha)\exp(-2\pi U)\Phi^*, \\
\exp(-\pi i r/4)P(e_\alpha)\exp(-2\pi U)\Phi^* &= (i)^k \tilde{J}_1 P(e_\alpha)\exp(-2\pi U)\Phi^*, \\
P(e_\alpha)\exp(-2\pi U)\Phi^* &= (i)^k \mathcal{F}(P(e_\alpha)\exp(-2\pi U)\Phi^*)
\end{aligned}
$$

which is the stated equality.

The Fourier transform of Hermite functions ($q = 0$). We consider symplectic spinors of the form

$$\exp(2\pi U)P(e_{\alpha^*})\exp(-4\pi U)\Phi^*$$

where P is homogeneous of degree k,

$$
\begin{aligned}
\exp(2\pi U)P(e_{\alpha^*})\exp(-2\pi U) &= P(e_{\alpha^*} - ie_\alpha) \\
&= P(-i(e_\alpha + ie_{\alpha^*})) \\
&= (-i\sqrt{2})^k P(\epsilon_{\alpha^*}).
\end{aligned}
\tag{1}
$$

Multiplying by $\tilde{J}_1$,

$$
\begin{aligned}
(-i\sqrt{2})^k \tilde{J}_1 P(\epsilon_{\alpha^*})\exp(-2\pi U)\Phi^* &= (-i\sqrt{2})^k \tilde{J}_1 P(\epsilon_{\alpha^*})\tilde{J}_1^{-1}\tilde{J}_1 \exp(-2\pi U)\Phi^* \\
&= (-i\sqrt{2})^k (-i)^k P(\epsilon_{\alpha^*})\exp(-2\pi U)\Phi^* \\
&= (-i)^k P(e_{\alpha^*} - ie_\alpha)\exp(-2\pi U)\Phi^* \\
&= (-i)^k P(e_{\alpha^*} - ie_\alpha)\exp(2\pi U)\exp(-4\pi U)\Phi^* \\
&= (-i)^k \exp(2\pi U)P(e_{\alpha^*})\exp(-4\pi U)\Phi^*.
\end{aligned}
$$

hence

$$\mathcal{F}(\exp(2\pi U)P(e_{\alpha^*})\exp(-4\pi U)\Phi^*) = (-i)^k \exp(2\pi U)P(e_{\alpha^*})\exp(-4\pi U)\Phi^*,$$

which yields, in accordance to the classical result,

$$\mathcal{F}(h_k) = (-1)^k h_k,$$

where h_k is the kth Hermite function.

Taking into account (1) and the expression of our symplectic spinors, we easily deduce that the only symplectic spinors which are invariant under the Fourier transform are of the form $\exp(2\pi U)Q(e_{\alpha^*})\exp(-4\pi U)\Phi^*$, where $Q(e_{\alpha^*})$ is a formal series (which may converge) consisting of even terms only : the set of even Hermite functions (i.e. k is even) therefore 'generates' the complex space of invariant functions under the Fourier transform.

Clearly, if k is odd, we obtain functions which undergo a sign change (this is the spectral decomposition of $\mathcal{F}$, given that $\mathcal{F}^2 = \text{Id}$).

In the notations of the remarks in section 5,

$$\mathcal{P}(x) = e^{2\pi(x \cdot x)} P\left(\frac{-1}{2\pi i} \frac{\partial}{\partial q^\alpha}\right) e^{-2\pi(x \cdot x)},$$

$x = (q^1, \ldots, q^r)$. Following J. Dieudonné ([2]), we set

$$\mathbf{w} P(x) = e^{-\pi(x \cdot x)} \mathcal{P}(x)$$

and, in our notations,

$$\mathcal{K}(\exp(-2\pi U) \exp(4\pi U) P(e_{\alpha \bullet}) \exp(-4\pi U) \Phi^*,$$
$$\exp(-2\pi U) \exp(4\pi U) Q(e_{\alpha \bullet}) \exp(-4\pi U) \Phi^*)$$

equals $\mathcal{K}(P(e_{\alpha \bullet}) \exp(-4\pi U) \Phi, Q(e_{\alpha \bullet}) \exp(-4\pi U) \Phi^*)$, relying on the results of section 6; if we take $q = 0$, for instance, we obtain

$$\mathcal{K}(\tilde{J}_1 P(e_\alpha) \tilde{J}_1^{-1} \exp(-4\pi U) \Phi^*, \tilde{J}_1 Q(e_\alpha) \tilde{J}_1^{-1} \exp(-4\pi U) \Phi^*)$$

and

$$\mathcal{K}(P(e_\alpha) \exp(-\pi U) \Phi^*, Q(e_\alpha) \exp(-\pi U) \Phi^*),$$

now since $\mathcal{F}(\exp(-4\pi U) \Phi^*) = 2^{-r/2} \exp(-\pi U) \Phi^*$, we get

$$2^{-r} \mathcal{K}(P(e_\alpha) \Phi^*, Q(e_\alpha) \Phi^*),$$

or $\mathcal{K}(\mathbf{w} P, \mathbf{w} Q) = 2^{-r} \mathcal{K}(P, Q)$ (cf. [2, p. 116]).

Many other examples could be given.

Conclusion. Now we have clarified the close link between the Fourier transform and the complex structure adapted to the symplectic structure in arbitrary signature, it is possible to algebrize and geometrize a number of computations which are usually carried out in a purely analytic context.

The Fourier transform defined in this way, acting on the symplectic spinor spaces can, under some conditions (the vanishing of the second Stiefel-Whitney class of the symplectic manifold, the existence of a pair of supplementary lagrangian fields), be extended over a manifold with a symplectic spinor fiber bundle.

The set of these global Fourier transforms is then parametrized by a Maslov index (or class).

The study of problems involving Poisson algebra deformations, geometric quantization and differential operators will then be much easier.

22.9   Selected references.

1. A. Crumeyrolle

(a) *Algèbres de Clifford symplectique et spineurs symplectiques*, J. Math. pures et appl., t. 56, 1977.

(b) *Classes de Maslov*, J. Math. pures et appl., t. 58, 1979.

(c) *Déformations d'algèbres associées à une variété symplectique*, Ann. de l'I.H.P., Vol. 35, n. 3.

(d) *Constante de Planck et géométrie symplectique.*, Lect. Notes in Math. n. 1165, Springer-Verlag, 1985.

2. J. Dieudonné, *Eléments d'analyse, tome 6*, Gauthier-Villars, Paris.

3. R. Howe, *On the rôle of the Heisenberg group in harmonic analysis*, Bull. of the Am. Math. Soc., Vol. 3, n. 2, Sept. 1980.

4. J. Leray, *The meaning of Maslov's asymptotic method : the need of Planck's constant in mathematics*, Bull. of the Am. Math. Soc., Vol. 5, n. 1, July 1981.

5. J. Sniatycki, *Geometric quantization and Quantum Mechanics*, Springer-Verlag, 1980.

6. M. Vergne, G. Lion, *The Weil representation, Maslov index and theta series*, Progress in math., Birkhaüser, 1980.

Chapter 23

BIBLIOGRAPHY.

23.1 Clifford algebras and orthogonal spinors.

- Ablamowicz R.

    - *Structure of spin group associated with degenerate Clifford algebras*, J. Math. Phys. 27(1), Jan. 1986—pp. 1–6.
    - *Deformation and contraction in Clifford Algebras, ibid.* 27(2), Feb. 1986—pp. 423–427.

- Ablamowicz R. and Salingaros N.

    - *On the Relationship between Twistors and Clifford Algebras*, Letters in Math. Physics 9 (1985) pp. 149–155.

- Ablamowicz R., Oziewicz Z., Rzewuski J.

    - *Clifford algebra approach to twistors*, J. Math. Phys. 23(2), Feb. 1982, pp. 231–242.

- Ahlfors L.

    - *Möbius transformations and Clifford numbers*, Diff. geom. and compl. analysis, Springer-Verlag, 1985.

- Atiyah M. F., Bott R., Shapiro A.

    - *Clifford modules*, Topology, Vol. 3, Supp. 1, 1964, pp. 3–38.

- Benn I. M., Tucker R. W.

    - *Fermions without spinors*, Comm. math. Phys. 89, pp. 341–362, 1983.
    - *The Dirac equation in exterior form*, Comm. in math. Phys., Vol. 98, n. 1, 1985.

BIBLIOGRAPHY.   333

- Bourbaki N.

  - *Algèbre, Chapitre 9*, Hermann, Paris, 1959.

- Brackx F., Delanghe R., Sommen F.

  - *Clifford analysis*, Research Notes in Math., Pitman, 1982.

- Brauer R., Weyl H.

  - *Spinors in n dimensions*, Amer. Jour. Math. 57, pp. 425–449.

- Brooke J. A.

  - *A Galilean formulation of spin I : Clifford algebras and spin groups*, J. Math. Phys. 19, pp. 952–959, 1978.

- Bugajska K.

  - *On geometrical properties of spinor structures*, J. Math. Phys. 21(8), Aug. 1980.

  - *Geometrical interpretations of spinors*, private communication (1979).

  - *Internal structure of space-time*, Journées Relativistes de Toulouse 1986.

- Cartan E.

  - *Leçons sur la théorie des spineurs*, Hermann, Paris, 1938.

- Chevalley C.

  - *The algebraic theory of spinors*, Columbia University Press, 1954.

  - *The construction and study of certain important algebras*, Publ. of Math. Soc. of Japan, 1955.

- Chisholm J. S. R., Common A. K. (editors)

  - *Clifford algebras and their applications in Math. Phys.*, NATO and SERC Workshop, Canterbury 1985, 592pp., Reidel 1986.

- Clifford W. K.

  - *Applications of Grassmann's extensive algebra*, Am. J. Math. 1 350 (1878).

- Crumeyrolle A.

  - *Structures spinorielles*, Ann. Inst. H. Poincaré, Vol. XI, n. 1, pp. 19–55, 1969. *Groupes de spinorialité, ibid.* , Vol. XIV, n. 4, 1971.

- *Dérivations, formes et opérateurs usuels sur les champs spinoriels des variétés différentiables de dimension paire, ibid.*, Vol. XVI, n. 3, 1972.

- *Spin fibrations over manifolds and generalized twistors*, Proceed. Symp. A.M.S. Stanford 1973, Vol. XXVII, Part 1.

- *Fibrations spinorielles et twisteurs généralisés*, Period. math. Hungarica, Vol. 6(2), pp. 143–171, 1975.

- *Théorie d'Einstein-Dirac en spin maximum 1*, Ann. Inst. H. Poincaré, Vol. XXII, n. 1, pp. 43–61, 1975.

- *Schémas d'Einstein-Dirac en spin 1/2*, *ibid.*, Vol. XXIII, pp. 259–274, 1975.

- *Les spineurs sur les espaces courbes et le problème de la quantification des champs*, Colloque centenaire Einstein, Publ. CNRS, Paris, 1979.

- *Twisteurs sans twisteurs*, Geometrodynamic Proceed., 1983, Bologna, 1984.

- *Construction d'algèbres de Lie graduées orthosymplectiques et conformo-symplectiques minkowskiennes*, Lecture Notes in Math. n. 1165, Springer-Verlag.

- *Conjugation in spinor spaces, Majorana and Weyl spinors*, Proceed. winter school 1986, Suppl. Rend. Circ. Mat. Palermo Serie II, n. 14.

- *Régularisation cliffordienne du problème de Kepler en dimension quelconque et symétrie conforme*, C. R. Acad. Sci. Paris, t. 299, Série I, n. 18, 1984. *Physical symmetry groups and associated bundles in field theory*, Reports on math. phys. n. 1, Vol. 23, 1986.

- Dabrowski L., Trautman A.

  - *Spinor structures on spheres and projective spaces*, J. Math. Phys. 27(8), pp. 2022–2028, August 1986.

- Delanghe R.

  - cf. Brackx F.

- Geroch R.

  - *Spinor structure of space-times in General Relativity*, Journal of Math. Phys. (9) n. 11, pp. 1739–1744, Nov. 1968.

- Graf W.

  - *Differential forms as spinors*, Ann. Inst. H. Poincaré, Vol. XXIV, n. 1, 1978.

- Haefliger A.

  - *Sur l'extension du groupe structural d'un espace fibré*, C. R. Acad. Sci. Paris, Aug. 1956.

- Hestenes D., Sobczyk G.

  - *Clifford algebra to geometric calculus*, Reidel 1984.

- Imaeda K.

  - *Quaternionic formulation of classical electrodynamics and theory of functions of a biquaternion variable*, Publ. Okayama Univ. of Science, 1983.

- Kähler E.

  - Rendiconti di Mat. (3.4)21 – 425 (1962).

- Karoubi M.

  - *Algèbres de Clifford et K-théorie*, Ann. Scient. Eco. Norm. Sup., 4e série, t. 1, pp. 161–270, 1968.

- Karrer G.

  - *Darstellung von Clifford bündeln*, Ann. Acad. Sci. Fennicae, series A, 1, n. 521, 1973.

- Kastler D.

  - *Introduction à l'électrodynamique quantique*, Dunod, Paris, 1961.

- Kosmann Y.

  - *Dérivées de Lie des spineurs*, thesis, Paris 1970. Annal. di mat. pura et applicata IV, Vol. XLI, pp. 317–395.

- Kwasniewski A. K.

  - *Clifford and Grassmann like algebras—old and new*, J. Math. Phys. 26(9), pp. 2234–2238, Sept. 1985.

- Lam T. Y.

  - *The algebraic theory of quadratic forms*, Benjamin 1973.

- Latvamaa E., Lounesto P.

- *Conformal transf. and Clifford Algebras*, Proceed. Am. Math. Soc., Vol. 79, n. 4, Aug. 1980.

- Lawrynowicz J., Rembielinski J.

  - *Hurwitz pairs equipped with complex structures*, Lecture Notes in Math. n. 1165, Springer-Verlag, 1985.

- Lichnerowicz A.

  - *Champs spinoriels et propagateurs en relativité générale*, Bull. Soc. Math. France, t. 92, Fasc. 1, pp. 11–100, 1984.

  - *Spineurs harmoniques*, C. R. Acad. Sci. Paris Série A.B. 257, pp. 7–9, 1963.

- Lounesto P.

  - *On primitive idempotents of Clifford algebras*, Reprot Math. Inst. Tek. Hog. Helsinki, 1977.

  - *Spinor-valued regular functions in hypercomplex analysis*, thesis, Helsinki, 1979.

  - *Sur les idéaux à gauche des algèbres de Clifford et les produits scalaires des spineurs*, Ann. Inst. H. Poincaré 33, pp. 53–61, 1980.

- Magneron B.

  - *Spineurs purs et description cohomologique des groupes spinoriels*, Actes Journées Relativistes de Toulouse, pp. 141–154, 1986.

- Maia M. D.

  - *Isospinors*, J. math. Phys., Vol. 14, n. 7, July 1973.

  - *Conformal spinor fields in general Relativity*, *ibid.*, Vol. 15, n. 14, April 1974.

- Micali A. and Villamayor O. E.

  - Sur les algèbres de Clifford, Ann. Sc. Eco. Norm. Sup., 4e série 1, pp. 271–304, 1968.

  - *Sur les algèbres de Clifford II*, J. für die Reine und Angew. Math., 242, pp. 61–90, 1970.

- Michelson A.M.

  - *Clifford and Spinor cohomology*, American Journ. of Math., December 1980.

- Morris A. O.

  - *On a generalized Clifford algebra*, Quart. J. math. Oxford, ser. 2, (18) pp. 7–12, 1967 and *ibid.*, (19) pp. 289–299, 1968.

- Oziewicz Z.

  - cf. Ablamowicz R.

- Penrose R., Ward R. S.

  - *Twistors for flat and curved space-times*, General Relativity and Gravitation 2, pp. 283–328, Plenum Press 1980.

- Popovici J.

  - *Remarques sur l'existence des structures spinorielles*, C. R. Acad. Sci. Paris, série A., t. 279, 1974.

- Popovici J. and Turtoi A.

  - *Prolongement des structures spinorielles*, Ann Inst. H. Poincaré, Vol. XX, n. 1, 1974.

- Porteous I.

  - *Topological Geometry*, Van Nostrand (1969).

- Rashevskii P. K.

  - *The theory of spinors*, American Math. Soc. Translations, series 2, 6, p. 1–110, 1957.

- Riesz M.

  - *Clifford numbers and spinors*, Lecture Series 38, Inst. for Phys. Sciences and Technology, Univ. of Maryland, 1958.

- Rzewuski J.

  - *On some geometrical consequences of physical symmetries*, Reports on Math. Physics, Vol. 22, n. 2, pp. 235–256 (1985).

- Salingaros N.

  - cf. Ablamowicz R.

- Sewerynski M.

– *On the Rashewskii approach to spinors*, Preprint, Inst. Chalmers Göteborg, 1985.

• Sobczyk G.

– cf. Hestenes D.

• Sommen F.

– cf. Brackx F.

• Timbeau J, Crumeyrolle A.

– *Quantification des systèmes classiques sur les variétés pseudo-riemanniennes à structure spinorielle*, C. R. Acad. Sci. Paris, t. 280, série A, June 1975.

• Timbeau J.

– *Structures torogonales et quantification sur des variétés pseudo-riemanniennes*, thesis, Toulouse, 1986.

• Thomas E.

– *A generalization of Clifford algebras*, Glasgow Math. J. (15), pp. 74–78 (1974).

• Trautman A.

– cf. Dabrowski L.

• Tucker R. W.

– cf. Benn I. M.

• Turtoi A.

– cf. Popovici J.

• Villamayor O. E.

– cf. Micali A.

• Ward R. S.

– cf. Penrose R.

• Weyl H.

– cf. Brauer R.

23.2   Symplectic algebras and symplectic spinors.

● Bayen F., Flato M., Fronsdal C., Lichnerowicz A., Sternheimer D.

  – *Deformation theory and quantization I*, Annals of Physics, t. 111, pp. 61–110, 1978.

● Crumeyrolle A.

  – *Algèbres de Clifford symplectiques et spineurs symplectiques*, J. Math. pures et appliquées, t. 56, pp. 205–230, 1977.

  – *Classes de Maslov, fibrations spinorielles symplectiques et transformation de Fourier*, ibid., t. 58, pp. 111–120, 1979.

  – *Indice d'inertie trilatère et classes de Chern*, St. Sc. Math. Hung. 13, pp. 259–271, 1978.

  – *Déformations d'algèbres associées a une variété symplectiques, une construction effective*, Ann. Inst. H. Poincaré, Vol. 35, n. 3, pp. 175–194, 1981.

  – *Constante de Planck et géometrie symplectique*, Lecture Notes in Math. n. 1165, Springer-Verlag, 1985.

● De Wilde M., Lecomte P. B. A.

  – *Existence of star-products and of formal deformations of the Poisson algebra of arbitrary symplectic manifolds*, Letters in Math. Physics 7 (1983), pp. 487–496.

● Howe R.

  – *On the rôle of the Heisenberg group in harmonic analysis*, Bull. Am. Math. Soc., Vol. 3, n. 2, Sept. 1980.

● Kostant B.

  – *Symplectic spinors*, Symposia Mathematica, Vol. 14, pp. 139–152, Academic Press, London.

● Leray J.

  – *Analyse lagrangienne et Mécanique quantique*, Cours du Collège de France, 1976-77.

  – *The meaning of Maslov's asymptotic method : the need of Planck's constant in mathematics*, Bull. Am. Soc., Vol. 5, n. 1, July 1985.

● Lichnerowicz A.

– *Les variétés de Poisson et leurs algèbres de Lie associées*, J. Diff. Geometry (12), 1977, pp. 253–300.

– *Déformations d'algèbres associées à une variété symplectique*, Ann. Inst. Fourier, t. 32, pp. 157–205, 1982.

– *Differential geometry and deformations*, Actes du colloque 'Symplectic Geometry', Toulouse 1981, Pitman publ. 1983.

• Moyal J. E.

– *Quantum mechanics as a statistical theory*, Proc. Cambridge Phil. Soc. 45, 99 (1949).

• Moreno C. and Ortega-Navarro P.

– *Deformations of the algebra of functions on hermitian symmetric spaces resulting from quantization*, Ann. Inst. H. Poincaré, Vol. 38, n. 3, 1983.

• Vey J.

– *Déformation du crochet de Poisson sur une variété symplectique*, Comment. Math. Helvetici, t. 50, 1975.

Notation index.

Symbols frequently used, together with the chapter and section where they are defined.

$\mathcal{A}$	10.2, 10.3
α	3.1, 17.1
B	1.1
$\mathcal{B}, \tilde{\mathcal{B}}$	8.2, 9.1
$\mathcal{B}_1$	10.4
β	3.1, 17.1
$\mathbf{C}$	1.1
$\mathcal{C}$	10.2, 10.3, 10.5
$\mathrm{Clif}(V), \mathrm{Clif}_V(Q), \mathrm{Clif}_x(Q)$	14.1
$\mathrm{Cl}(p,q)$	12.1
$C'(r)$	3.2
$C(Q), C(E,Q), C^+(Q), C^-(Q)$	3.1
$C'(Q), C(Q')$	3.1
$C_n(p,q), C_n^0(p,q)$	12.1, 12.2
$C(r,s)$	3.2
$C_S(F), C_S(hF), \check{C}_S(hF), \check{C}_S(hF)\Phi^*$	17.2
$D^1(V)$	14.1
e	5.1
e_N	3.3, 13.2
ϵ	8.2
ϵ'	10.3
$f = y_1 y_2 \dots y_r$	3.3
f_α	14.2
G, G^+	4.1
G_e	14.4

Subject index.

344